SCIENTIFIC INFORMATION SYSTEMS

This book is dedicated
to
Elsbeth Monika Holt

Scientific Information Systems

ROBERT W. HOLT
George Mason University

Routledge
Taylor & Francis Group

LONDON AND NEW YORK

First published 2001 by Ashgate Publishing

Published 2016 by Routledge
2 Park Square, Milton Park, Abingdon, Oxon OX14 4RN
711 Third Avenue, New York, NY 10017, USA

Routledge is an imprint of the Taylor & Francis Group, an informa business

British Library Cataloguing in Publication Data
Holt, Robert W.
  Scientific information systems
  1.Information storage and retrieval systems
  I.Title
  003

Library of Congress Control Number: 00-109530

ISBN 13: 978-0-7546-1116-5 (hbk)

Contents

List of Figures

List of Tables

xiii

Acknowledgements

I would like to thank Dr. Eleana Edens and Dr. Thomas Longridge of the Federal Aviation Administration for guiding and funding the research from which this book was developed and supporting the writing of the book. I would also like to thank the Office of the Provost at George Mason University for funding a sabbatical to write this book. Most of all, I would like to thank Elspeth Holt, my wife, whose continued support and editing assistance made this book possible.

1   Overview and Introduction

Overview

This book discusses a synthesis that blends scientific methods with the broad capabilities of computer database information systems. This synthesis should aid anyone who is trying to explain, predict, or change the behavior of a complex system involving humans. This unique blend has been developed from research on information systems in the aviation industry, but these principles and methods can be broadly applied to other areas. Since the original work was in the aviation domain, many of the specific examples in this book will have aviation content. The scientific issues addressed by these examples will, however, be universal. Each chapter will cover different aspects of this approach, and the examples should clarify the advantages of this synthesis for understanding a complex system completely and accurately.

Chapter 1 introduces the concept of a Scientific Information System (SIS) and discusses why it is necessary to use scientific principles and methods in an information system. After distinguishing a SIS from a traditional Management Information System (MIS), this chapter discusses the interaction of theory and data in a SIS. The process of using scientific methods to refine both theory and data over time will be reviewed. An example of a SIS focused on job performance will be discussed in detail and used to clarify important theoretical and practical issues.

Chapter 2 covers issues of the design of the databases for a SIS and methods for ensuring the basic quality and integrity of the data in the system. Each measure in a SIS must be evaluated for basic data quality. The design of a SIS requires a specification of relevant sets of measures plus the linkages among the databases containing these measures that are required for effective use. Finally, Chapter 2 discusses methods for ensuring basic data integrity, which is a prerequisite to evaluating the scientific quality of the SIS data.

Chapter 3 focuses on the three major criteria for ensuring the scientific quality of the data in the SIS: sensitivity, reliability, and validity. Sensitivity is the precision of measurement or how much a measure changes when the thing being measured changes. Reliability is the stability, consistency or lack of error in measurement. Validity is the accuracy or correctness of measurement. Chapter 3 covers appropriate methods for establishing each aspect of high quality scientific data.

Chapter 4 covers basic methods for evaluating and modifying the theories on which the SIS is based. Each SIS theory should help explain, predict, and control events, processes, and outcomes of a complex system. A careful conceptual and empirical evaluation of the theoretical basis of the SIS is necessary to evaluate the extent of empirical validity of each theoretical viewpoint used in the SIS. Theoretical validity has practical implications. The SIS user can be more confident when applying or interpreting SIS results that are part of validated theory and more cautious when applying or interpreting results for which the theoretical validation is weaker. Evaluation should also involve competitively assessing different theories and thereby improving the theoretical basis of the SIS over time. The co-evolution of theory and data improves the usability of SIS information for solving complex system problems.

Chapter 5 covers the application of the information in the SIS to identify and solve problems. Chapter 5 discusses the stages or steps in a problem solution cycle. Practical use of a SIS starts with problem exploration or elaboration. Both qualitative and quantitative methods can be used with SIS data to find or define a problem. Next, the SIS theoretical viewpoint and information databases are used to elaborate problem origins and potential solutions. Finally, the SIS is arranged for collecting appropriate data to evaluate the problem solution. The utility of the SIS for solving problems will depend on the precision, completeness, and validity of the SIS theories. Developing better SIS theories will often require the use of more advanced analytical methods on the data which are covered in succeeding chapters.

Chapter 6 focuses on analyzing structure for systems that are in a steady state or have negligible dynamic processes. Systems that do not change or change very slowly or infrequently can be described by theories and methods that focus on static states. Chapter 6 covers methods for evaluating this type of theory such as factor analysis, structural equation modeling, and multidimensional scaling.

Chapter 7 focuses on advanced analyses for dynamic systems that are mainly driven by the occurrence of events or transitions among distinct states of the system. For these systems, change is the focal point and different methods must be used to understand and model change. State-based modeling covers the systems that change among a small set of possible states. The transitions among the system states may be predictable or deterministic as well as unpredictable, stochastic, or probabilistic. Event-based modeling methods focus on the entities of a system such as persons or teams or focus on simple processes that affect the entities in different ways.

Chapter 8 focuses on dynamic systems with more complex processes. These processes are often analyzed as functions. For systems with a quantitative focus, these functions may be the traditional numerical functions described in algebra and calculus. For systems with a qualitative focus, these functions may be non-numerical or symbolic. Symbolic production systems, for example, have been used to model human thought processes. In a general, functional models can be used to explain, predict, and control other systems with very complex processes.

The final chapters provide examples of how these methods can be used in a SIS focused on job performance to provide practically useful information. In each chapter, the use of the SIS theoretical viewpoint and data for dealing with practical domains will be illustrated. Chapter 9 covers the domain of personnel selection. The focus will be on the identification and measurement of personal qualities that affect job performance. The use of the SIS to do job analysis and design and evaluate personnel selection instruments will be discussed. The advantage of the SIS for long-term as well as short-term evaluation of selection instruments will be covered.

Chapter 10 covers the role of the SIS in developing and evaluating training. The focus will be on the development of facet, stage/phase, and process theories of training. The use of SIS information to make training changes such as changing training materials or content, and rational re-allocation of training resources from over-trained areas to under-trained areas will be described. The pros and cons of different scientific evaluation methods will be covered.

Chapter 11 covers the effects of the job and job context on performance. The focus is on determining the effects of the person, the workplace, and the person workplace interface on performance in an integrated multi-level model. The elaboration of the SIS to systematically include team and organizational level effects on performance will be illustrated in detail.

Chapter 12 integrates the material in previous chapters from a corporate or business point of view. Three major roles of the SIS in the business enterprise are discussed. The first role is to give scientifically-validated information about the system. This information can be used to diagnose problems as well as to extrapolate the effects of changes or interventions. The second role of the SIS is a source for feedback for all information stakeholders. The third and ultimate role of the SIS is to effectively understand, predict, and manage all aspects of the system. The use of scientific methods and theories in the construction of the SIS helps ensure that it will fulfill these three roles.

Introduction

A Scientific Information System (SIS) is a computerized management information system (MIS) that takes a systems viewpoint (De Greene, 1970) and is based on scientific principles of theory construction, data quality, and systematic data analysis. The emphasis on scientific principles in each step of the construction, use, and evolution of such a system distinguishes it from a typical management information system, data warehouse, or similar systems. The use of scientific principles requires a more careful and thoughtful approach to the collection and use of data in a SIS. This cost is balanced against the gain in the quality of information in the system and the knowledge that can be derived from this information.

Comparison to Management Information Systems

Although part of a SIS is information expressed in a set of computer databases similar to a Management Information System, the SIS is distinct from a MIS in both content and use. First, the data content of a SIS has an explicit scientific and theoretical grounding that the MIS lacks. Second, the knowledge of the SIS is succinctly summarized in the set of underlying theories that are not present in a MIS. Third, the use of a SIS will typically include scientific inferential or descriptive methods for gaining information over and above the direct reports on system results used in a MIS.

Data Basis The essential distinction between a SIS and MIS is that the SIS establishes and refines a theoretical or conceptual basis for the data residing in the databases. In contrast, a MIS does not have a theoretical basis. Although this conceptual structure may be minimal at the outset of the construction of the SIS, it is refined in an evolutionary manner using accepted scientific methods. The ultimate goal is to be able to explain, predict, and alter the performance of a complex system. This utility must achieve a level that meets the information and control needs of SIS users.

Another distinction between a SIS and a MIS is that typically a SIS will be as focused on the process as on the product. In contrast, a MIS is often product or outcome-oriented with little or no emphasis on causal processes. Where applicable, the process emphasis gives a SIS the leverage to clarify the causal processes and to specify how changes in the causal processes will lead to changes in outcomes. This increases the ability to alter the system in desired directions.

Knowledge To the extent that the SIS theories are empirically validated, they serve as a concise and effective summary of what is known about the inputs, processes, and outcomes of the target system. This abstract knowledge of the system goes beyond the data contained in the databases. Validated theories may be used to extrapolate and integrate knowledge to areas where data have not been collected. The theoretical knowledge creates a wider potential for using the SIS to explain, predict, and alter system behavior.

Scientific Information Methods Although the databases in a SIS can be used in the same way for routine reports as a MIS, the SIS has the advantage that the scientific quality of the information can be established. Establishing the sensitivity, reliability, and validity of the information in the system allows the legitimate use of a variety of scientific methods for investigating the data. Methods ranging from classical statistical methods to sophisticated modeling methods can be used to answer empirical questions and solve practical problems.

Although these methods can also be used on the unqualified data that may be in a MIS, the quality of the result will be correspondingly poor. Because there is no accurate knowledge of the quality of the data and how well the data fit a particular type of analysis, there can be no accurate knowledge of the quality of the result. This lack of determination of the quality of a result makes the systematic accumulation of knowledge about the system virtually impossible. Thus, the use of scientific analysis methods in a SIS is much more likely to result in cumulative knowledge about a complex system.

Appropriate domains for a SIS Developing the theoretical view of a SIS and applying scientific methods to extract knowledge about a complex system can be justified wherever the costs of doing so are outweighed by the benefits of complete, accurate, and valid knowledge. Although both the costs and benefits of a SIS may be extremely difficult to assess quantitatively, a very rough qualitative judgment can be made of both costs and benefits. Establishing a MIS requires information input and people with the computer skills to develop and maintain the databases. The SIS requires at least one additional person with extensive scientific training to develop the theoretical view and guide the use of scientific inference methods. Additional collateral costs include the acquisition of scientific data analysis software, access to scientific journals relevant to the domain, and other minor costs. Thus the costs, while real, are not extensive.

The potential benefits of better knowledge must offset these costs. There are many potential benefits of better knowledge about a complex system. Potential benefits can be in increased efficiency of training, better performance of routine tasks, increased morale or satisfaction, or in avoiding catastrophically bad outcomes such as accidents or lawsuits. Domains where the system is so simple that the knowledge is complete and accurate do not require a SIS. A fully automated production line of simple widgets would be an example. In such a case, the issue is to effectively *apply* the appropriate engineering and mechanical knowledge to address system problems and obtain performance benefits.

Domains where the system is complex and not completely known can be appropriate for development of a SIS because for these domains the benefits outweigh the development costs. In general, the performance of humans on complex, real-world tasks is critical to the effective functioning of many modern systems, but the human aspect of performance is only partly understood. Therefore, many domains that require individuals or teams to complete complex tasks would be appropriate for a SIS.

The aviation domain is one example of a complex domain appropriate for a SIS. The system of a crew flying an aircraft is inherently complex and dynamic. The potential benefits of better system knowledge are large. Training costs are high. The total cost of operation of large aircraft is quite high and efficiency of operation has a large impact on the profitability of the airline. The morale and satisfaction of pilots and other groups is important to avoid crippling strikes. Finally, the cost of poor performance such as errors can be catastrophic. For example, airlines ranging from Air Florida to Pan American have gone out of business in part due to accidents. Since the aviation domain is appropriate for a SIS and since some of the methods were developed in that domain, examples from the aviation domain will be used for critical SIS topics in this book.

However, there are many other potential domains appropriate for a SIS, and the principles of construction and use of a SIS are quite general. Prime domains for developing a SIS occur wherever the situation is complex enough to require a complex theoretical and empirical description, and the cost of poor performance or errors is high enough to justify the cost of the SIS. Surface transportation domains such as ship or rail have the same high cost of catastrophic errors as aviation and are complex, team-dependent tasks. Similarly, team-based control of complex systems like nuclear power plants, oil refineries, or hospital operating rooms, are also examples of domains with complex, dynamic tasks and a high cost of catastrophic errors.

In other domains, critical costs may be accumulated over time rather than attributable to catastrophic accidents. For example, poor planning and decision-making by mid-level managers in a corporation may ultimately impact on the competitiveness and viability of the organization, but this effect may be quite gradual over time. Inherently, the tasks of middle and upper-level management across a variety of domains are complex and yet critical to the organization. Establishing a SIS focused on management processes and decisions may be appropriate in these domains.

In still other domains, the function of critical committees, teams, task forces, or groups may strongly determine organizational outcomes. In such cases, a SIS focused on the processes and performance of these entities may have benefits that justify the costs. To illustrate the wide variety of potential domains, chapters of this book will also contain simple examples chosen from a variety of potential domains as well as the continuing example of the aviation domain. These varied examples will help emphasize the generality of the techniques and procedures covered in this book.

Gaining knowledge about a system

There are different possible sources of non-scientific knowledge such as intuition, insight, expert opinion, or dogma. Each of these knowledge sources has potential flaws that inhibit the accumulation of accurate knowledge about a system. Intuition can be unreliable. Insight can be undependable. Expert opinions can be inconsistent. Dogma can be unalterable in the face of changing conditions. To minimize potential knowledge flaws as much as possible, the development of knowledge in a SIS is empirical knowledge based on scientific procedures. In the long run, scientific approaches and procedures offer more reliable, consistent, verifiable, cumulative and adaptive knowledge than non-scientific approaches.

Scientific knowledge is not the same as data or information. Statistical abstracts are full of numerical facts and figures, but this data may not lead to knowledge because the magnitude of data can confuse as well as illuminate the issue. Even if the magnitude of data is not overwhelming, data representation can interfere with gaining accurate knowledge about a system. Data or raw numerical results can be represented in the form of numbers, tables, or graphs. Different forms of representing data are possible, and each form can have different implications to the observer (e.g. Tufte, 1983). However, it is possible to distort the truth or mislead the observer, even with data or basic graphs or tables based on the data. Thus, raw data or data representations are not infallible guides to knowledge (e.g. Huff, 1954).

In general, knowledge is not the same as information. The Internet, for example, has a great deal of raw information, but much of this information is of dubious or unknown quality that makes knowledge acquisition difficult. Scientific methods for gaining information from data typically require reducing the data to a summary conclusion or estimate. Many sophisticated methods can be used to extract information from data (e.g. Hays 1981, Tabachnik & Fidell,1996). Each method has its own set of assumptions for reducing data to a conclusion or estimate, and a corresponding set of advantages and disadvantages. No matter how sophisticated the technique, however, the information cannot be any better than the data upon which the information is based. This is also known as the "Garbage In, Garbage Out" principle. Even sophisticated methods of data analysis will give little or no information if the underlying data is of poor quality.

Since knowledge is quite distinct from the data or information upon which it is based, we must try to define the additional "something" that results in knowledge. Gaining knowledge about a system implies obtaining general conceptual information about that system with some degree of certainty that the information is correct. These validated conceptual structures are the theories of the SIS. A SIS uses scientific methods to offer a degree of certainty about the SIS theories on the one hand, and the quality and meaning of the database information on the other.

Good information is an essential building block for accurate knowledge about a system. Once scientific methods have been used to ensure the quality and relevance of the SIS data, then obtaining good information requires a good question and an appropriate method for analyzing the data. Figure 1 shows the triad for good information: quality data, a well-formed question, and an appropriate analysis. The lack of any component jeopardizes the quality of the information. Poor data can produce misleading results and conclusions from any method. A poorly framed or ambiguous question can make it difficult or impossible to select relevant data and appropriate analysis methods. Poorly chosen analysis methods may give either irrelevant or possibly very misleading answers. Obtaining good information requires the combination of good quality data with an appropriate analysis method for answering a clear-cut question.

Good information on a specific question is a key part of accumulating accurate, empirically based knowledge about a system. Systematic accumulation of general knowledge in any domain requires the synthesis of relevant pieces of information. Kuhn (1970) has described this process as the normal progress of knowledge accumulation in science. This information is summarized and organized by the scientific theory or theories for a given domain. These theories may evolve gradually as new information is integrated or shift abruptly when a scientific revolution occurs (see Scientific American, 1999 for examples).

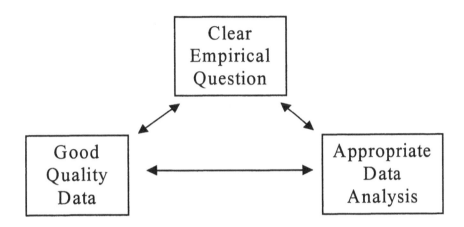

Figure 1.1 Three requirements for good empirical information

Scientific revolutions are a large, qualitative change in the theory or viewpoint that accounts for the empirical results in a field. An example is the adoption of the theory of "Continental Drift" and plate tectonics by the field of geology in the 1950s and 60s. This new viewpoint was qualitatively very distinct from the preceding theoretical view of stable continents and resulted in a very different way of viewing geological phenomena.

Both the normal and revolutionary phases of scientific progress rely on a theory or conceptual viewpoint. Similarly, the construction, appropriate use, and evolution of a SIS rely on one or more theories that embody a conceptual viewpoint. As in science, however, this viewpoint is not static but rather changes and evolves as the SIS is constructed, used, and altered to meet changing informational needs or goals.

Theory guides the initial development of the SIS by defining data that are relevant to explaining, predicting, or changing the performance of the system. In addition, the theory may specify the appropriate methods for measuring these data and necessary links among the data sets. Understanding how the theory specifies the contents of the SIS requires carefully defining a scientific theory.

Definition of theory A theory is the set of relationships among variables and processes which characterizes a system. All theories use variables to specify the basic elements of the theory. Variables are any single, unitary thing such as height, weight, or length. Variables may represent an abstract construct such as velocity or mass. Weight, for example, is a measure that indexes the more abstract construct of mass.

One way of looking at this is that constructs are the essential conceptual properties dealt with by the theory and variables or measures are the concrete way to get at these constructs. Constructs run the gamut from relatively simple to complex. Physical constructs range from the simple constructs of mass and velocity to complex constructs such as space/time and relativity. Similarly, constructs applicable to persons range from relatively simple constructs such as age and gender to such complex constructs as ability or personality. Constructs form a major part of the theories underlying a SIS and their associated measures form the heart of the data in the SIS.

Typically, a theory will specify the relationships among two or more constructs in a system. The relationships among the constructs in the theory suggest the ways that the data in the SIS ought to be related. Having a guide to how data ought to be related is particularly important for complex systems represented by complex sets of data. For complex data, the theoretical guide helps the user define and understand the set of plausible or likely relationships from among the almost infinite ways the data could be related. To the extent the theories are correct, this guidance will efficiently optimize the use of the SIS data.

One of the most interesting and important ways that variables may be related is by the processes of the system. Processes are the dynamic components of a system. Processes may underlie changes in the system over time, adaptive responses to changing conditions, positive and negative feedback loops, growth or decay, and so forth. The extent to which a theory emphasizes a static system of variables and relationships vs. a system with dynamic processes has important implications for the data required for a SIS. Different sorts of data will be required if the theory is mainly static compared to a theory that has many dynamic components.

Static or steady-state theories Static theories specify a set of relationships among constructs that is stable. The Big Five theory of personality structure, for example, specifies a set of thirty elementary personality variables organized into a set of five stable factors (McRae and Costa, 1990). This personality structure is hypothesized to be stable over time and relatively unchanged by life events. A theory emphasizing a static structure in this manner has very different requirements for measurement and verification than a dynamic theory of personality such as Freudian psycho-dynamics.

The stable set of relationships among variables or constructs is the focal point of a static theory. The validation of theories that emphasize only these structural relationships will typically require less data than the validation of dynamic theories. Due to the presumed stability, the data can be gathered at one time rather than being tracked across several time intervals. These static-focus theories can also be analyzed with simpler statistical methods than dynamic theories, such as the methods discussed in Chapter 6.

Dynamic theories Dynamic theories emphasize the change in a system over time, conditions, or some other critical variable. The processes underlying these changes may be described in greater or lesser detail. The simplest description is the specification of the change process as a "black box" which produces the changes in either a random or systematic fashion. Small, random fluctuations in a variable or system state are often assumed to be due to some random shock or perturbation of the system. These minor perturbations may make a system change over time and affect the precision of predictions about system behavior. Unless these minor perturbations result in major changes, such as in some chaotic systems, they are not usually the focus of the dynamic theory.

A dynamic theory will typically focus on parameters or processes that result in major changes in system variables or system states over time. These parameters or processes represent the key elements of the dynamic theory. Focusing on key elements gives the maximum power for a theory of a certain size and complexity to explain, predict, and control outcomes in the target system. One important distinction for dynamic theories is the extent to which predicted change in the system is a smooth, continuous change vs. an abrupt, discontinuous change that may be more difficult to predict. Chapter 7 covers methods for modeling systems where the dynamics can be described as shifts among a small set of states or as changes due to a discrete set of events. Chapter 8 covers methods for modeling systems with more complex dynamic processes.

Constructing the SIS Although theory is a critical component of the development and use of a SIS, the theory will not usually be specified before the initial establishment of the databases in the SIS. The construction of both the SIS and the integrated theory or theories for this information will generally take time. Commonly, the information in the SIS databases will change and co-evolve with the theoretical viewpoints that give this information a systematic, coherent meaning. The sequence should be viewed as a bootstrapping procedure in which the initial the theoretical viewpoints about the nature of the system are confirmed or disconfirmed with the data in the SIS. Both confirmation and disconfirmation are crucial to refining the viewpoint into a better theory and refining the type and amount of data that is collected and analyzed in the SIS.

Definition and example of a job performance SIS   The detailed definition of a SIS must include the goal of the development of the system as well as a general statement about the scientific theory and methods and relevant information databases used to construct the system. Therefore, SISs defined for different domains or for different goals will have different specific definitions. The detailed definition of a SIS may change over time as the domain or goals for the system alter. For analyzing performance of individuals or teams in the workplace, the definition of a SIS necessarily involves the development of a theory of job performance together with the relevant database information. Therefore, the definition of a SIS focused on job performance is

> A job-performance Scientific Information System is the combination of the information databases related to job performance together with relevant theories that are developed by scientific methods to explain, predict and alter job performance.

The information databases required for this type of SIS will initially depend on what should predict job performance given the initial viewpoint or theory. The initial set of information may have some components that are later proved to be relevant to job performance and other components that are not. The information databases are empirically refined by augmenting relevant information and deleting irrelevant information. If new theoretical views are examined in the SIS, these views may dictate the addition of supplemental information in the SIS. For individual and team job performance, different theories may require a variety of possible relevant databases. The information databases in the SIS allow a strong, competitive evaluation of the different theories.

The information databases for a job performance SIS will certainly include one or more databases containing performance appraisal information. Additional relevant databases will depend on the theoretical views initially adopted for the SIS, but in general these databases may involve the person, the organization in general, the team or work group in particular, the job content and context, and the appraisal of job performance as focal points. Figure 1.2 represents a possible theoretical viewpoint for which the information in each type of databases is relevant for explaining job performance.

Personal factors A wide range of data could be included in the personal database. These factors should include any known job-relevant experience, training, knowledge, skills, or abilities. Additionally, job-relevant aspects of attitude, motivation, or personality could be included. Almost always, the information included in these databases will be directly or indirectly relevant to the explanation, prediction, or alteration of performance. Under some specific situations, however, information could be included that would NOT theoretically be related to performance.

Personal factors, for example, may include factors that should NOT be relevant to job performance evaluations such as age, marital status, race, and ethnicity. The reason for including this information would be to empirically check that they are in fact irrelevant. For example, for legal reasons it may be necessary to show that there is no discrimination against protected groups in the workplace. In such a case, the personal information on these factors would be analyzed and hopefully show no significant relationship to performance evaluation. If, however, age, gender, or ethnicity did predict job performance over and above the other job-relevant factors, this might be evidence of workplace bias or discrimination and require management action.

Organizational factors Organizational factors include the overall culture of an organization as well as specific aspects of organizational climate. Cultural norms may influence the functioning of individuals or teams in the organization. Norms may influence both the content as well as the process or dynamics of an individual, team, or section of the organization. At the individual level, the relative importance of different goals as well as strategies for solving problems or attaining goals may be influenced by these norms. At the team level, the nature of the relationships among team members, the exercise of leadership, and the process of team interaction may be influenced by cultural norms.

Aspects of climate such as the safety climate of an organization can influence both individual and team decisions related to performance. A climate emphasizing "Better safe than sorry" may have very different performance outcomes than a climate emphasizing "Nothing ventured, nothing gained". Additionally, attitudes toward organizational policies and procedures can increase or decrease compliance with these procedures.

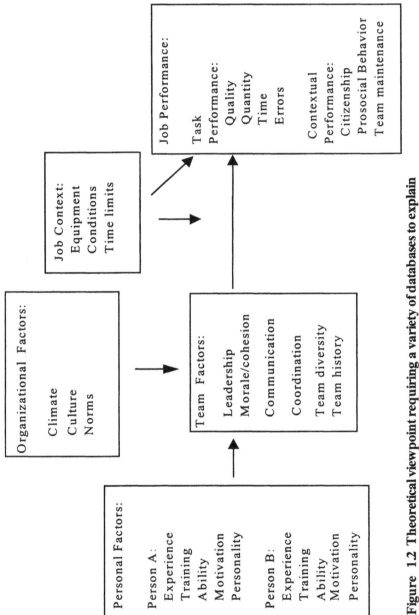

Figure 1.2 Theoretical viewpoint requiring a variety of databases to explain job performance

Attitudes toward specific sections of the organization can influence morale or motivation of people or teams in these sections.

Team factors   Since the modern workplace is increasingly team-oriented, team factors can play a critical role in job performance.   Effective leadership, morale, cohesion, communication, and coordination of activities could be relevant for a wide variety of teams.   Subsidiary factors such as team diversity may affect communication, coordination, creativity, or adaptability of the team.   Other factors such as the team history may play a role for only some types of teams which have well-established processes based on tradition or custom.

Job contexts will moderate additional subsets of relevant team factors. For example, a job context like crews piloting an aircraft has inherently high workload, strong time constraints, and potentially high costs for making errors. For such a job context, workload management may be a critical team factor.

Job context   There are at least two distinct levels of job context that could be relevant for performance.   At the general level, the overall conditions for performing job tasks include general working conditions, the quality of the tools or equipment for doing the job, time limits, and the like.   These situational factors can either directly impact on job performance or moderate the impact of team or personal factors on job performance.

At a more detailed level, the exact interaction of the person with tools or equipment in the workplace can become critical.   This is the domain of human factors psychology.   From the extensive research in human factors, aspects of the tools or equipment that either facilitate or inhibit performance can be selected for measurement.   For simple physical tools, the shape, location, and required motions are relevant.   For complex tools such as computers, the cognitive aspects of the task are relevant in addition to the physical aspects.   For complex tools used by a team, physical and cognitive aspects of the tool as well as social aspects of the team are all relevant to the task situation.

Job performance information   One focal point of a SIS devoted to job performance would be the job performance information. A global evaluation of job performance would be, for example, a yearly evaluation by a superior. A more detailed analysis of job performance may be based on components determined, for example, by a task analysis.   For any particular task, performance may be broken down further into specific performance criteria such as quality of work products, quantity of work products, time required for the task, or errors of omission or commission.

Contextual job performance as defined by Borman and Motowidlo (1993) includes work-related behavior such as social interaction and prosocial behavior, but distinguished from tasks directly related to formal job requirements. This component of performance can also be measured at either a global or a more detailed level. For the detailed level, the contextual tasks and expected levels of performance can be determined by adapting methods used for defining and measuring job task performance that are discussed in Chapter 9.

Information granularity

Information granularity is the level of detail of the information in an SIS. The measurements for data in any of the databases in a SIS may range from global to detailed, depending on the level of granularity required by the underlying theory and the required use of the SIS. Figure 1.3 presents different levels of granularity for information about "intelligence" in a personal database.

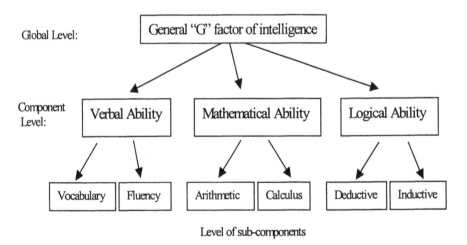

Figure 1.3 Possible levels of measurement for intelligence

The most global or coarse-grained measure of intelligence would be a single summary index such as an intelligence quotient or standardized score from a Stanford-Binet test or Wechsler Adult Intelligence Scale. At the next level of detail, different scores in the database could represent the distinct but related verbal, mathematical, and logical abilities as separate scores. At the finest level of detail, the database would contain a set of scores such as the six sub-component scores listed in Figure 1.3.

Clearly more levels exist than are depicted in this figure. A more global construct than "intelligence" would be overall capability to perform on the job. Such a general view of ability would subsume intelligence as measured by IQ tests plus social intelligence, past experience and skills, and other factors. Conversely, a more detailed level of analysis for intelligence may be necessary in which vocabulary, fluency, arithmetic, calculus, inductive logic, and deductive logic are broken down by different task domains or further subdivided into ability sub-components or specific skills.

This raises the issue of how to specify the granularity of measurement for each construct in a SIS. The granularity can be set a priori by the theory or by pragmatic reasons of cost, practicality or time limits on measurement. The theory may specify, for example, that verbal, mathematical, and physical ability have qualitatively distinct contributions to job performance, but that further subdivisions yield no more useful information. In this case, the theory would require measurement of the three ability components.

Conversely, if only a global intelligence was desired, the distinct contributions of the three abilities might not be distinguished, measured, or tracked in the SIS. Measuring the six ability sub-components would also be unnecessary from this viewpoint because they would give no more useful information.

On the other hand, the theory may specify that each individual sub component is a distinct ability with different contributions to job performance. In this case the sub components would be measured, tracked, and validated against job performance in the SIS. Therefore, theoretical considerations are one major determinant of the granularity of the information in the SIS.

Pragmatic constraints also play a role. Typically, the more fine-grained the level of measurement, the more time-consuming and costly the measurement process. Therefore, there may be a conflict between a theory that specifies a very fine grain of measurement and the practical limits of the measurement situation.

In aviation, for example, a scenario-based flight test is limited to about two to four hours. If the theory specifies over 50 teamwork skills and each skill takes 10 minutes to observe, clearly not all skills can be assessed during a single flight test. In practice, one airline coped with this challenge by turning to "topics" which represented clusters or aggregations of relevant skills. Assessing skilled performance at the topic level was successful, but some of the fine-grained analysis of separate skills was no longer possible. The acceptability of such a compromise depends on the ultimate goals and use of the SIS.

Where such compromise is necessary, the important thing is to recognize and define as carefully as possible the limits of the information in the SIS. Knowing the limits of information granularity for a construct allows the analyst to gauge how the lack of finer levels of measurement may impact on the quality of prediction and use of the other information in the SIS. For example, certain questions about the effects of the fine-grained variables simply cannot be answered, which is a qualitative loss of information.

In addition to the loss of qualitative information, the lack of finer levels of measurement may degrade the accuracy of prediction in the ultimate use of the SIS. This would be a quantitative loss of information. The qualitative and quantitative losses of information are costs that should be weighed against the cost of more fine-grained measurement. Depending on the information requirements for the SIS, this loss of accuracy may be acceptable or unacceptable. The information requirements of the users of a SIS also dictate the level of required completeness or thoroughness of the SIS information. Requirements for absolutely precise prediction, for example, may require a more complete theory and set of information in the database.

Sparse vs. complete theory

A "sparse" theory is a theory that uses a minimal number of constructs to provide a "good enough" account of a given phenomenon. A "complete" theory is a theory that uses all necessary constructs to account for as much of the reliable variance in a phenomenon as possible. Each form of theory represents distinct goals for explaining or predicting key constructs.

In Figure 1.4, the key construct is the number of widgets produced by a person operating a complex device such as a computer terminal under certain conditions. A sparse theory may involve just the predictors of the operator's speed, agility, verbal, mathematical and logical abilities. A complete theory, on the other hand, would attempt to account for all of the systematic differences in widget production by also measuring the more minor predictors of adequacy of the lighting, the quality of the equipment the worker is using, and the working conditions.

The information requirements of the SIS should specify the goal for explaining reliable variance in key variables. If the goal is to predict stock market prices, for example, a goal of explaining only 2 or 3% of the variability in movements of the Dow Jones average may be good enough to yield profitability when using the SIS. If the goal is to predict aircraft incidents or accidents, a goal of explaining or predicting 90% of such incidents might be far more desirable than only two or three percent.

Narrow vs. broad scope of theory

The information requirements of a SIS also dictate the initial scope of the theory, which may be narrow or broad. Initially, the scope of the theory may be narrow, while increasing information demands require a subsequent broader scope. For example, the initial construction of a SIS for job performance might focus narrowly on total widget production as described above. Later on, the scope may be broadened to include other relevant aspects of job performance such as widget quality, contextual variables of job performance such as maintaining team cohesion and morale, and important collateral variables such as training, job satisfaction, likelihood of attrition, and so forth.

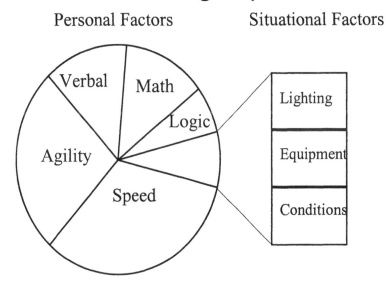

Figure 1.4 Factors affecting widget production

Narrow or broad scoped theories can be either sparse or complete. Table 1.1 gives an example of each kind of theory. Each cell in this table represents the amount of the target variable (i.e., the column variable) that could be predicted by a cause (i.e., the row variable). For this simple illustration, a broad scope theory has more target variables than a narrow scope theory (i.e., more columns). Including more contributory causes (i.e., more rows) gives a more complete version of either a broad or narrow scope theory.

Table 1.1 Illustration of sparse and complete, narrow and broad-scope theories

	Narrow-scope theory:	Broad-Scope theory targets the entire set of variables related to job performance: ⟵――――⟶			
Causal Variables:	Number of Widgets	Quality of Widgets	Job Satis-faction	Likely to Quit	Aids Team Morale
Speed	30%				
Agility	20%				
Verbal	**15%**	10%			
Math	**15%**	10%			
Logic	**11%**	10%			
Lighting	**3%**	10%	15%		
Equipment	**3%**	15%	15%		
Conditions	**3%**	15%	15%		
Motivation		**30%**	25%	30%	25%
Extroversion					50%
Education			25%	30%	25%

Predicting the number of widgets just using speed and agility would be a sparse theory of a narrow scope since these variables only predict 50% of the variability in widgets produced. Predicting the number of widgets using all personal factors and situational factors would require adding the six predictors with bold entries in the second column. This extended theory would still be narrowly focused on widget production, but it would be able to account for 100% of the variability in the number of widgets produced. In consequence, it would be a narrow scope but complete theory.

A broader theory of widget production could try to predict the quality of widgets produced, the job satisfaction of the employee, likelihood of quitting, and contextual performance such as aiding team morale. Broadening the scope of the theory may require the additional measurements of motivation, extroversion, and education level. These additional variables would be necessary to have any prediction of the likelihood of quitting or aiding team morale outcomes (bold entries in last two columns). Adding these variables would also obtain a more complete prediction of the quality of widget production and job satisfaction (bold entries in third and fourth columns). In general, the number of required variables will increase as a theory is developed to have broader scope or more complete prediction of each key variable. The completeness of the SIS information and theoretical structures is discussed in Chapter 3.

However, the granularity of the data also has a very strong effect on the number of measured variables. Fine-grained theories will typically require more measured variables. If the theory addresses a narrow scope of behavior but analyzes it at a fine grain, it may require more information than a theory addressing a broader scope of behavior at a much coarser grain. For example, detailed models of human cognition that specify activation of each separate memory element and the distinct steps of each cognitive process may require measurement of key variables across millisecond time intervals for complete validation. In contrast, a coarse-grained theory of job performance may require measurement of key variables on a yearly basis. The issue of granularity and integrating across levels of prediction is further discussed in Chapter 11.

Clearly scope and granularity are aspects of the theoretical view guiding the SIS that can change over time. In part, the change in scope or granularity may result from the natural refinement and elaboration of the theoretical view. Just as likely, however, is the possibility that the development of expanded information needs on the part of users drives an increasing scope or finer granularity of the theory and information in the SIS. For example, a job performance SIS initially focused on individual production may be forced to a broader scope by the conversion of a company to team-based production. Similarly, the requirement for more efficient and effective training may necessitate the development of a finer-grained account of the processes underlying human performance. For many potential applications, the expansion of information needs could be the major determinant of SIS growth. The growth and integration of both the theoretical and data components of a SIS is further discussed in Chapter 12.

Chapter summary

A Scientific Information System is an information system built on scientific principles and methods. Using these methods with the information capabilities of computer databases and scientific analysis methods, the SIS can evolve a comprehensive and accurate picture of the processes and products of the targeted system. The scope and granularity of the theoretical view and corresponding data represented in the SIS can be tailored to the information needs of the users, which may change over time. The scientific aspects of a SIS increase the confidence in the information about the system and ensure that the knowledge about the system will generally improve as the SIS is developed.

This approach is particularly important and useful when the targeted system is complex or has dynamic processes. Performance of complex jobs by teams in a dynamic environment is one example of such a situation. To obtain a complete picture of job performance in such a context would require this type of information system. The SIS approach offers a feasible method to examine the way multiple factors interact to produce job performance at the individual, team, and organizational levels. The SIS information can be used to select personnel (Chapter 9), design and evaluate training (Chapter 10) or alter the physical or social aspects of the workplace (Chapter 11). Discussion of different aspects of the construction, evaluation, and use of a SIS is the focus of this book.

2 Basic Data Quality

Overview

The basic data quality requirements for a Scientific Information System involve the source and type of data in each database plus the linkages among the databases that are required for effective use. Additionally, data security or special data handling may be required for legal or contractual reasons. Finally, the basic integrity of the data must be ensured in preparation for scientific analyses of data quality.

Data sources

Data for a SIS can come from many different sources ranging from questionnaires or scales requiring an overt response from a person to unobtrusive measures such as archival or trace measures (Webb, Campbell, Schwartz, Sechrest, & Grove 1981). Each source of measurement has a set of practical and scientific advantages and disadvantages associated with it. Practical advantages and disadvantages have to do with costs of measurement such as time, money, and the requirement of trained personnel. Scientific advantages and disadvantages have to do with the aspects of measurement that tend to increase scientific indexes of data quality such as sensitivity, reliability, and validity.

Generally, each measure is intended to produce data that reflect one or more *constructs*. A construct is the theoretically or practically meaningful target of measurement such as "intelligence". Different specific measures such as a Stanford-Binet or Wechsler IQ test can be used to measure the intelligence construct. Measurement methods may vary from very direct assessments of the construct to very indirect assessments and may vary in the measurement procedure.

To some extent, all measures rely on a theory of measurement to establish the relevance of the measure to the target construct, but the theoretical dependence ranges from very minimal to extensive. The more indirect the measure is, the more involved this theoretical link must be, and the stronger the measurement assumption. The evaluation of this theoretical link is part of establishing the validity of the overall SIS. Further, the measurement theory should be considered one component of the total theoretical viewpoint expressed by the SIS. If these measurement theories involve key variables in the SIS, they may be central and important parts of the total theoretical viewpoint. An example of such a key variable is the measurement of productivity in a job performance SIS. In this case, a failure in the validity of the measurement theory would impact the analyses of job performance with respect to many other variables and processes in the SIS. The measurement assumptions of very direct as well as very indirect key measures must be carefully assessed.

Very direct measures

Very direct measures include direct behavioral observations, direct trace measures and job sample tests. These measures are very closely tied to the construct they are supposed to reflect. The conversion of the response to the measurement requires little or no interpretation or processing of the measure, and minimal measurement assumptions beyond the direct relevance of the measure to the construct.

For example, if producing widgets is the job performance criterion, the direct observation and count of the number of widgets produced per hour by an employee is a very direct measure. The simple widget count requires no interpretation on the part of the observer and can be automated. The relevance of the number of widgets produced to job performance defined in this manner is clear and relies on little or no theory. Similarly, systems that track the number of calls answered by telephone operators are direct *trace* measures of this aspect of their job performance.

Direct observations or trace measures can also, however, have multiple indicators of a construct. For example, the performance of people with a new type of computer interface can be very directly indexed by the time and total number of key presses required to do typical tasks. Better performance would be indicated by less time and fewer key presses. These aspects of performance can be recorded as trace measures by a background computer program that captures and time-stamps each key press or mouse action by the users.

The same measure may be a very direct measure of one construct but only a moderately direct or even indirect measure of other constructs. Recording the direction of a person's gaze with an eye-tracking system, for example, gives a very direct measure of whether a person has looked at a given object or part of the visual field (Figure 2.1). This measure may only be a moderately direct measure of attention, however, since gaze may only imperfectly index attention. The person could have been attending to something else while their gaze wanders, for example. Conversely, the person could direct their gaze at something by mistake due to a momentary distraction, and immediately stop attending to the information. Finally, direction of gaze may be an indirect measure of understanding, mental workload or other underlying cognitive processes. Understanding a perceived object may require, for example, accessing long-term memory knowledge about the object. This process may occur for some attended objects but not others, and therefore be only partially and indirectly indexed by eye gaze.

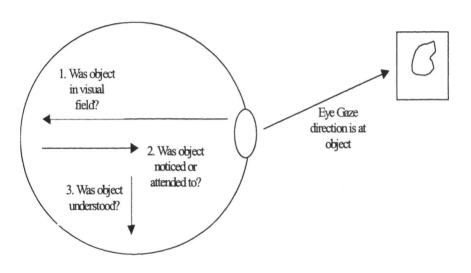

**Figure 2.1 Eye gaze direction as a very direct (1),
moderately direct (2), and indirect measure (3)**

Moderately direct measures

Interviews and objective tests like "paper-and-pencil" tests are examples of moderately direct measures. These measures typically use testing content that is fairly directly related to the theoretical construct but a testing context that is different from the actual context. The first assumption is the relevance of the measure to the construct. For a multi-item test, responses to all the test items should reflect the construct. For most of these measures, the relationship of the items to the measured construct should be plausible and objectively confirmable by some sort of analyses such as those discussed in the data quality section below. A further assumption is that the measured construct is both stable and broadly applicable or general.

Situational context generalization   One common assumption of these tests is that the entity measured is stable and generalizes from the measurement context to other contexts. Specifically, these measures typically assume that the measurement in the testing context generalizes to the actual system context.   For example, an intelligence test is generally given under controlled conditions in a separate testing context. The assumption is that this level of intelligence is also active in other contexts such as the work or job context. The basic assumption for these types of measures is that the underlying construct is context independent. Many paper-and-pencil tests of attitudes, abilities, and related personal attributes which make this assumption have been devised (e.g. Buros' Mental Measurements Yearbook, 1978).

    Such an assumption is not always justifiable. The generalization of even a basic construct such as intelligence across domains and contexts can be questioned.   Gardner (1983) proposed multiple distinct intelligences specialized for different domains: linguistic, musical, logical-mathematical, spatial, bodily-kinesthetic, and personal. Assessing intelligence in each domain would require appropriate corresponding items or stimuli. Another distinction is intelligence associated with appropriate pragmatic action in particular settings or situations, often labeled "practical intelligence".

    The extent of generality of measurement should be a theoretical part of the SIS, and the generalization should be empirically tested where possible. Generalizability theory (Cronbach, Gleser, Nanda, & Rajaratnam, 1972). includes specific methods for empirically evaluating generalization. If a construct is hypothesized to be context specific or data analysis shows that the measurement is influenced by different contexts, then the construct must be measured in an appropriate context.

Other generalization   In a similar manner, most current measurement techniques make the nomothetic assumptions that the construct exists to some extent in all people and that the same measurement items can be used to index the construct for all people. The measurement assumptions may be incorrect.   The alternative view is the idiographic view that people are qualitatively distinct and that particular constructs may not occur in all people. Kelly (1955), for example, strongly advocated such a position in his theory of personal constructs. He proposed qualitatively distinct cognitive structures in each person.

Idiographic measurement uses a different approach than nomothetic measurement. Idiographic measurement methods allow the person to choose relevant stimuli and structure the meaning of the responses to these stimuli. In Kelly's Role Construct Repertory Test, for example, the respondent designates examples of other persons with a specified relationship. The dimensions that characterize these people are elicited by comparing sets of people for similarities and dissimilarities (Table 2.1). Thus, in idiographic measurement each person may respond to a different set of relevant stimuli in qualitatively distinct ways.

Table 2.1 Example of idiographic measurement of personal constructs

Person A			Person B		
Similar set	**Contrast set**	**Personal Construct**	**Similar set**	**Contrast set**	**Personal Construct**
Father Brother Grandpa Uncle Bill	Mother Sister Grandma Aunt Bea	Male vs. Female gender	Sister Brother Best friend	Father Mother Grandpa Grandma	School vs. Home setting
Grandpa Grandma	Sister Brother	Young vs. Old age	Father Mother	Grandpa Grandma	Working vs. Retired
Mother Father	Sister Brother	Large vs. Small size	Mother Grandma	Sister Brother Best friend	Cook vs. Wash Dishes

In this table, for example, the sets of constructs distinguishing family members for person A and B are qualitatively distinct. Person A considers the nuclear family as the essential set of persons and uses gender, age, and size to discriminate among them. Person B includes the best friend in the essential set of persons and uses school, work, and household duties to discriminate among them. Both the relevant stimuli and the underlying constructs are qualitatively distinct.

One difficulty with idiographic measures is obtaining comparable scores across a set of people. Despite the differences in the detailed measurement content across people, certain higher-order properties of the entire set of responses such as the cohesiveness, complexity, or structure may be comparable. In some cases, analytic methods such as multi-dimensional scaling (Torgerson, 1958) or Pathfinder (Schvaneveldt, Durso, & Dearholt, 1989) can give comparable higher-order scores.

It is important in the documentation of a SIS to clarify the measurement assumptions for at least the measures of the key variables in the system. This is particularly important as the assumptions of the measurement theory become more involved for more indirect measures.

Indirect measures

Indirect measures require a strong theoretical bridge from the basic observation or datum to the final measurement or variable. These measures typically assume both a generalization from the test context to the real system context, and the validity of a theory that gives meaning to the responses. For indirect measures, the link between the responses of the person and the measured construct may not be obvious and may be difficult or impossible to objectively confirm. For example, such measures as projective tests require strong, theoretically guided scoring keys to derive the measure of the construct from the responses of the person. The quality of the measurement depends both on the quality of the individual's initial responses and the quality of the conversion of those responses to the scores stored in the SIS.

Projective tests Most projective tests are indirect and rely on the validity of the underlying theory to justify the measurement. In the Thematic Apperception Test (TAT), for example, the person must tell stories about ambiguous pictures of scenes. The content of the story can be classified for themes reflecting underlying motivations such as Need for Achievement, Need for Affiliation, and so forth (e.g. McClelland, 1980). The final scores for these motivations are very indirectly related to the basic task of telling a story. As a result, the scientific quality of those scores depends on the validity of the theory upon which the TAT is based as well as on the accuracy of the transcription and interpretation of the contents of the stories. Unlike moderately direct measures, the validity of the scoring key for the TAT cannot be clearly established by simple confirmatory analyses. Therefore, the validity of the resulting score strongly depends on the assumed validity of the underlying theory.

Other indirect measures   Examples of indirect measures that are not projective tests would include measures of implicit memory such as free association to words that are parts of sets of previously trained pairs of words. In this case, the implicit memory for the trained word associations is indexed by the number of trained word pairs that are verbalized during the free association recall. This score is justified by the underlying theory that specifies free recall as one manifestation of implicit memory. Similarly, the use of eye-gaze tracking information to index importance or preference for various stimuli relies on the theory that connects gaze to these constructs. If these measures index key constructs in the SIS, the measurement theories are important components of the SIS theoretical viewpoint.

If, however, the measurement theories concern minor variables in the SIS, they may be peripheral parts of the theoretical viewpoint. For example, in a job performance SIS, cohabiting status may be expected to play a minor role in determining job performance by influencing the amount of work-family conflict. The measurement theory may assume that self-reports of cohabiting status are veridical. However, suppose some self-reports of cohabitation would be inaccurate due to potential legal complications or personal reasons. In this case, measurement problems would affect analyses of the effects of cohabitation, but would not greatly alter the overall validity and utility of the SIS.

Multi-component measures

Multi-component measures are assessment situations that mix direct and indirect measures. An example is assessments of the performance of pilot crews flying a test scenario in a simulator. Some performance indexes may be very directly indexed by the occurrence of aircraft stalls, tail strikes, landing bounces, or other physical motions or conditions of the simulated aircraft. Other performance indexes may be moderately direct assessments of performance by trained and calibrated evaluators. Other moderately direct assessments could include an oral examination of aircraft systems knowledge, knowledge of relevant FAA rules and regulations, and so forth.

An example of an indirect measure might be post-flight recall of relevant information. For example, an indirect measure of a construct such as "situation awareness" might be a post-flight free recall of all Air Traffic Control (ATC) communications during approach and landing.

For multi-component assessments it is important to carefully examine the assumptions underlying each distinct facet of assessment. Direct measures should be examined to ensure that they are in fact relevant to the construct and appropriate for the inference purposes of the SIS. For more indirect measures, the measurement theory should be clarified and examined for plausibility, empirical support, and compatibility with the other theories involved in the SIS. The appropriate empirical analyses for any of these purposes depend, however, on the basic type of data that measures the construct.

Data types

Answering a question scientifically requires relevant data combined with an appropriate data analysis (Figure 2.2). There are several different basic types of data that can be derived from measurements. These types of data vary in the amount as well as the type of information they may contain to answer questions. The type of data is also important in determining the sensible and appropriate use of scientific analysis methods for the data. Ultimately, the net gain in information from an analysis depends on correctly matching the question or information goal with an appropriate type of data and appropriate scientific analysis. The connecting arrows in Figure 2.2 illustrate these interdependencies. To develop appropriate data for analysis it is important to be aware of the meaning and limitations of fundamentally different types of data.

S.S. Stevens (1951,1959) described data as being nominal or categorical, ordinal, interval, or ratio. Although other types of data have been defined for specific measurement situations, these four basic types of data cover the goal of measurement for most common scientific measures and will be used to illustrate scientific data issues in a SIS. Based on Stevens' work, the differences among these types of data are summarized in Table 2.2.

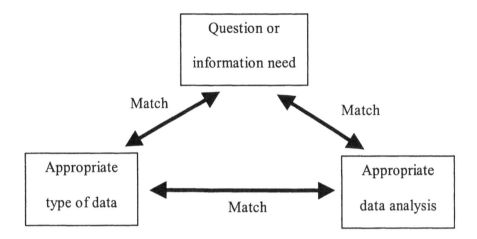

Figure 2.2 Matching information needs to appropriate data and analyses

Nominal or categorical data

Nominal or categorical data includes any form of data for which the basic information is membership in a category. For most analyses, these categories should be mutually exclusive and exhaustive. An example is categories of religious membership. Protestant, Roman Catholic, Jewish, and Other may be four mutually exclusive and exhaustive religious categories. Membership in a category can be symbolized by unique characters (Group P, Group C, Group J, Group O) or numbers (Group 1, Group 2, Group 3, Group 4). Since the only meaningful aspect of these codes is the designated category, any 1:1 transformation to a different set of unique identifiers preserves this information and is legitimate. That is, any four distinct codes could be used to replace the original group codes and accurately represent these four religious categories.

The use of numbers for category membership can be very misleading if the user mistakenly uses the ordinal or interval qualities of the numbers when interpreting category membership. In this example, Catholics are not "greater than" Protestants and Jews are not "greater than" Catholics. Further, despite the implication of the numerical codes, the difference between Catholics and Protestants ($2 - 1 = 1$) is *not* really the same difference as the difference between Jews and Catholics ($3 - 2 = 1$). Any statistical method that uses the ordinal or interval properties of these numbers will give very misleading results and must be avoided.

Table 2.2 Comparison of four types of data

Data Type:	Nominal	Ordinal	Interval	Ratio
Definition	Set of unordered exclusive and exhaustive categories	Set of rank-ordered scores on a continuum	Scores on a continuum with meaningful intervals	Interval scale scores on a continuum with a natural zero point
Amount of Information	Category membership only	Relative place on a continuum	Relative place & relative distance on a continuum	Relative place, relative distance & distance from natural zero
Allowable transformations	1:1 mapping to new set of labels (isomorphic)	Any monotonic transformation $Y = f_{monotone}(x)$	Any linear transformation $Y = aX + b$	Multiply or divide by a constant $Y = aX$
Measure of central tendency	Mode	Median	Mean	Mean
Types of statistics	Non-parametric	Non-parametric	Parametric	Parametric

Ordinal or ranked data

Ordinal or rank data presumes a single underlying dimension of measurement. An example is the relative hardness of rocks. Relative hardness between each pair of rocks can be ascertained by finding which rock scratches the others. The scratched rock is softer and given a lower ranking number. For example, diamond is harder than quartz which is harder than granite which is harder than sandstone. These ranks can be represented by the numbers 4, 3, 2, 1, but any four numbers that expresses these ordinal differences preserves this information and is a legitimate recoding. That is, any other set of numbers that correctly expresses the ordinal differences (e.g. 23, 17, 13, 11) can be used to represent these four ranks. Ordinal qualities of the original data are preserved under any monotonic transformation on the original set of rank numbers. Transformations that do not preserve the rank order are not allowed.

Although the existence of a difference in hardness is represented by these numbers, the size of the difference is not. That is, the 4, 3, 2, 1 coding scheme would imply that difference in hardness between diamond and quartz is the same difference as between granite and sandstone. That implication may be quite incorrect. For ordinal data, any method that uses the interval properties of the numbers could give very misleading results and should be avoided.

Interval data

Interval data also presumes a single underlying dimension of measurement. In this case, however, not only the difference between measured levels but also the relative size of those differences is reflected in the scale numbers. An example is the measurement of temperature. A temperature of 100 degrees is hotter than 80 degrees which in turn is hotter than 60 degrees. However, measured temperatures also give information on the relative magnitude of the differences. That is, the difference between 100 degrees and 80 degrees is the same difference as between 80 degrees and 60 degrees and twice as big a difference as the difference between 50 degrees and 40 degrees.

Legitimate transformations for interval data include any linear transformation. That is, interval scale data can be multiplied by a constant or have a constant added and still preserve the interval qualities. For example, Celsius temperatures can be converted into Fahrenheit temperatures by the formula: $F = 32 + C * (180/100)$. The interval properties are preserved because the transformation is only multiplying by a constant and adding a constant. Any other form of transformation will change the interval properties of this type of data.

Most quantitative techniques that use procedures like taking an average are implicitly assuming the interval qualities of the data. Interval data allows a wide variety of univariate and multivariate statistical techniques to be used for gaining information from a SIS. For many techniques, the information in an interval scale measure is summarized as variances around the average and covariances with other variables. The average value itself may not, however, be meaningful for interval data as arbitrary constants can be added to the scores. Many psychological measures strive to be interval quality or close enough that no serious distortions occur in the data analysis.

Ratio data

Ratio data is interval data with a natural zero point for the scale. For temperature, the Kelvin and Rankine temperature scales are ratio scales since they are based on a true absolute zero temperature, while Fahrenheit and Celsius scales are not. The average of a set of ratio scale scores is an inherently meaningful number that can be compared to the zero point of the scale. All statistical techniques that are appropriate for interval data can be used on ratio data.

The only appropriate transformation for ratio data is multiplying by a constant. For example, Kelvin temperatures can be converted to Rankine temperatures by multiplying by the constant 180/100 or 9/5. Adding a constant would change the zero point and is NOT allowed. That is, adding or subtracting a constant from absolute measures of temperature distorts the natural zero point and destroys the ratio qualities of the scale.

The presence of the natural zero point and meaningful averages makes certain additional tests of results meaningful. For example, the mean temperature of a set of measures can be compared to see if it is above absolute zero. In certain other types of multivariate analyses, the use of sums of squares and cross products can be a meaningful variation on the usual analysis of variances and covariances.

Impact of data types on scientific results

Comparing the types of data, there is more information contained in a score as the type of data changes from nominal to ordinal to interval to ratio data (see Table 2.1). Since many sophisticated statistical techniques are parametric and are only sensibly used on interval data, there is a strong tendency to assume interval data without carefully considering the quality of the measurement. The most common error in human judgment or ratings is to presume that ordinal data are really interval in quality or that scale numbers represent equal intervals. For example, judgments on a 1-5 scale such as a Likert scale (Likert, 1932) may be presumed to be equal-interval judgments as represented by the numbers 1, 2, 3, 4, and 5 (Table 2.3). If the data are "close" to being interval quality, this assumption is not that wrong and the results of the analyses will not be too distorted (see Cliff, 1993 for more details). The Likert scale anchors are typically assumed to be evenly spaced and to represent an interval response scale.

Table 2.3 Five-point scales with different anchors

Likert Scale anchors: (equal interval)	Coding scheme	Frequency Scale anchors: (not equal interval)
Strongly Agree	1	Never (0%)
Agree	2	Occasionally (10%)
Neutral	3	Sometimes (30%)
Disagree	4	Frequently (70%)
Strongly Disagree	5	Always (100%)

However, if the data are really far from the specified intervals, derived information could be extremely distorted. The right-hand column of the Table 2.3 gives a possible set of response anchors for a frequency scale. Assuming the listed percentages correctly reflect the verbal labels, this set of judgment anchors does NOT represent an interval scale. The interval from "Never" to "Occasionally" is 10% while the difference from "Sometimes" to "Frequently" is 40%. If the equal interval 1-5 coding scheme were used for these responses, the resulting analyses could be distorted.

Coding of "don't know" or non-responses must also be carefully considered. If, for example, on a Likert scale the "don't know" response is coded as a "0", this code would imply that the lack of response was more extreme agreement than "Strongly Agree". Since that is probably not correct, a better choice for coding "don't know" responses could be the "3" or Neutral coding. This would be tacitly assuming that a person's "don't know" response is equivalent to a neutral response. Alternatively, the "don't know" response could be treated as missing data that is discussed in a subsequent section. These types of decisions should be recorded in a data codebook.

The equal-interval assumption cannot be made for all scales. Scales that have not been scientifically evaluated should be examined for the quality of measurement as part of their evaluation in the SIS. In particular, new scales should be checked for interval quality during their initial development or use in a SIS. It is important in the SIS to specify the presumed or confirmed data type for each measure so that users of the information will not inadvertently use inappropriate analyses or derive incorrect conclusions. This information should be kept in a data codebook that summarizes the properties of the data at a detailed level for each type of measure.

Data codebook The data codebook is a coherent reference for the type of data, allowable scores, and meaning of each measure in a set of related measures. For example, the recurrent yearly evaluation of air transport pilots may involve knowledge exams about FAA regulations and aircraft systems, maneuver exams for assessing physical control and maneuvering of the aircraft, and a scenario-based exam of the pilot's performance in a crew for a typical flight. The data codebook for each exam would detail each measurement item, the type of scoring scale and allowable scores, and meaning of possible scores for the item.

This information will help authorized users make appropriate use of the SIS and should be stored with the data either physically or electronically. As the measures in the SIS are changed or revised, the data codebook must be consistently revised. The user of older versions of measures in the SIS must be able to consult appropriate archival versions of the data codebooks for correct use of old as well as current data.

Data storage, security, and linkage

In the design of the SIS, the issues of data storage, data security, and data linkage must be resolved so that the informational goals of the SIS can be achieved with minimal time, cost, and effort. Data storage must be efficient, error-resistant, and flexible. Data security must be sufficient to safeguard the interests and allay the concerns of different stakeholder groups such as management, unions, regulatory agencies, and so forth concerning the unauthorized access to the SIS data. Data linkage must be sufficient to connect the data required for both the initial set of planned analyses and for future developments in information needs for the SIS. Successfully addressing storage, security, and linkage issues will facilitate the maintenance of basic data integrity, which is essential to the practical and scientific value of a SIS.

Data storage

Currently, the most common way of storing data is in *reduced normal form* in a relational database (e.g. Ullman, 1982). This method is efficient and is typically designed to represent natural aggregates or clusters of data. A relation is simply a set of related pieces of information. A relation is often visually represented as rows in a table where the column headers represent the data fields. Each qualitatively distinct set of information that is organized on a different basis should be represented as a different relation. Each separate relation is typically depicted as a different table. Figure 2.3 represents a set of interlinked tables.

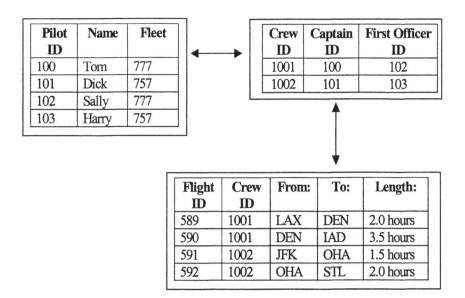

Pilot ID	Name	Fleet
100	Tom	777
101	Dick	757
102	Sally	777
103	Harry	757

Crew ID	Captain ID	First Officer ID
1001	100	102
1002	101	103

Flight ID	Crew ID	From:	To:	Length:
589	1001	LAX	DEN	2.0 hours
590	1001	DEN	IAD	3.5 hours
591	1002	JFK	OHA	1.5 hours
592	1002	OHA	STL	2.0 hours

Figure 2.3 Tables representing a linked relational database

Each table in this aviation example has a different basis. The basis of the first table is the individual pilot; the basis of the second table is the crew; the basis of the third table is the flight. Each table has the information that is naturally associated with its basis. Each table also has unique ID numbers or keys that allow the information to be combined across tables. This would be necessary, for example, to keep track of the number of hours that each pilot has flown in a month. The hours for each flight can be associated back to a crew and in turn associated back to each individual pilot to keep track of his or her total flight hours for the month. This total can be compared to the legal maximum number of hours to ensure that the pilot does not exceed that limit.

Clearly the SIS may have a great deal of interlinked information. Information in particular fields or tables may be confidential, sensitive, or subject to limitations in access or distribution. In general the data security for the SIS information must be ensured.

Data security

If SIS information includes identifiable personal information, the privacy of the individuals must be considered in constructing the SIS and authorizing use of the information in the SIS. Furthermore, SIS information may have to be protected from unauthorized access for legal or contractual reasons. For all of these reasons, safeguards may be implemented in either the construction or use of the SIS. In construction of the databases, for example, individuals may be identified by a Personal Identification Number (PIN) rather than by names or social security numbers.

PIN numbers Different methods can be used to establish PIN numbers, and each method has its advantages and disadvantages. An informal system in which each person picks his or her own PIN has the advantage that if the PIN number is reliably recallable it is not necessary to keep a master list of names and PINs. The lack of a master list may help people feel their data is confidential. They may also feel more in control of their data since they may use a different PIN at any time. The key feature of the informal system is that the PIN number be unique and easily recallable. In our aviation work at one carrier, the pilots were advised to pick a PIN number representing their mother's birthday or some other easily recallable date. The disadvantage of this informal system is that users forget their PINs and that the naturally-selected PINs may occasionally be duplicated. The former problem can be severe. In our sample, over half the pilots had difficulty remembering their PINs over a period of three years. The latter problem is more minor; in our sample of 600 pilots, only one or two pairs of pilots chose the same 6-digit date for their PINs. A PIN composed of more digits should alleviate the latter problem, but may exacerbate the former problem.

Institutionally-determined PINs can eliminate both the PIN recall problem and the PIN duplication problem. A further advantage is that the algorithm generating the PIN numbers can generate quasi-random numbers that are totally unrelated to any background information of the person. This helps prevent any "guessing" of the PIN on the part of a potential hacker or unauthorized user and gives added protection to the privacy of the personal information.

However, PINs determined by an institution have the disadvantage that somewhere in the institution a master list of names and associated PINs will exist. This list is required if the institution wants to be able to remind users of their correct PINs, and will almost certainly exist in some archival form somewhere in the institutional records. In a practical, day-to-day sense the data in the SIS may be de-identified, but if pressed in a legal or regulatory proceeding the data for a person could conceivably be traced. This means that the data are not absolutely, positively, and permanently de-identified. Therefore, a level of trust between the institution and the persons who give information for the SIS must be established for the institutional assignment of PINs to be effective.

Establishing this trust may be facilitated by detailing the measures taken for securing the use of the SIS for authorized users and purposes only. One aviation organization, for example, established institutional PINs with a firewall such that only one person in the entire organization was privy to the master list linking names and PINs. This person necessarily had the jobs of creating new PINs for newly-hired pilots and of reminding the pilots of their PINs if they forgot them. Trust had to be established, however, in the fact that no one else would have access to the master list and that the person in this job would not make unauthorized disclosures of the PINs. Similarly, the relevant stakeholders must be convinced that the SIS is a safe and secure repository for information. This is critically important if information that is private or sensitive is stored in the SIS, which will often be the case.

General SIS security To establish security in the use of the SIS, several methods can be employed. First, the authorized users may be strictly limited and regulated. Second, physical access to the system may be limited and monitored. Third, guidelines may be established for proper use of the data and enforced by legal or contractual measures. Fourth, actual system use may be monitored in real time or by trace measures such as activity logs to ensure that unauthorized use is not occurring. Striking a proper balance between security and usability of the SIS may be difficult. Low levels of security may result in unauthorized use and potential legal or contractual difficulties. These difficulties may be very costly if, for example, the information in the SIS could expose the organization to legal liability claims, precipitate a union strike, or endanger the business basis of the company.

On the other hand, high levels of security may increase the time, effort, and overall cost of obtaining information from the SIS. This can induce potential users to simply do without the information or obtain lower-cost information elsewhere if feasible. The costs to the organization of decisions or actions based on inadequate knowledge must be weighed against the costs of unauthorized use to determine an appropriate level of security. Once the appropriate level of security is set, the system can be designed to achieve that level of security. A combination of appropriate software methods ranging from passwords to data encryption and appropriate hardware methods can be used to limit physical access to the system and/or use of the data in the system.

One part of data security that should be considered is restricting the copying or dissemination of the information in either physical or electronic form. Such simple acts as copying a file to a disk, for example, make it difficult or impossible to control the dissemination of the data. This becomes a particularly critical issue if legal or contractual requirements require information to be universally removed or expunged from the database. For example, if a person has background information with a criminal conviction that is later overturned on appeal, the court may order that all records of the original conviction be expunged. If the data file has been copied, it may be impossible to guarantee that ALL copies of the database have been correctly expunged, resulting in grave legal difficulties. Data security issues must be carefully considered in the initial design as well as later use of a SIS. The implemented security measures must not, of course, interfere with the necessary links among the data sets that allow the data to be selected and reformatted for all the required SIS analyses. For example, encryption techniques must not interfere with the retrieval and connection of information from different databases or different tables within each database.

Data linkage

To be useful for systematic analyses, the information in many different tables will have to be linked. As an example, the performance SIS example in Chapter 1 has been recast in Figure 2.4 to illustrate the linked set of relational tables that could characterize an aviation performance SIS.

Using this example, the Crew Factors relational table should contain all the information relating to the crew and have the necessary links to appropriately connect this information to personal factors, organizational factors, job context factors, and job performance results. These links will usually be expressed as identification numbers (IDs) or key numbers that designate unique rows in the related database.

In this example, the Crew ID uniquely identifies each row or relation in the Crew Factors table (see Table 2.4). Fleet ID is the key that can be used to access the organizational factors such as climate and culture for the specific organizational division, which in this case is the fleet. Job ID is a key describing the distinct nature of the operational conditions for the team's job performance. In this example, suppose fleet aircraft are being used either in shuttle operations from DC to NY or in regular hub-to-hub operations from DC to Chicago. These different flying jobs are identified by different Job IDs and have a set of related characteristics. Shuttle operations may have shorter flights, stricter time limits, reduced meal service, and a reduced number of flight attendants compared to regular hub-to-hub flights. Similarly, the person information under Captain and Copilot are ID numbers (possibly PINs) that point to appropriate rows in the Personal Factors table.

The crew data on Morale, Communication, Coordination, and History are the data linked to a particular crew. These data may come from different measures, but they are all targeted at the crew as the focal point rather than individuals, jobs, or tasks and are therefore represented in this table. Crew morale, communication, and coordination may be judged by expert evaluators, and crew history may be the simple archival record of how long or how many flights these crewmembers have flown together. Reduced normal form puts data together that naturally fit together in a relation. Notice in Table 2.4 that the same person, Person F in this case, can be a member of two different crews or teams. To avoid confusion, it is critical to keep the related measurements at the team level distinct from related measurements at the individual level.

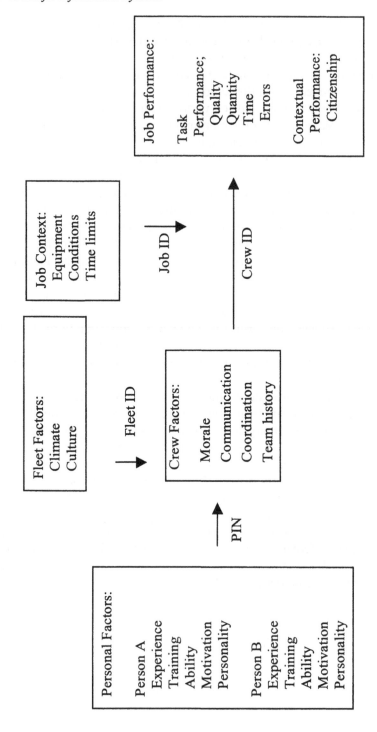

Figure 2.4   Links among relational tables in an aviation job performance SIS

Table 2.4 Table representing the relation for the Crew Factors part of an aviation SIS

Crew ID	Fleet ID	Job ID	Capt	F.O.	Commun-ication	Coordin-ation	Long-evity
001	Boeing 737	DC-NY shuttle	A	B	[team 1 score]	[team 1 score]	1 week
002	Boeing 737	DC-Chicago	C	D	[team 2 score]	[team 2 score]	3 days
003	Boeing 747	DC-Los Angeles	E	F	[team 3 score]	[team 3 score]	1 month
004	Boeing 747	DC-Frankfurt	G	F	[team 4 score]	[team 4 score]	2 months

Reduced normal form also has the important advantage that it guarantees a unique piece of data or datum is represented only once in the database. In the above example, the team communication and coordination scores are stored just once for this crew in the team database. Multiple representations of the same datum make modification difficult and error prone because the user must locate and change every representation of the datum.

For example, if crew longevity scores were stored with each pilot record rather than with the crew record, changes would require that the analyst track down each pilot's record and change the longevity score. If one representation would be missed in this change, the data would become inconsistent and data integrity would be endangered (discussed below). Representing each piece of data only once in the database allows any change to be made only once and has the desirable consequence that this modified value is correctly represented in any possible use of the database.

At least two types of data linkage are important for a SIS: 1) linkage among different types of data and 2) linkage of the same type of data across time. Linkage among different types of data is required for uniquely identifying and appropriately connecting and combining the data required for informational analyses. Unique keys or identifying numbers in the relational database usually provide this type of linkage as discussed above. The keys are essential for converting the reduced normal form of the database to the forms used in the information analyses.

Linkage of the same type of data across time is important for cumulative and longitudinal analyses. For example, the cumulative number of pilots trained or flights flown during the specific time interval may be important aspects of the analysis. In addition, linkage of data across time allows precise analyses of key system variables such as trends in performance over time. Similarly, any other important information that may dynamically change over time such as fleet culture or crew morale can be systematically tracked and analyzed with the SIS data.

Statistical normal form   One particularly important form of SIS data is a tabular form expected by several statistical packages in which the rows represent the cases or units to be analyzed and the columns represent the data variables. For some analyses, this "statistical normal form" may be the same as the reduced normal form of a database table, but for many if not most analyses this form will be quite different. In particular where an analysis requires the melding of information from different reduced normal form tables, the statistical normal form will be created from these source tables by a suitable query and will not be a direct representation of any source table.

For example, a statistical analysis of the extent to which crew factors influence job performance could require the synthesis of several data tables in the SIS. Suppose the question was how much the age and experience of the crew would add to the observed communication and coordination of the crew in producing superior performance. This information would be assembled by a query from the personal, team, and job performance data tables, and might look like the example in Table 2.5. In this example, the Personal Factors data table would be accessed using pilot IDs or PINs to determine the age and experience of each pilot. The Crew data table would be accessed to find the communication and coordination scores for the crew. The Job Performance database would be accessed to obtain an overall evaluation of crew performance in a standard evaluation. This information would be assembled into a table like Table 2.5 for analysis within the database or exported to a statistical package for analysis.

Linkage over time   For any SIS that must handle system development or changes over time, data should also be linked and identified over time as well as interlinked among the basic data tables. Data linkage over time would be necessary for any SIS that has a goal of tracking the state of a system over time or modeling a process of change. In most complex, human-centered systems, dynamic change may play a key role in the explanation, prediction and alteration of many aspects of the system. For some SISs, in fact, the scientific understanding of change may be the central focus of the entire enterprise. For example, a SIS may be designed to evaluate the effects of changes in training or personnel qualifications on the subsequent performance of the target system. In such a case, the appropriate linkage of the information over time is critical.

Table 2.5 Statistical Normal Form of information for examining the effect of crew factors on job performance

Crew ID	Capt. Age	F.O. Age	Capt. Exp.	F.O. Exp.	Commun- ication	Coordin- ation	Job Perf.
001	50	30	20,000 hours	4,000 hours	Crew 001 score	Crew 001 score	Perf. 001
002	54	44	25,000 hours	10,000 hours	Crew 002 score]	Crew 002 score	Perf. 002
003	53	52	22,000 hours	15,000 hours	Crew 003 score]	Crew 003 score	Perf. 003
004	58	38	28,000 hours	7,000 hours	Crew 004 score	Crew 004 score	Perf. 004

If appropriate links over time cannot be established, the ability to precisely analyze changes in the system over time will be severely compromised. Aggregate cross-sectional analyses that give a "snap shot" of system functioning at different times can still be created, but the precise carry-over of the values and processes in the system across time cannot be ascertained. Lacking this kind of precise information would preclude the SIS user from some of the more powerful forms of dynamic modeling of system processes discussed in later chapters.

Appropriate linkages over time for persons would necessarily require stable PIN numbers or the equivalent as discussed above. In addition, however, other aspects of the system that are uniquely identifiable but may change over time may require unique identification numbers. For example, the contents of a training curriculum will typically change over time as it is revised. In such a case, each training curriculum should be identified with a unique version number in the SIS so that the distinct effects of different versions of the training on individual or team performance can be tracked over time. The information about the structure and content of the data in a

SIS is critical to efficient and correct use of the data. For this reason, this information should be collected in a coherent document called the metadata guide that explains the general data structure and content for the SIS.

Metadata guide

The metadata guide is a roadmap to the information contained in the SIS. The concept of metadata for a SIS is an augmented version of metadata information for a data warehouse (Barquin & Edelstein, 1997). Metadata information is required in some data warehouse applications due to the complexity of information in those systems. Similarly, metadata is necessary in a SIS to guide the user through the overall meaning, structure, and general content of the sets of measures included in the SIS.

A data warehouse is typically constructed from available, convenient information sources and does not have specific scientific or informational goals. Unlike a data warehouse, a SIS is constructed from carefully constructed and scientifically evaluated measures, and is designed to satisfy both informational and scientific goals. Therefore, the metadata guide for a SIS must also convey the purpose and scientific quality of the measures as well as basic data structure and content. To do this efficiently, the SIS metadata guide is connected with the data codebooks discussed earlier.

System overview The first component of a SIS metadata guide is an overview of the system being described and analyzed in the SIS. In a system focused on training and performance, for example, the metadata guide should contain a map, schedule, or layout of the basic phases or events in both the training and evaluation process. This map must effectively communicate the sequencing, structure, and dependencies of the set of training and evaluation events as well as the basic meaning of the measures for each event.

The system map should not, however, be limited only to events that currently result in data, but rather be a general overview of the big picture. In aviation, for example, there may be training events or types of exams such as oral exams that are part of the sequence but do not currently result in SIS data. The overview is valuable for the user to see what is possible as well as what is currently measured or assessed as part of the SIS. One way of communicating this big picture is by a formalized graph such as a flow diagram, event calendar, PERT chart, or other device that depicts the objects, events, and processes of the system described by the SIS.

This graph should help the user clarify any dependencies among the events in the system, particularly the dependencies that influence data measured in the SIS. In either training or evaluation contexts, for example, there is often a sequential dependency from exam results to future testing or training. Failing an exam may lead to retraining and reevaluation, while passing an exam may lead to more advanced training.   Similarly, a promotion or change in job can trigger training or changes in assigned tasks, appropriate performance measures, and the like. Sequential dependencies as well as any other important conditional information on the system should be clarified by appropriate figures or graphs and the accompanying text.

Database map The second part of a metadata guide is a map of the databases and links among them. This part of the guide should describe the basic tables that contain all the relevant SIS data, and the content and basic structure of each dataset should be clarified. Additionally, the form of the data structure in the database should be depicted verbally and graphically.

   For a relational database, the form depicted might be the reduced normal form of the database that shows the linked structure of the data tables in the system. This part of the metadata guide gives an overview of the data obtained from each of these events, a general, high-level explanation of the meaning of the set of measures producing the data, and references to the data codebooks that give the detailed meaning of each specific item or measure.

Relation of system map to database map The final part of the metadata guide describes the correspondence of the system map to the database map. This correspondence must be sufficiently clear so that the user can work in both directions. That is, the user must be able to clearly and unambiguously identify how a given aspect of the system is measured by one or more databases and tables in these databases. Conversely, if the user starts from the map of the databases (or representative codebooks), he or she must be able to determine clearly and unambiguously which system events or processes in the system map produce that data. The metadata guide should also refer to documents or appendices describing the procedures for ensuring basic data integrity and scientific data quality.

Data integrity

Data integrity is defined as the completeness and correctness of the values for each variable in the data tables. Sometimes these values will be text or alphanumeric strings that represent variables. For example, "comments" variables typically have text entries.   Since scientific explanations and methods tend to emphasize quantitative data, the large majority of the values in a SIS may be numbers. In this case, data integrity is ensuring the completeness and correctness of the numbers entered for measured variables.

Data integrity can either be ensured prior to and during data entry or ascertained by preliminary data analyses after data entry. In general the up front protection of data integrity by security measures is preferable to attempts to fix data integrity errors after entry. After-entry correction leaves inaccurate data in the SIS for at least some period of time, which complicates systematic analysis and use of the data. Further, sometimes it may not even be possible to correct the data error, as in cases where the old measurement tool is no longer available, the evaluation context cannot be recreated, etc. Therefore, both *a priori* and *a posteriori* methods for ensuring data integrity should be considered for a SIS.

Data completeness

Completeness of a data record is essential to avoid problems with missing data. For cases where variant records are recorded containing different information depending on some variable, completeness still requires that the variant records be complete. For example, in recording personal biographical data the initial determination of a person being unmarried precludes asking any questions about the age, job, and income level of a spouse. This creates a variant record in which the data fields for spouse's age, job and income are either not present or left blank, depending on the design of the database. Even in this case, however, all the other fields for biographical information about the person must be complete. The type of missing data caused by the design of variant records is systematic and absolutely predictable, so checking for completeness can still be objectively done up front or after data collection.

Up front checks for data completeness involve checking each field for a legitimate entry. These checks for completeness should be designed into the system. A check can be as simple as having the data-collection person scan each form for completeness as it is submitted and immediately return the incomplete forms to the evaluator for completion. If data entry is directly through computer-based entry forms that feed into the databases, this type of check can be automated. In this case, the entry program would check all fields for acceptable values before accepting the values and transmitting the set of data to the database. This type of editing program for data entry can be extended to cover the other integrity checks discussed below.

Completeness checks after data entry require a person or, more typically, a program to inspect the fields of the database tables and report cases with missing data. This leaves the problem of how to fix the missing data. A person can use his or her expert knowledge to provide best-guess values for the missing data. In entering questionnaire data, for example, the person entering the data may be able to use written comments of the respondent on the questionnaire to complete missing data. However, these estimates will be subject to all the possible errors of human judgment, so the trade-off of more complete but possibly less accurate data must be carefully considered.

Fixing incomplete data   If the measured attribute is not expected to change, the best solution may be to simply re-measure and record the missing data. For example, if an ability measure is missing for a person and the ability is assumed to be stable, simply re-measuring the missing data by re-administering the ability test is justified.   The usefulness of this method depends on the presumed or demonstrated stability of the underlying construct and the time elapsed since the original measurement attempt. For example, some aspects of persons such as ability and personality are presumed to be very stable, others such as skill may change slowly with experience, and still others such as mood may change quite quickly.

Unfortunately, in a system with any degree of quick dynamic change, re-measurement may be impossible.   Re-measurement may also be impossible if the measurement context cannot be accurately re-created. For a scenario test such as a Line Operational Evaluation (LOE) in the aviation domain, for example, experiencing the scenario a second time may result in quite different behavior than the first time due to memory and learning from the first testing experience. Other significant changes in the testing situation may shift or bias the measurements away from what the original values would have been. For example, if the pilot is paired with "fill-in" crewmembers for re-testing rather than regular line pilots, the interaction of the pilot with the new crewmembers may not be exactly the same as with the crewmembers in the original test.

Alternatively, the missing value can be estimated from other information in the SIS. To estimate the precise value of missing data requires analyses to determine the best method of approximating the missing value and implementation of the best method in the fix-up program. The precise approximation of a missing data point depends on how much systematic variance is in the variable and how well the current data sets and theories developed for the SIS can predict that variance. At the initial stages of construction of a SIS, the data sets are likely to be limited and the demonstrated theoretical connections rather sparse, so the approximation of missing data will be imperfect.

A simple, cautious variation of estimating missing values is to use only the data for that measure and substitute the mean from the sample for the missing values. This method is widely used for interval or ratio data because it does not systematically bias the mean or create artificial empirical relationships with other variables. However, such substitutions reduce the normal variability of the measure. Further, mean substitution biases the observed correlations with other variables to be smaller in size, or closer to 0 than they would otherwise be.

In any case, the approach of replacing missing values with mean values is NOT appropriate for nominal or ordinal data. For nominal data, only the mode or most frequent category of the distribution could be used to replace missing values. For ordinal data, the median rank could be used as the replacement value. These replacements preserve the central tendencies of each respective distribution. All methods of replacement make strong assumptions about the values that would have been measured for each case, and these assumptions may be incorrect. Therefore, these methods should not be used when there is any large percentage of missing data. Rather, the source of the missing data should be isolated and corrected.

Deleting incomplete data   The final approach to missing data is to delete it from the analyses. The entire case and all its data may be deleted from the analysis (sometimes called "listwise" deletion), or the missing data can be deleted only from the particular analysis that directly involve it (sometimes called "pairwise" deletion). The likely effects of both deletion methods should be carefully considered before choosing one.

Typically, missing data is randomly scattered across variables. For an analysis using a large set of variables, the sample size will be strongly decreased using listwise deletion. Decreases of 20-30% of the sample can easily occur. This loss of sample size will decrease the statistical power for making inferences from the data (Cohen, 1977).

The accuracy and stability of statistical description and estimation are similarly adversely affected by the loss of sample size. In particular, the results of complex analyses on small samples of data may produce a very unstable and untrustworthy result. Even if a reasonable sample size is left for completing the analysis, the omission of the segment of the sample with missing data may bias the results, particularly if there were systematic reasons for the missing data. This type of sample bias can severely bias the results of the analyses.

Using pairwise deletion, the other data in a case record are used to a maximum extent. Unfortunately, this may result in situations that violate basic mathematical assumptions about the possible interrelationships in the data, making certain techniques unusable. The data in Table 2.6 below illustrates this problem. Age, Weight, and Self Esteem are measured for nine persons. The Age data are missing for persons 7, 8, and 9; the Weight data are missing for persons 4, 5, and 6; and the Self Esteem data (rated on a 7 point scale) are missing for persons 1, 2, and 3. Using pairwise deletion when analyzing the relationship among age, weight and self esteem generates inconsistent relationships and a real problem in interpretation, which is illustrated by the graphs in Figure 2.5.

Table 2.6   Data that create inconsistent results with pairwise deletion

Person:	Age	Weight	Self-Esteem
1	25	110	
2	30	120	
3	35	130	
4	40		1
5	45		2
6	50		3
7		140	6
8		150	5
9		160	4

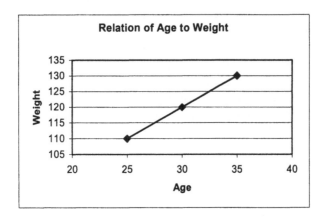

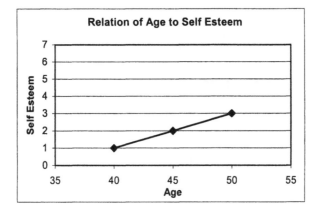

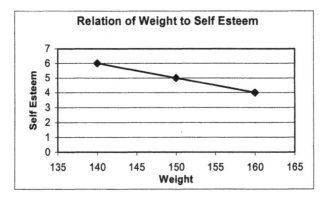

Figure 2.5 Graphs of inconsistent correlations caused by missing data

In Figure 2.5, the graphs of the relationships using pairwise deletion show that age is strongly positively related to weight, that age is strongly positively related to self esteem, but that weight is strongly *negatively* related to self esteem! Taken as a set, these three relationships are logically impossible. Statistically, these relationships form a mutually inconsistent set of relationships that violates basic statistical assumptions. Thus, neither listwise nor pairwise deletion really solves the issue of missing data. Therefore, every effort should be made first to ensure complete data at the time of measurement, and then to ensure correct data entry.

Correctness

Ensuring that data are correct requires checking the data against known constraints. The constraints may be value constraints on a single variable or constraints on a set of values from interrelated variables. Value constraints for each variable depend in part on the type of data that is being encoded and how values are assigned to each particular measure.

Single variable value constraints All types of data should have a defined upper bound and lower bound for permissible values. In addition, data may have additional limits on values in between the upper and lower bounds. Examples of limits are given in Table 2.7 below.

For nominal data, for example, the upper and lower bounds are determined by the maximum and minimum values used as category labels, and the values between these limits must be one of the exact values used as a category label. One of the thorniest problems with nominal data is responses indicating multiple categories. If the respondent checks off two different religions, for example, the problem is how to classify that person into categories that are supposed to be mutually exclusive. If the data for each response alternative are coded as separate variables, this type of multiple category response can be accommodated in the data, but at the expense of exclusiveness in the nominal scale.

For ordinal data, typically the ranks are expressed as whole numbers where each number corresponds to the rank. In this case, data reflecting ranks should be in the range of 1 (highest rank) to the number of ranked objects. In this coding scheme, values in between these bounds would be limited to whole numbers.

One problem with ordinal data is tied ranks. For example, the respondent may refuse to rank order two or more objects and rather give them all the same rank. Most ordinal data analyses can accommodate tied ranks, but the power of these techniques to answer questions is limited to the extent that tied ranks are widespread in the data.

Table 2.7 Checking correctness for four types of data

	Nominal (numerical codes)	Ordinal ("1" is high rank)	Interval	Ratio
Upper bound	Highest number designating a category	Maximum of number of ranked objects	Maximum of Response scale or known system limits	Maximum of Response scale or known system limits
Lower bound	Lowest number designating a category	"1" if rank 1 is basis for all ranks	Minimum of Response scale or known system limits	Minimum of Response scale or known system limits
Intermed-iate values	Each number must equal a category label	Whole numbers from 1 to max rank	Response scale intervals	Response scale intervals
Potential problems	Multiple category responses	Tied ranks	Fractional or decimal responses	Responses below an absolute zero

For interval or ratio data, upper and lower bound values depend on the scale of measurement adopted. For example, Kelvin temperatures should all be positive values whereas Celsius temperatures can be negative or positive. Special scales adopted for research, such as a 5-point Likert scale for attitudes (Likert, 1932), also set upper and lower bound values as well as constraining the data values in between those limits.

The nature of the target system itself can set limits on values for certain variables. One type of limit is physical constraints on the system. If the target systems are families, for example, the number of children is a whole number—neither negative numbers nor fractions of children are acceptable values. In aviation, the altitude of an operating aircraft cannot be below ground level so a report that exceeds that limit is incorrect. Similarly, velocities of subsonic commercial aircraft are limited to Mach numbers less than 1.0. All of these are examples of physical limits on measured variables that should be checked during data entry.

One other important database check is a scan of unique table keys to ensure there is no duplication of values for these keys. Duplicate records or fragments of records can play havoc with the combination process required to convert data from the reduced normal form to the statistical normal form. The typical error is that an undetected duplicate record in a basic relation table produces 2^n duplicate records in the statistical normal form, where n is the number of variables in the join. For combinations of many variables, this tremendous number of duplicate records will strongly bias any statistical analysis, so the error of duplicate records must be avoided.

Constraints based on a set of variables Constraints among a set of variables can be logical, numerical, or constraints due to known properties of a target system. The essential check is that the data is consistent with the constraint. An automated data collection system may include such consistency checks in the data entry process. Alternatively, these constraints can be checked after the data is entered into the system by a "clean up" program.

An example of logical constraints could be as simple as checking the logical consistency between the person answering that they have a spouse on a questionnaire and the person filling out the spouse's age and income. A logical error occurs if the person says they have a spouse but do not enter an age or income, or if the person says they do not have a spouse but do enter an age and income for a spouse. In either case, one of the set of values is incorrect.

An example of numerical constraints would be checking the consistency between total number of children listed for a family and the number of male and female children. The sum of the number of male and female children must exactly equal the total number of children. Numerical inconsistency implies that one of the three numbers in the set is incorrect.

Some constraints are due to the measurement process. Measurements such as "ipsative" scores can also constrain a set of values. Ipsative measures for a set of variables must sum to be a constant or some known value. In the Cattell 16 Personality Factor test (Cattell, 1965), for example, the total score across all 16 personality dimensions must be a constant. A similar constraint is true for the dimensions of the Myers-Briggs Type Indicator. These types of constraint on measured values must, of course, be checked.

A hierarchical constraint is a system constraint due to known properties of set of categories and sub-categories in the measurement process. For example, if illegal drugs are seized in a police raid, the listed drug must correspond to the known categories of proscribed drugs such as marijuana, cocaine, etc. In this case the total value for all the seized drugs should also match the sum of the listed values of each type of seized drug, which would be a numerical constraint.

Just as for data completeness, data correctness can be ensured at data entry or checked after data entry. Ensuring data correctness at data collection is only feasible with an automated data entry system because some of these constraints may be subtle and too difficult for a human to check in a quick scan of a data sheet. At the very least, data correctness should be checked after data entry by a "clean up" program that inspects all the data for a database or table in the SIS.

Fixing incorrect data   At a minimum the clean up program will, of course, check upper bounds, lower bounds, and acceptable values for the data for each variable. Procedures should be created to systematically handle the incorrect data found by this program. These procedures should be consistently enacted by either the clean up program or the data manager who finds and corrects the incorrect data. Correction procedures include the conversion of the incorrect values to missing values or an attempt to replace the incorrect values with more correct values.

Conversion of the incorrect value or values to missing data values is the default option when it is not possible to estimate a more correct value. The procedure is simple and objective, but the net result is to exacerbate the problems of missing values in the SIS. Although this causes a loss of information, allowing the incorrect values into the SIS is a much more serious error because a statistical analysis cannot generally cope or adjust for an incorrect value. Therefore, the results of such analyses can be extremely distorted and the information consumer may have no clue that this is occurring.

If, however, the data can be estimated in some manner and replaced with a more correct value, the problem can be avoided. If upper or lower bounds are exceeded, for example, the clean up program can convert the value to the corresponding maximum or minimum permissible value. The alternative "trimming" approach converts extremely high values to the highest legitimate value observed in the sample and extremely low values to the lowest observed legitimate value in the sample. In a similar vein, fractional numbers entered where whole numbers are expected, such as 2.5 children, can be converted by either rounding or truncation procedures.

Ensuring basic data quality is critical to the success of all subsequent analyses. Although the required checks are tedious and should be programmed wherever possible, some of the decisions about how to handle the data problems uncovered must be made by a human. Making these decisions carefully and with a full appreciation of their effects on the subsequent possible analyses is important. The exact decisions made and the rationale for those decisions should be attached to the metadata guide. One method is to store this information in files linked to the metadata guide so that later users can understand the resulting data completely.

The problems encountered with a specific measure should also be referenced in the data codebook. One method is to link the appropriate file to the definition of that measure in the data codebook. The file becomes a permanent record of problems encountered for each measure and actions taken for those problems. The user can refer to this file to check that the SIS data were appropriately inspected for basic data problems. Storing the decisions made also permits a re-examination of these decisions at a later date and the possible "undoing" of the decisions. For example, specific transformations might be reversed or alternative approaches to replacing missing data or coping with the various forms of measurement error adopted. Metadata guide and codebook information allows re-examination of decisions made and evaluation of different approaches to data problems.

Measurement error

All forms of measurement can have error. More indirect measures may have a greater potential for error or confounding during the measurement process, but even very direct forms of measurement can have error. In counting widgets produced, for example, a partially completed widget at the end of the measurement period can either be counted or not counted by the observer. This variability in counting can give variation in the scores representing productivity and is a source of measurement errors.

Measurement errors may be systematic biases in the measurement process, random fluctuations in the measurement process, or any combination of these. An instructor evaluating aircrew performance during a simulated scenario may have random errors due to not observing critical behaviors at specific moments. The evaluator may also have a systematic bias due to using his or her personal standards of performance as a baseline for ratings rather than the specified standards. Both systematic and random measurement errors strongly impact on the scientific use of the data in a SIS.

Random error

Random error is chance, non-systematic fluctuations in the result of a measurement process due to "noise" in the system or any one of a variety of chance factors. Chance factors are influences that operate occasionally in an unpredictable fashion. Examples of chance factors are memory lapses, pressing an unintended key, using the wrong word in a sentence, misinterpreting an oral or written communication, marking the wrong box on a response form, and the like.

Random errors reduce data quality, specifically the reliability of measurement. Decreasing reliability creates fuzziness in the information in a SIS. Reducing the reliability of measurement has serious consequences such as decreasing the precision of prediction from a SIS and requiring larger sample sizes to make clear statistical decisions. However, random error will not systematically bias the predictions or inferences from a SIS in a particular direction and is therefore less dangerous for scientific inferences than systematic error. The major inference problem for unreliable measures is that the loss of statistical power may result in concluding for no relationship or difference when one really exists, a Type II error. Since in many cases this loss of statistical power can be compensated for by increasing the sample size of the analysis, a mildly unreliable measure can still give useful results. Increasing the sample size will not, however, remove systematic errors.

Systematic error

Systematic error is a systematic bias in the result of the measurement process that is unrelated to the measured construct. This bias or slant in the measurement process will typically cause values that are generally too high or too low to be accurately indexing the construct. For example, performance evaluations given by supervisors to subordinates may be biased too high in order to avoid the unpleasant task of communicating negative evaluations. This may result in a situation in which everyone is evaluated to be above average, which we know cannot be the case.

Systematic measurement biases can severely impact the SIS. A leniency bias such as in supervisor-subordinate performance evaluations can result in a "ceiling effect" in which all the measurements cluster at the top of the scale and there is insufficient variability to establish relationships or test relevant questions. The contrary form of "floor effect" where scores pile up at the bottom of the scale is also possible if there is a systemic bias toward harsh evaluations. These biases can hamstring the use of a SIS for the respective constructs.

Confounding of measurement More important, however, is a confounding of the measurement of the target construct by some other construct. Essentially, the measure of the target construct is contaminated by the presence of variability due to some other construct. In the case of performance appraisals, suppose that the goal is to have a clean measure of job performance. But if a supervisor rating of performance is used, the measure may be contaminated by the level of ingratiation skills on the part of the employee.

One consequence of this confounding is that any attempt to predict this measure of job performance by other information in the SIS is now hindered by the contaminating influence of the ingratiation behavior. Further, when job performance is used to predict other outcomes such as job satisfaction, employee turnover, and so forth, a systematically inaccurate picture will emerge due to the unrecognized presence of ingratiation skill effects in the job performance measure. Reduction of measurement errors and confounds is important for the data in a SIS to address clearly and correctly the information needs of the users.

Reducing measurement errors

Reducing measurement errors requires a careful inspection of all aspects of the measurement technique (i.e., the measurement instrument, context, and process) to identify the likely source of error. It also requires an inspection of the measurement assumptions. More indirect measures will have more complex measurement assumptions and more possible sources of error among these assumptions.

Reducing errors in very direct measures Very direct measures rely on a direct measurement technique and assume that the direct measure is relevant to the targeted construct. Therefore, the only two sources of error are the measurement technique and the assumed relevance to the construct. Suppose that direct, automated recordings of performance such as widgets produced or phones calls answered was the performance measure. The only source of error is either the technology for making the trace recording or the relevance of the trace recording to the desired construct of job performance. Technology for counting widgets or tracking completed calls may be quite accurate and error-free, but technology for other trace measures such as tracking eye movements may be prone to errors under certain conditions. If errors occur, the distinct components of the measurement process and the conditions of measurement must be examined in detail to find and correct the sources of error in the measurement technique.

Direct measurement techniques may result in systematic or random errors. A systematic error would be a consistent or predictable error. Such errors systematically bias the score or result in a predictable way. The most common predictable errors are constant or proportion errors. An example of a constant error would be using a scale that is known to consistently weigh two pounds light. An example of a proportional error would be using a scale that gives a weight 10% higher than the true weight.

Random errors occur in an unpredictable manner. Such errors typically do *not* systematically bias the resulting score. Even for physical direct measures some amount of random error may be unavoidable. For example, the sensors for tracking values of a system variable may have a certain amount of irreducible "fuzz" or noise in their output. When observers are used to rate variables, they may also have a certain amount of irreducible individual differences among them that result in slightly different scores.

The relevance of a direct measure to the desired construct must also be considered as a potential source of error. Even if the recording technology gives a precise count of widgets produced or calls completed, the *quality* of performance may not be indexed and be a source of measurement error. Certain employees, for example, may emphasize quantity of production at the expense of quality while other employees emphasize quality of production at the expense of quantity. These differences would make the direct count of objects produced an imprecise measure of "real" productivity.

Similarly in the aviation domain, a direct recording of the aircraft's position, altitude, heading, airspeed, angle of bank, and so forth during a landing may be technologically perfect but still have error when used to index crew performance. For example, crew performance may be scored by the number of times the aircraft exceeds certain critical limits such as a limit on airspeed. Several noticeable but minor airspeed deviations may, however, not index poor performance as much as one large deviation on airspeed that could cause an aircraft to stall. A similar difficulty could occur in scoring deviations from a course heading by the number of degrees off course. A large deviation that is quickly corrected may be better performance than a more moderate deviation that lasts for a long period of time. Thus, even very direct measures may be susceptible to systematic and random error as well as potential problems with the relevance of the measure to the target construct. Relevance problems may be due to either raw data that are partially irrelevant or to inaccurate coding or transformations of the raw responses so that the scores do not correctly reflect the target construct.

Reducing errors in moderately direct measures   For moderately direct measures, the sources of error may lie in the measurement technique or in the theory linking the item or items to the desired construct. Clearly the sources of error for very direct measures are still possible for moderately direct measures. Additionally, the assumption that the results of the measurement context will generalize to the real system can be a source of errors as well as the basic theory specifying the relevance of the items to the construct. Violations of the context-generalization assumption can produce random or systematic errors.

Anything in the measurement context that affects the measure but is NOT present to the same degree in the natural system can cause errors. Random errors can be caused by factors such as mood or motivation of the persons measured. To the extent that their mood or motivation impacts the measurement and varies randomly, the measurements will be discrepant from what is true of these persons in the real system. Paper and pencil tests, for example, may be influenced by a person's mood on that day or by the desire to achieve a high test score. These influences will vary across persons and cause random measurement errors. Since the impact for some persons will be positive and other persons negative, the random errors in measurement will tend to cancel out over large samples of persons and not systematically bias results from the SIS.

Differences between the measurement context and the real system can also cause systematic errors. In such a case, the scores in the SIS will be systematically higher, lower, or different from what they would be without the error. One example of such biases is a social desirability bias in which persons present themselves in the most positive or socially desirable fashion. Due to this bias, self reports of the number of traffic tickets one receives or cheating on income tax reports may be systematically lower than reality. Similarly, reports of attending religious services or giving money to charity may be systematically higher.

These systematic distortions of the measures may be intentional or unintentional. Lying, faking, or attempting to look socially desirable are examples of intentional systematic errors. Examples of unintentional errors would be systematic errors due to memory biases, use of heuristics to make judgments, and so forth (Tversky & Kahneman, 1982). Using observers is one way to avoid the distortions of self reports, but observers are also part of the measurement technique that may introduce errors.

When an observer is used to measure a construct, the observer becomes a possible source of both systematic and random measurement errors. In evaluating pilot performance, for example, systematic differences in evaluations have been documented and training has been designed to reduce these errors (Williams, Holt, & Boehm-Davis, 1997). This training used statistical methods to document the occurrence of observer errors. Training feedback and establishing common grading standards among groups of observers worked to reduce systematic observer errors.

The observation and evaluation process can also produce random errors. The clearest example of these errors is when an observer is asked to rate exactly the same set of stimuli in the same context but at different times. The occurrence of spontaneous changes in the observer ratings is one indication of random errors in the measurement process. Reducing this kind of error may involve externalizing as much of the evaluation process as possible. Using explicit, step-by-step rating forms so that the observer does not have to rely on memory recall for the relevant evaluation steps or judgment criteria may help reduce these errors.

The relevance of the measurement items or content for measuring the target construct must also be considered as a source of errors for moderately direct measurement. Even if the theoretical justification for these measures is fairly simple, it must be checked wherever possible. For example, if the measurement theory states that ten items should form a consistent scale measuring a construct, this should be checked before the information is used in the SIS. If in fact only nine of the ten items consistently measure the construct and the tenth item is unrelated, the tenth item may contribute random error to the measurement.

Reducing errors for indirect measures For indirect measures, the theory tying the measurement items to the target construct is more complex. The potential errors of indirect measures include all the types of errors for direct and moderately direct measures. However, to the extent that the theory tying the measurement to the construct is more complex, there are correspondingly more potential sources of error in the theoretical linkages than for the direct or moderately direct measures. Further, unlike the moderately direct measures, the theoretical links may be so complex that it is not possible to empirically evaluate the plausibility of the links.

A Rorschach Inkblot test for measuring psychopathology, for example, may require several theoretical steps to tie the responses of the person to the measured construct (Figure 2.6). The first assumption may be that ambiguous stimuli rather than content-related stimuli are the best vehicle for eliciting psychopathology. The second assumption may be that a specific form of psychopathology systematically will influence perception of an ambiguous stimulus in a consistent and predictable way. The third assumption may be that the person will precisely and accurately report these perceptions. The fourth assumption may be that the coding scheme employed to classify the responses for particular psychopathologies is complete and accurate. Any step in the theoretical justification of the measurement technique can be a potential source of random or systematic error, so indirect measures have more potential sources of error.

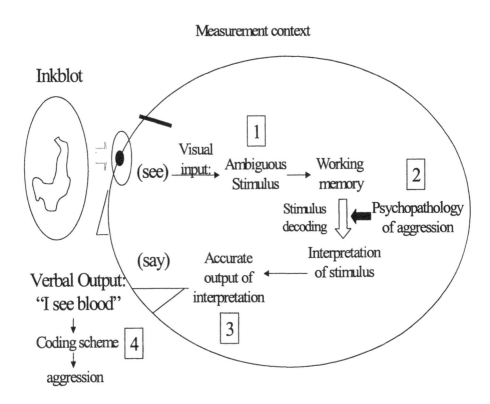

Figure 2.6 Example of measurement assumptions

Clearly the extent to which a measure is indirect is a matter of degree, and the vulnerability to error depends on the degree of complexity of the assumptions. More specifically, if the theoretical steps in the justification of the measurement technique can be enumerated, the degree of vulnerability to theoretically based measurement error can be examined and roughly estimated. If the steps are sequentially dependent as in the above example, error vulnerability is the functional concatenation of the error vulnerability of each step. In the above example, the theoretical vulnerability of ink-blot measures of psychopathology is the concatenation of possible errors for theoretical assumptions one through four as depicted in the figure.

Measurement techniques for which the steps in the theoretical justification *cannot* be enumerated probably should not be considered scientific and should not be used in a SIS. Similarly, any measurement technique that has enumerable steps but also has extreme vulnerability in one or more steps should not be considered for a SIS. Consider the measurement method illustrated in Figure 2.6 for a sample of violent prisoners applying for parole status. The theoretical steps in the measurement process can be clearly enumerated, but step 3 may be vulnerable. The accurate output of the aggressive interpretation of the stimulus may be suppressed by the prisoners due to their goal of gaining parole. Since this step may be very vulnerable to systematic reporting error, the Rorschach measurement of aggression should not be used for this population.

If practical considerations prevent the reduction in measurement errors by changes in the technology, context, or process of measurement, the overall quality of measurement can still be improved by two methods. The first method is using multiple or additional forms of measuring the target construct. The second method is attempting to measure the error itself and trying to adjust the scores for the construct using these error measures.

Multiple measures

Multiple measures can improve the quality of measurement of a construct. Multiple measures will generally be more expensive than single measures, and thus may only be considered for key constructs in the SIS. In a performance SIS for example, multiple measures might be considered for job performance since it is a critical variable in the system.

Multiple measures can either use the same basic method or different methods. For example, multiple measures of intelligence could all be paper-and-pencil methods. Alternatively, intelligence can be measured using different forms of response such as problem-solving, oral debate, and so forth.

Multiple measures with same method Multiple uses of the same or a similar method to measure a construct can add to the stability or reliability of measurement. Multiple measures should be checked to ensure that they converge on measuring the target construct. If these measures converge, a combined measure can be constructed by averaging the multiple measures. This combined measure is more stable than any single measure and has the advantage that the measurement of the construct is not totally dependent on any single measurement.

The disadvantage of adding multiple measures with the same method is that the multiple measures are still subject to any confounding or contamination caused by the particular measurement method. Multiple paper-and-pencil measures of intelligence may be contaminated by the fact that some persons have dyslexia and tend to score low on all paper measures, while others have more experience or dexterity with paper tests and score higher. Therefore, the aggregate measure of intelligence based on multiple paper-and-pencil measures would enhance measurement reliability, but still be subject to contamination by confounding personal factors which might reduce measurement validity.

Multiple measures with different methods Multiple measures using different methods reduce or avoid contamination by the method of measurement. Campbell & Fiske (1959) called this the multimethod approach. Any factors that affect a person's score on a particular measure are less likely to also confound the person's score on a different type of measure. Reading problems, for example, would affect performance on a paper-and-pencil intelligence test but would not affect an oral test. A second advantage is that the convergence of multiple measures gives evidence for the validity of the underlying construct.

If the multiple measurements do converge, an aggregate score based on these multiple measures is much less contaminated by variance due to a particular method. If multiple measurement methods do not converge, the existence of the construct is questionable (Campbell & Fiske, 1959). Thus, while expensive to implement, multiple converging measures can increase the scientific quality of measurement such as the reliability and validity of measurement, which are discussed in the next chapter. In situations for which there is only one available measure for a key SIS construct and multiple measures cannot be added, an alternative approach would be to attempt to directly estimate and correct for the error in the measure.

Measuring the error

If the error component of a measure can be estimated with some degree of certainty, the score can be adjusted to correct for the known amount of error or bias. For example, if the only available scale is known to weigh two pounds light, the user can simply add two pounds to the indicated weight and arrive at a more correct score for weight. If the error is constant and stable, this simple correction may suffice. Estimating and correcting for errors will depend on whether the source of the error is 1) the technology or process of measurement, or 2) the measured system itself such as the persons or teams being measured.

Error caused by the measurement technique If the source of the error is the technology or process of the measurement technique, the estimation and correction of error naturally focuses on the critical aspects of the technique. Some of these errors may be systematic as in the example of the biased weighing scale given above. If a system required multiple scales to measure weight, the known errors of each scale could be used to correct all the measured weights and make them more accurate. The example of observers being systematically more harsh or lenient in their grading of performance is another example that would allow a similar approach to error correction. The known bias of each observer could be added or subtracted from their judgments to correct for the known systematic errors.

Random errors in the technology or process are more difficult to detect and correct due to the fact that by their very nature they occur in an unpredictable fashion. Nevertheless, these errors may be indexed by either the collateral occurrence of some other measurement in the system or by identifiable values of the scores that indicate error. An example of the collateral occurrence of another measurement indicating error would be the occurrence of a number of "missed observation" ratings for an observer evaluating job performance in a structured setting. A high number of missed observations may indicate that the resulting observations and judgements have random errors.

Correcting for such a random error in the measurement processes depends on whether the direction and extent of the error can be estimated. If direction and extent of error can be estimated, the original score can be appropriately adjusted. If the direction and extent of the error cannot be estimated, the real score may still be estimated using an approximation based on other information in the SIS. For an aviation example, suppose an evaluator misses numerous observations during a particular phase of flight for a crew. In this case, the performance of the crew on other phases of flight in the session can be used to estimate their likely performance on the particular phase of flight. Although certainly not as good as initially obtaining a good measurement, this method can be a decent second-best if the degree of predictability of the measure from other SIS information is high.

Error caused by persons, teams, etc. The entities measured in a SIS can also be a source of systematic or random error. That is, the persons, teams, organizational divisions, and so forth can influence measurement. The systematic influence of persons on measurements is often a focal point of scientific investigation in itself and is called the study of individual differences. Individual differences may systematically contribute to higher or lower measurements on certain variables. For a test of performance, individuals with higher intelligence and motivation may score higher while individuals with higher test anxiety or evaluation apprehension may score lower. If these systematic biases due to individual differences are known and can be accurately measured, this source of error can be adjusted for.

Random errors in the measurement can also be due to individual factors. For example, momentary variations in mood, energy, memory capacity, or other personal factors can impact the measurement of performance. Estimating these random errors with separate tests may be possible if the factor is stable enough to have the same value during an auxiliary measurement. An auxiliary measurement will probably not work if the error factor is extremely unstable. However, if the basic measure of the construct is carefully designed, the influence of some of these random errors can be reflected in the measure itself. Knowing that an error of some type has occurred allows the possibility to adjust the data analyses for that error.

An example of the collateral occurrence of another measurement indicating error would be the occurrence of a large score for the "lie" scale on the Minnesota Multi-phasic Personality Inventory (MMPI) personality measure (Hathaway & McKinley, 1943). The lie scale is composed of items that have high social desirability but which are not true for most people. The occurrence of a high lie scale score means that the person was not answering the entire set of questions honestly and the obtained personality profile may be seriously inaccurate. Knowing that the measured personality profile may be inaccurate for this person allows the data analyst to omit that person's record or possibly use other information instead of the MMPI profile for the data analyses.

Adjusting for person or team-based errors When the differences among individuals, teams, or departments being measured are the source of errors, indexing the extent of error and correcting for it may be possible in the data analysis step of using the SIS. Statistical methods can be used to adjust or covary for the influence of some types of error. However, these methods typically require that a very accurate measure of the error-producing factor or factors can been made. If test anxiety is contaminating performance scores, for example, the removal of its influence requires an extremely accurate measure of test anxiety. If such a measure is available, it can be used to covary and adjust the scores for the focal variables in the analysis.

Detecting errors in data Even if the precise source of errors cannot be identified, heuristic methods can be used to minimize the impact of error on the SIS. One heuristic is that extreme values of a measurement may indicate error. An extreme value is sometimes called an "outlier" value. In measuring simple reaction times for tasks like pushing a button, for example, the typical range of scores is less than one second. A response taking longer than one second is extreme and may indicate that some error occurred in the measurement process. The person may not have seen the stimulus, been momentarily distracted, and so forth.

For variables that are measured on a continuum, either extremely high or low scores can indicate an error of measurement has occurred. Such scores can be identified statistically and can either be deleted from the data or corrected by replacement with a "best guess" value. One rule is that scores lying above 3 standard deviations above the mean or below 3 standard deviations below the mean are sufficiently extreme that they should be replaced. In such a case, these scores can be replaced with the nearest acceptable value ("trimming"), or a base rate value such as the overall mean for that variable. If the scores can be predicted or estimated from other SIS information, a best guess can be calculated and substituted for the extreme value.

The use of error detection and correction methods depends on defining and measuring the known errors for a measure. Initially in the construction of a SIS, the amount and types of errors for each measure may not be known. Each measure must be subjected to careful scientific scrutiny to establish as objectively as possible its measurement qualities including error.

Once the basic data quality is ensured, the scientific quality of the data can be further assessed, which is the focus of the next chapter. Checking scientific quality may also uncover basic data problems and require revisiting the issues of this chapter. For example, the assessment of scientific reliability may indicate that a measure has more error than was expected. That finding might prompt a re-examination of the measurement process and a search for collateral indications of error. If this re-examination causes a change in the basic data, this change may require re-doing the original analyses. This type of backup costs time and effort and should be avoided wherever possible by paying close attention to basic data quality.

Chapter summary

The data in a SIS should represent meaningful scientific constructs. These constructs can be presumed to be universal for all cases (nomothetic), applicable to certain types or subgroups of cases, or unique to each particular case (idiographic). Initial scientific development may focus on universals as this approach is more parsimonious, but a complete theoretical account may ultimately require idiographic constructs.

Measures of scientific constructs will vary in how directly they tap each particular construct, and on the type of data they produce. The type of data will limit certain transformations and analyses, and should therefore be described in the data code book in the SIS. The overall structure and linkage of the entire set of SIS measures should be described in the metadata guide for the system. This guide should clarify the storage format (e.g. reduced normal form), the level and type of each set of measures, and the keys that allow linking the information.

Basic data integrity includes the completeness and correctness of the information in the SIS. Random and systematic errors of measurement should be reduced or eliminated wherever possible. Ensuring basic data quality is an important preliminary step to checking the scientific quality of the data that is the focus of the next chapter.

3 Scientific Data Quality

Overview

The three basic aspects of scientific data quality are the sensitivity, reliability, and validity of measuring a construct. A construct is any unitary thing defined by a theory or by common usage. Constructs range from concrete things such as height or weight to abstract things like intelligence or motivation. The sensitivity, reliability, and validity aspects of construct measurement are informally defined in Table 3.1. Although related, each aspect is a distinct part of scientific data quality.

All three aspects of quality involve the variability of the measure. Adequate sensitivity ensures that the scientific measure will have some variability that reflects the variability of the construct in the real system. The criterion of sensitivity is the variability of the construct that is faithfully preserved in the measure. That is, the measure should neither artificially reduce nor enhance the amount of real variance in the construct. For example, classroom tests that are much too easy or much too hard will result in all students passing or all students failing. Either of these results would show much less variance in student learning outcomes than might really exist for the class.

Adequate reliability ensures that the systematic variability of the measure will not be overwhelmed by random variability. The criterion of reliability helps ensure the consistency of measurement results if alternate forms of a measure are used such as the Stanford-Binet Form A and Form B measures of intelligence. Reliability should also be reflected in the stability of the measurements over suitable time intervals. Unreliable measures will give inconsistent or unstable measurements.

Adequate validity ensures that the variability of the measure reflects the intended construct and nothing but that construct. The criterion of validity helps ensure that the measurement is not biased by systematic confounds. Possible confounds include variance caused by the measurement process, social desirability of responses, habitual response sets, and so forth. Invalid measures do not reflect the intended construct and will give misleading or completely erroneous results.

71

Table 3.1 Comparison of sensitivity, reliability, and validity definitions

	Sensitivity	Reliability	Validity
Informal Definition	The extent to which small variations in the actual construct result in different measurement values	The consistency, stability, or lack of random error in measurement	The extent to which a measure reflects the intended construct and only that construct
Legal analogy with witness testimony	"the whole truth"	"the consistent truth"	"nothing but the truth"
Radio analogy	"Gain" How well does a radio pick up and amplify weak stations?	"Noise" How well does the radio suppress or reduce random noise or static?	"Selectivity" How well does the radio receive the tuned station and reject all other stations?
Example of the evaluation of crew performance	Small differences in crew performance result in different evaluations	Evaluations are performed without random or idiosyncratic rater errors	The measurements reflect only real differences in crew performance and not something else

The interrelationships among these criteria are illustrated in Figure 3.1 for an example of measuring differences in the construct of intelligence. Figure 1 shows the real variability of intelligence on the left, and three different components of variability of the IQ measure: random variability, unrelated systematic variability (e.g. method variance), and valid variability. This figure illustrates that sensitivity is the amount of real variability in intelligence that is ultimately reflected in valid variability in the IQ measure. Reliability is the proportion of systematic (non-random) variability in the IQ measure, which combines systematic method variance as well as valid variance reflecting differences in intelligence. Validity is the proportion of the total variance in the IQ measure that really reflects differences in intelligence.

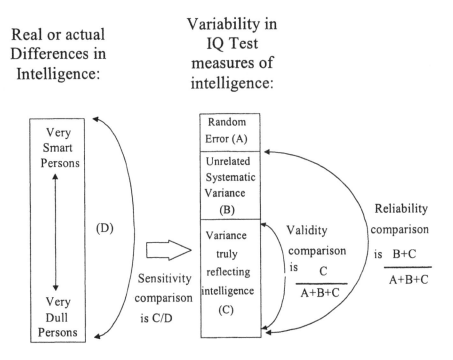

Figure 3.1 Interrelationship of reliability, sensitivity, and validity

Clearly, a good scientific measure must have all three aspects of data quality. If the measure is not sensitive, there will either be no differences in the measured scores or these differences will bear little relationship to the differences in the real system, either of which is unacceptable. If a measure is unreliable, the random variability correspondingly reduces the possible validity. Even if a measure is reliable, the variability in the measure must still be checked to ensure that it reflects the target construct rather than things such as measurement confounds, biases, or different constructs. To slightly paraphrase the legal requirements of a witness, a measure must be "the whole truth, the consistent truth, and nothing but the truth", which are respectively the issues of sensitivity, reliability, and validity. Just as we require all three things for a legal witness, we require all three aspects of quality for scientific data.

Measurement sensitivity

Definition of sensitivity

Informally, sensitivity is the extent to which a measure varies as the construct measured is varying (Figure 3.1). We want measures that preserve the variability of the construct as completely as possible. Measures that have less variability than the construct are under-sensitive and can understate differences or relationships. Conversely, measures that are exaggerations of the construct can also be problematic since the observed variability will be larger than it really should be. Further, the observed differences or relationships using the measure can also seem larger than they really are. The definition of sensitivity used here is the extent to which the real differences among cases on the construct are preserved by the measured differences.

Accurately indexing the degree to which real distinctions are preserved in the measurement process depends on the type of data. Table 3.2 gives a definition of sensitivity for each basic type of data. For nominal/categorical data take the example of measuring religious affiliations for a sample of people. A category set of Protestant, Catholic, Jewish, Moslem, and Other would have less sensitivity than a category set which would recognize the major subdivisions within each of the major religions. Similarly, extending the measured religious affiliations to Hindu and other non-western religions will increase the degree to which the measured affiliations reflect real differences in the religions of a broad sample of people.

Clearly there are practical limits to how fine the categorical system can be for measuring a particular construct like religious affiliation. Having all possible subdivisions of the major religions included may make for an unwieldy measure. Typically a balance must be struck between sensitivity and practicality of measurement in any particular situation. However, the important point is to consider the sensitivity of measurement when constructing measures for the SIS and when evaluating results from the analysis of SIS data.

For ordinal data, take the example of a pecking order among a flock of chickens. Having people subjectively judge which chickens are more dominant may lead to tied ranks where chickens are perceived as equally dominant, or inconsistent rankings among the judges. The alternative measurement procedures of having each pair of chickens compete for a limited food source would resolve the exact dominance differences for each pair of chickens in the flock, and would therefore be more sensitive.

Table 3.2 Definitions of sensitivity for different types of data

	Type of Data		
	Nominal or Categorical	**Ordinal or rank-ordered**	**Interval or Ratio**
Definition of sensitivity:	Distinctions in real or natural categories are preserved in measured categories.	Real differences in ranks are preserved in the rank order of the measurements.	Real variability among cases on the construct is preserved in measurement variability.

For interval/ratio data, take the example of evaluating crew performance that is a critical issue in the aviation domain. Typically crew performance is evaluated by more experienced pilots observing crew actions and outcomes. Sensitivity is the extent to which the real variability in crew performance is picked up by variability in the evaluators' assessments of crew performance. An assessment scale such as "Pass/Fail" would not allow the evaluator to describe the different levels of performance with any degree of sensitivity. A scale with more judgment points (e.g. a 5 or 7 point scale) may allow the evaluator to assess the different levels of crew performance more sensitively.

Accurately preserving the real variability of the measured construct is important for obtaining good information from the SIS using statistical methods. Basic information such as group differences or relationships among measured constructs will be distorted by insensitive measures. For example, the differences in average scores for different groups or conditions will be understated for insensitive measures. Similarly, a relationship between two variables that is estimated by insensitive measures will be incorrect; the real relationship between the underlying constructs will typically be understated. Finally, many analytic techniques depend on partitioning the variance of a measure in some way. For insensitive measures, these techniques will be partitioning a reduced amount of variance that may produce inaccurate results and estimates.

Under-sensitivity Most typically, some of the fine-grained details about the construct are lost in the measurement process. In this case, the resulting measure has less variability than the construct and is under-sensitive. For interval or ratio measures, variability in the measure is less than the real variability. In practice, we do not have an accurate measure of the real variability, so sensitivity must be estimated by one of the methods discussed below.

In rating performance of aircrews on a 9 point scale, for example, certain evaluators might tend to over-use the "standard" category of 5.0 and conservatively evaluate extremely good or bad crew performance (Figure 3.2). These evaluations would have under-sensitivity.

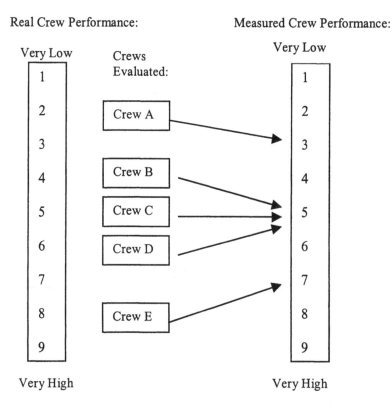

Figure 3.2 Under-sensitive evaluation of crew performance

Over-sensitivity   Much more rarely, the measurement process can exaggerate real differences in the construct.   In rating performance of aircrews, for example, certain evaluators may tend to exaggerate ratings for performance that is slightly above standard or slightly below standard (Figure 3.3).   That is, evaluations of below-standard performance (e.g. below 5) would be lower than they really should be, while evaluations of above-standard performance would be higher than they really should be. These evaluations would have over-sensitivity.

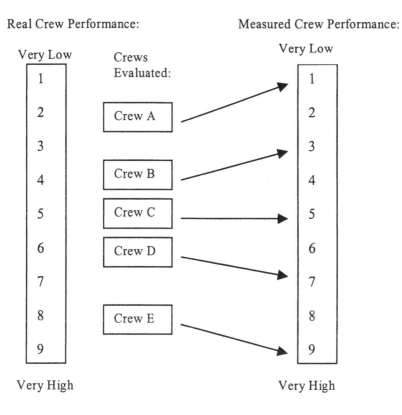

Figure 3.3 Oversensitive evaluation of crew performance

Measurement units   Analogous to the different grain size of categorical measurements of religion discussed earlier, the measurement units for interval/ratio scales may differ in precision.   These different measurement units can give different basic levels of sensitivity.   For example, weighing a person in kilograms is less sensitive than weighing them in pounds, which is less sensitive than weighing them in ounces.

However, sensitivity cannot be increased indefinitely by simply using a finer unit of measurement because of practical limitations (such as those discussed under categorical measures) and because the error of measurement typically increases for finer measurements. Since the error of measurement decreases reliability and validity, increases in sensitivity must be weighed against possible decreases in reliability and validity. The trade-off that may occur between sensitivity and error as the unit of measurement is changed is illustrated in Figure 3.4. In Figure 3.4, sensitivity increases with a more fine-grained scale. However, random measurement error also increases as the scale has more fine-grained response options.

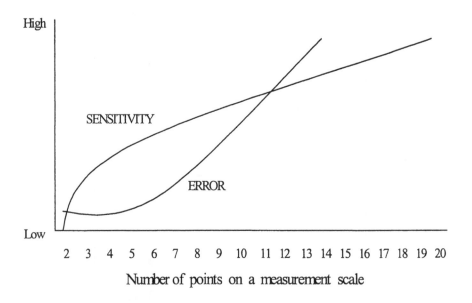

Figure 3.4 Example of the trade-off between sensitivity and error of measurement

If the trade-off for measuring a particular construct follows this example, the optimal point for the measurement unit may be found by comparing the relative height of the two curves. The best trade-off point is one that maximizes sensitivity while at the same time not inducing large amounts of error. For this example, the trade-off point of between 4 and 6 scale points would give the best balance of sensitivity to error for this measurement.

The exact value chosen could, however, change with the relative value placed on sensitivity vs. error in measurement. In this case, increasing the relative value on sensitivity will shift the trade-off point toward finer units of measurement. That would mean using a 6 or 7 point scale in this example. Increasing the relative value on keeping error at a minimum will shift the trade-off point toward larger units of measurement. That would mean using a 4 or 5 point scale in this example. In the aviation domain, for example, several airlines have adopted a four or five point evaluation scale to increase sensitivity while keeping the error of the evaluation process to a minimum.

Empirical methods to estimate sensitivity

Several methods can be used to estimate the sensitivity of measurement of interval or ratio data. Each method has different requirements, focal points, and potential problems or biases. Table 3.3 summarizes three different approaches to estimating sensitivity for interval or ratio data.

Range ratio The range ratio is the ratio of the obtained range of measurements for a sample to the real or expected range of measurements for that sample. Using the data in Figure 3.2 as an example, the obtained range of evaluations for crews was from 3 to 7 for a range of 5 intervals. The real range of crew performance was, however, from 2 to 8 for a range of 7 intervals. The ratio of the obtained to real ranges for Figure 3.2 data would be = 5/7 or .71. The value less than 1.0 indicates under-sensitivity.

Conversely, using the data in Figure 3.3 as an example, the obtained range of evaluations was from 1 to 9 for a range of 9 intervals. Since the range of real differences in performance is the same as in Figure 3.2, the ratio of the obtained to real ranges is 9/7 or 1.28. The value greater than 1.0 indicates over-sensitivity. Alternatively, if obtained range is precisely the same as the real or expected range, the range ratio has value of 1.0 which would indicate perfect sensitivity of measurement.

Although these examples have used a real range of performance as the basis for computing the range ratio, in practice the real range of performance will have to be estimated by some method. One common method is the use of Subject-Matter Experts (SMEs) to estimate the range of expected values of measures of a construct for a defined population. The maximum expected value and minimum expected value are then used to compute an expected range which can be used in place of the real range for estimating sensitivity.

Table 3.3 Comparison of basic methods for estimating sensitivity

Empirical estimates of Sensitivity:	Requirements:	Assumed to be important:	Potential confounds, biases, or problems:
Range ratio	Good estimate of real range of the construct	Differentiation of measurement for high and low scores	High instability due to occurrence of outliers
Variance ratio	Good estimate of real variance of the construct	Differentiation of measurement across the whole distribution	Moderate instability due to occurrence of outliers
Omega-squared	Good examples of two or more levels of the construct	Differentiation of measurement for the specified levels	Choosing a non-representative or too extreme set of levels for the estimate
Intra-class correlation	Representative sample of the types of cases to be measured	Differentiation of measurement across the set of levels in the entire sample	Choosing a biased rather than representative sample of cases

For this example, if SMEs expect a range of crew performance from 2 to 8 but the data in the SIS only shows a range of observed assessments of crew performance from 4 to 6, under-sensitivity may be a problem. Using SME estimates of maximum and minimum values for the range ratio has the advantage of being easy to implement but the disadvantage of relying on the accuracy of SME estimates.

Due to its ease of implementation, the range ratio lends itself to inclusion in the analysis of data quality in the SIS. Measures that show either under-sensitivity or over-sensitivity can be flagged for further examination. For example, if crew performance is judged on a 4-point scale and SMEs expect performance to differ from 2 to 4 on this scale, any sample of evaluators that showed all "3"s would be clearly under-sensitive.

Variance ratio The variance ratio is very similar to the range ratio except that it uses the variance of a distribution rather than the range to index variability. This ratio compares the variance of the obtained distribution of scores is compared to the variance of the expected distribution of scores. The advantages of using the variance to index the expected and obtained variability are that the variance is a more complete reflection of the variability of all cases in a sample and is less affected by the presence of one or two extreme outliers in the distribution.

As for the range ratio, the expected distribution must be obtained from SMEs or from a theory that makes precise predictions of the distribution. The observed variance for a sample is the sum of squared deviations around the mean divided by the sample size (or sample size - 1). The variance ratio is formed by dividing the observed variance by the expected variance.

This ratio would have the value 1.0 if sensitivity is optimal, a value less than 1.0 if the measure is under-sensitive, and a value greater than 1.0 if a measure is over-sensitive. In the preceding example, the real variance in the performance of crews A, B, C, D, and E was 5.0. The under-sensitive evaluations of crew performance in Figure 3.2 would have a variance of 2.0, which would result in a variance ratio of 2/5 or .40. The over-sensitive evaluations of crew performance in Figure 3.3 would have a variance of 10.0, which would result in a variance ratio of 10/5 or 2.0. Although these values are different from the range ratios, they estimate under-sensitivity or over-sensitivity in the same manner. Since the variance is a more stable index of variability than the range, the variance ratio would generally be preferred to the range ratio if SMEs can accurately determine the entire shape of the expected distribution.

Since the range ratio and variance ratio are subject to the accuracy of the estimates of the SMEs and to the occurrence of extreme outliers that influence the range in any particular sample, these estimates of sensitivity should be augmented by other methods wherever possible. These other methods will rely less on the judgments of SMEs and be less susceptible to bias by extreme values, but will require multiple ratings from the judges or evaluators. Two methods that can be used for this purpose are omega-squared (Hays, 1981) and the intraclass correlation coefficient.

Unlike the range or variance ratio, these methods of estimating sensitivity require a sample of measured objects which either (1) have established values on the underlying interval or ratio scale, or (2) are a reasonably-sized random and representative sample of the possible values on the underlying scale. Both of these methods express the sensitivity as a value between 0 and 1, where 0 is no sensitivity and 1 is perfect sensitivity (similar to the range ratio). Having sensitivity indexes with values from 0 to 1 facilitates comparisons to the reliability estimation procedures which also result in indexes which range in value from 0 to 1.

Both these indexes estimate the relative percent of total variation in the obtained measure due to real differences in the stimuli. The two indexes differ depending on whether the real value of the stimuli on the underlying continuum can be established at *a priori* fixed levels or not. If fixed levels or values of the construct can be established by experts or some other a priori method for a subset of stimuli, then the omega squared statistic can be used to estimate sensitivity. Alternatively, if no *a priori* evaluation method is available but a representative random sample of the stimuli can be obtained, then the intra-class correlation coefficient can be used to estimate sensitivity.

Omega-squared Hay's (1981) omega-squared index is generally used to estimate the proportion of variance due to a small subset of known, fixed levels of a factor. For estimating sensitivity, the levels of the factor are groupings of test stimuli that all have the same value on the underlying construct. For example, if a 5 point scale is used to evaluate performance, the fixed levels may represent examples of 2, 3, or 4-level performance. The exact levels for the test stimuli should be carefully established by experts or an alternative evaluation method because the estimation of sensitivity critically depends on getting the correct estimates of real values for the test stimuli.

The essential logic of this use of the omega-squared index is to estimate how much variance in measurements is due to differences in the real values of the stimuli. This estimate uses the differences in mean judgments for stimuli at the 2 level, the 3 level, and the 4 level, for example, to estimate how much of the total variability is due to real differences in the stimuli. The omega-squared index compares the variability due to real differences among the stimuli in the numerator to the total variability of all judgments in the denominator.

When used in this way, the omega-squared index reflects the sensitivity of a set of judgments to real differences in the stimuli. More sensitive measurement methods will have relatively more variability due to the stimuli and less variability due to judgment errors or other causes. Therefore, these measurement methods will produce values of omega squared closer to 1.0. In the example in Figure 3.5, videotapes of crews "A" through "I" have been judged by SMEs to represent 2, 5, and 8 level performance.   The judgements of the assessors map these different performance levels quite well. The obtained value of omega squared would be near 1.0, indicating excellent sensitivity.

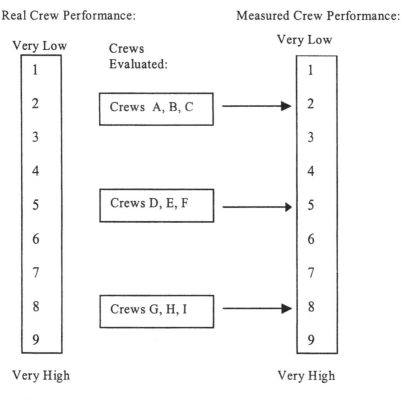

Figure 3.5 Example of good sensitivity using Omega-squared method

Less sensitive measurement methods will have less variability due to the stimuli and more extraneous variability. This will produce values of omega squared closer to 0. The example in Figure 3.6 below uses the same stimuli as before but shows a pattern of more modest sensitivity. Crews A, B, and C are evaluated as 1, 2, and 3 instead of all being evaluated at the real value of 2. Similarly, Crews D, E, and F are evaluated at 4, 5, and 6 instead of all being evaluated at the real value of 5. The omega- squared value for this set of evaluations would be well below 1.0, indicating a poorer sensitivity of judgment then in the previous example.

Real Crew Performance: Measured Crew Performance:

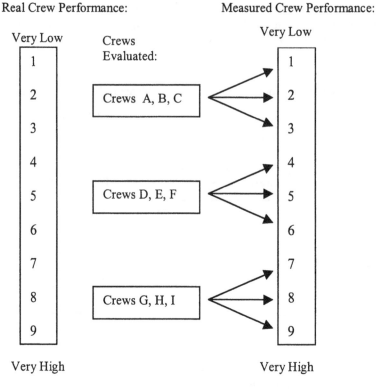

Figure 3.6 Example of modest sensitivity using Omega-squared method

Correcting under-sensitivity Based on experience in the aviation domain, typical problem with the initial assessments by evaluators may be under-sensitive measures. If under-sensitive measures are found, this problem can be addressed by changing the rating scale or by rater training. The changes to the evaluation scale must be designed to increase discrimination and reduce error in the evaluation process. Especially if training is minimal, it is important that each scale point has a qualitatively distinct, well-understood meaning to the evaluator. For performance assessments, each scale point should be a discrete, meaningful level of performance with an appropriate label or judgment anchor.   For example, when evaluating aircrew performance, the scale points should have explicit andqualitatively distinct meanings to the evaluators. An example of a 5-point scale for evaluating aircrews using this principle is given in Table 3.4.

Table 3.4 Example of a 5-point crew performance judgment scale

Judgement level	Performance Description	Required evaluation action
(1) Fail/ Unsafe	Simulator crash. Performance endangers passengers or aircraft	Crew fails and must be re-scheduled for additional training
(2) Repeat Required	Performance is unsatisfactory but not dangerous	Crew must repeat the item or maneuver until satisfactory
(3) debrief required	Performance is satisfactory but has significant weaknesses	Crew is debriefed on the performance weaknesses
(4) standard performance	Crew complies with company standards for performance in all respects	Crew passes with no repeats or debriefs required
(5) above standard performance	Crew exceeds company standards for performance in one or more critical areas	Crew is given positive feedback for good performance

These five judgment levels may constitute a natural scale for the aircrew evaluators since the levels correspond to qualitatively distinct required actions. When the judgment scale includes levels that cannot be directly tied to distinct actions, examples, or anchors, the level of rater training will probably have to be correspondingly increased. Training may not be able to overcome the inherent error in using judgment scales with an extremely high number of levels, such as a 100-point judgment scale.

For expert evaluators, training can allow them to use a more fine-grained, sensitive scale while still preserving adequate levels of reliability and validity. For experienced raters evaluating crew performance in aviation, sensitivity values of .3 to .4 were achievable with only modest amounts of training (Williams, Holt, and Boehm-Davis, 1997). The upper limit on the sensitivity of measurement in this case will depend on the extent of training and degree of experience of the evaluators. Even untrained evaluators may be able to distinguish an extremely coarse "good" vs. "bad" or "pass" vs. "fail" performance. Conversely, evaluating performance on an 11-point scale might require considerable training. Based on our experience, evaluators with minimal training can use a four or five-point scale with adequate reliability and validity. For this group of expert evaluators, the four or five-point scales may provide an adequate trade-off between sensitivity of measurement and the cost of extensive evaluator training.

Using the omega-squared estimate of sensitivity is particularly important when there is a critical area of the scale for which sensitivity must be established. In the above example, it may be critically important to distinguish unsafe crew performance (level 1) from inept but safe crew performance (level 2). In such a case, the estimate of sensitivity can be performed on a set of test stimuli that represent a 1 vs. 2 level performance. However, if the real level of the test stimuli cannot be precisely established, then using the omega-squared approach to estimate sensitivity is not possible. In these cases it may still be possible to use an alternate approach such as the intra-class correlation coefficient that only requires a representative sample of the possible stimuli.

Intra-class correlation coefficient The alternative approach to establishing sensitivity requires a representative sample of stimuli as a starting point. Generally, this in turn requires some type of enumeration of a stimulus domain and a random selection of stimuli from that domain. The sample of stimuli will be representative to the extent 1) that a reasonable sample of the stimuli can be obtained, and 2) each stimulus or case is randomly selected. For evaluating aircrew performance, for example, this approach would require an archive of representative videotapes of crew performance that could be randomly sampled.

To estimate the sensitivity of measurement for such a sample, the intra-class correlation coefficient can be used. The logic of the intra-class correlation coefficient is quite similar to omega-squared discussed above. The variability in mean judgments caused by a sample of stimuli is compared to the total variability of all judgments. This ratio has the same expected range of values from 0 to 1. As was the case for omega-squared values, values near 1.0 would indicate good sensitivity of judgment while values near zero would indicate poor sensitivity of judgment.

The interpretation of the values for this sensitivity index is similar to omega squared with one critical difference. Inherently, the omega-squared estimate was for a fixed set of stimuli with known values. In contrast, the intra-class correlation coefficient is the proportion of judgment variance caused by the random set of stimuli. Hence, the intra-class correlation coefficient is general for a stimulus domain. The potential advantage of using the representative sample is that the sensitivity estimate generalizes to all stimuli in the specified domain.

In practice, different methods of estimating sensitivity can be used at different points in the construction of a SIS. The range or variance ratio are heuristic methods that lend themselves to being included in the initial scan of the data input to the SIS. The range or variance of expected real values for a construct can be compared to the observed range for the corresponding measure and any discrepancies noted for further more specific analyses. If the real values of specific test cases can be determined reasonably well, the omega-squared method can be used to more precisely check on the sensitivity of a measure. Alternatively, if the population can be explicitly defined and a random sample of test cases drawn from the population, the intra-class correlation can be used for a more precise sensitivity estimate.

In addition to sensitivity, the reliability and validity of measurement must be evaluated to assess the scientific quality of the data in a SIS. Evaluating reliability requires an estimate of all the systematic variance in a measurement compared to total variance. Measuring reliability depends on having a precise definition of reliability and carrying out appropriate procedures to make reliability estimates.

Measurement reliability

Definition of reliability

Informally, reliability can be thought of as the consistency or stability of measurement. Formally, reliability is defined as the lack of random error in the measurement instrument (Nunnally, 1978). All sources of systematic (non-random) variance in a measure contribute to reliability (see Figure 3.1). Unfortunately, this systematic variance can include systematic variance due to the measurement method rather than the desired construct. Further, the measurement variance may reflect constructs other than the intended one. Therefore, reliability overlaps with, but is distinct from, validity. Systematic variance can reflect any stable aspect of the person, team, or system being measured, even if this variance is not related to the targeted construct. The coverage of the validity issue in the next section will focus on the extent to which systematic variance reflects the intended construct.

The basic methods of estimating reliability were developed primarily for paper-and-pencil measures of constructs. Paper and pencil measures are usually rather direct measures of the intended construct. However, measurement methods which rely on human evaluators are more indirect than paper and pencil methods since the additional process of human judgment is required. Therefore, additional methods for training and estimating reliability of human judges have been developed. These advanced methods are intended to increase reliability of the human evaluators and have important nuances of their own; therefore, they will be covered separately.

Basic methods of estimating reliability

Basic methods of estimating reliability include: test-retest reliability, alternate form reliability, and internal consistency reliability (see Nunally, 1978 or Pedhazur & Pedhazur Schmelkin, 1991 for more information). Each basic method makes different assumptions about what are the main source of error in measurement and what things can be ignored. The basic methods are compared in Table 3.5.

Table 3.5 Comparison of basic methods for estimating reliability

Type of Reliability	Requirements	Assumed to be important	Potential confounds, biases, or problems
Test-Retest	One sample is measured at two different times	Stability or similarity over time	Memory, carry-over, stable response set
Alternate Form	Two equivalent forms of the measure	Stability over different samples of items from the same domain	Learning or carryover effects across forms, stable response set
Inter-item consistency	Multiple items designed to measure the same construct	Similar responses to multiple items in a single sample	All items are answered at the same time, inter-item carryover or sensitization, stable response set

The exact form of estimate for reliability for each of these methods will depend on the basic type of data. The different reliability estimating methods are listed in Table 3.6 below. Since the interval/ratio measures are the most common type of data in research, the versions of test-retest, alternate form, and internal (inter-item) consistency that are appropriate for interval/ratio data (right-most column) will be covered below.

Table 3.6 Possible indexes for estimating reliability with different types of data

	Type of Data		
	Nominal or Categorical	**Ordinal/Rank-ordered**	**Interval/Ratio**
Test-retest index	Kappa concordance index across tests	Spearman's Rho for rank-ordered correlation	Pearson's r for product-moment correlation
Alternate-form index	Kappa concordance index across forms	Spearman's Rho for rank-ordered correlation	Pearson's r for product-moment correlation
Inter-item index	Kappa concordance index across item subsets	Guttman's Coefficient of Reproducibility	Coefficient Alpha

Test-retest reliability Test-retest reliability is used to assess the temporal stability of a measurement over time. The same measure is given at two or more times and the results are systematically compared (Figure 3.7). Internal consistency and alternate form reliability methods are also presented in this figure and will be discussed in subsequent sections. The comparison for test-retest reliability is typically a correlation index. The value of the correlation reflects the reliability of measurement in the sense of stability over time. The maximum correlation value of 1.0 would indicate perfect test-retest reliability, whereas values near 0 would indicate a total lack of test-retest reliability.

For some situations, test-retest reliability may be the only possible estimate of reliability. For example, if there is only a single item that can reflect performance, then the only feasible way to assess reliability is test-retest. An example might be measuring performance of shoe salesclerks by using the number of shoes sold. For that particular job, this single item may be the only relevant reflection of better or worse performance. To assess reliability of this performance measure, the shoe sales of a set of clerks could be correlated from time 1 to time 2.

Clearly, even in this simple case, the reliability estimate could be influenced by factors such as the degree of aggregation for the measure. Daily or weekly shoe sales figures may be influenced by a variety of factors such as the weather, holidays, and so forth. Basing the single-item measure on a longer time period such as months would help cancel out some of these fluctuations and could well give a higher estimate of reliability. That is, the correlation of shoe sales figures from month 1 to month 2 should be stronger than from week 1 to week 2, and the correlation of yearly shoe sales figures should be stronger still. Conversely, correlating shoe sales from day 1 to day 2 might emphasize the chance fluctuations and show very poor test-retest reliability.

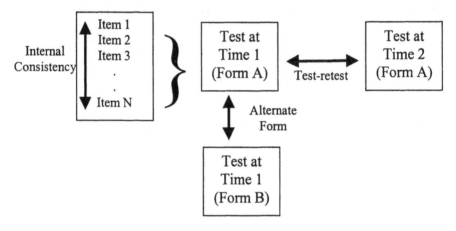

Figure 3.7 Illustration of three basic methods of estimating reliability

The level of aggregation issue in measurement has similar implications for the SIS as the level of granularity of the theory discussed in Chapter 1. That is, fewer data points are represented in the SIS if shoe sales are measured monthly or yearly than if they are measured weekly or daily. One consideration for setting the level of aggregation in measurement in a SIS is that it is relatively easily to aggregate lower-level data into high-level aggregates in computer analyses. However, if only the high-level aggregate is entered as data it may be impossible to subsequently decompose it into lower-level data. Therefore, a rule of thumb for measurement in a SIS is to measure an item at the most detailed level that is feasible. In this case, shoe sales could be recorded on a daily basis and aggregated as necessary for weekly, monthly, or yearly measures.   The reliability of performance measures at each level of aggregation could be assessed and only those performance indices with acceptable reliability used in further analyses in the SIS.

Entering the low-level basic performance measure may have a further advantage that it can be modified during the aggregation process to obtain a more precise measurement of the intended construct. For the example of shoe sales, the aggregation of daily sales performance into weekly and monthly sales figures can take into account the number of hours and days that each clerk was actually on the sales floor for that week or month. A more precise measure of sales performance can be made by using a corrected ratio of number of shoes sold per hour worked for the weekly or monthly interval. The test-retest reliability of this corrected ratio may be superior to the correlation of raw weekly or monthly sales figures since the fluctuations due to different number of hours worked are controlled for.

Similarly, the confounding influence of other important factors on performance can be adjusted for or if daily sales figures are recorded. Suppose shoe sales were larger on weekends than on weekdays. The average sales for weekends vs. weekdays can be used to adjust the average daily sales figures during aggregation. This adjusted average should be a better performance index for further analyses. Many other forms of adjustments are also possible if the basic performance measurement recorded in the SIS is daily sales.

Unfortunately, although test-retest reliability is simple and may be the only appropriate method for some measures, it may not be the most suitable method for all measures. Test-retest reliability assumes that all important sources of random error will be fairly reflected in the time interval from Time 1 to Time 2, and that the only factor causing stability in measurements is the intended construct. Each of these assumptions may be violated and bias the resulting reliability estimate.

The choice of the time interval for test-retest reliability must be carefully considered to see if all important sources of error have a reasonable chance to play a role during that interval. Take, for example, the influence of a person's emotional state on measures of variables like job satisfaction or marital satisfaction. Clearly the measure of satisfaction may be influenced in a positive or negative direction by the person's current emotional state. Some sources of influence on the emotional state, such as the mood of the respondent, could reasonably occur within a span of days. Other sources of influence, such as the respondent having a seasonal affective disorder, would require an interval of several months to a year in order to have a reasonable chance to affect the measurement.

The most common mistake is to choose a time interval that is too short to correctly estimate test-retest reliability. A short interval reduces the potential impact of the error sources with a longer time span and will generally lead to a positively biased estimate of reliability. A rule of thumb is to choose the longest feasible interval for estimating test-retest reliability.

Besides the constraint of practicality, the one important constraint on longer test-retest intervals is that the measured construct should be theoretically stable over the specified interval. Constructs such as ability and personality are typically theorized to be very stable over long time intervals, whereas constructs such as task motivation may only be stable for shorter time intervals.

Besides the measured construct, many other factors in the situation could cause similar responses from Time 1 to Time 2. Depending on the type of evaluation, these factors include being in the same testing situation, memory of the Time 1 set of responses at Time 2, and systematic differences in evaluators. Generally, these factors would positively bias the test retest estimate of reliability. For example, if a person remembers the response pattern from Time 1 while responding to the same items at Time 2, he or she may attempt to make the Time 2 answers the same to appear consistent rather than making a natural, independent response at Time 2.

Alternate form reliability Alternate form reliability uses a correlation of two separate tests of the construct in the same manner as test retest reliability. However, in this case each measure has a distinct, unique set of items (Nunnally, 1978). Potentially, both measures can be administered at the same time because the memory for the response to any particular item on form A will be unrelated to the response to the different items on form B. For example, the Stanford-Binet intelligence test has two equivalent forms that can be used to measure the construct of mental ability.

Although each form should have a unique, distinct set of items, the forms should otherwise be equivalent in difficulty, length, and so forth. Since memory should not bias this form of reliability estimate, it is usually preferred to test retest reliability. The disadvantage, however, is that two parallel forms of measurement instruments must be constructed and equated.

The conditions of administration of each form must still be carefully considered. If both forms are administered at the same time, the possibility of fatigue, sensitization, learning, or other sequential effects on the second measurement must be examined. Conversely, if the forms are administered at different times to avoid these problems, then the context of administration and the underlying stability of the construct itself should be considered. Any change in the context of administration may potentially impact on the measurement of the construct and reduce the reliability estimate. Similarly, a construct that may change over time may give a downward-biased estimate of reliability when assessed with alternate forms at two distinct time intervals.

Internal consistency reliability Internal consistency reliability refers to the internal coherence or intercorrelations of a set of items which are all measuring the same thing (Nunnally, 1978). The interrelationships among items in a set are summarized into a Coefficient Alpha index which ranges from 0 (no internal consistency reliability) to 1 (perfect internal consistency reliability). Several factors influence Coefficient Alpha, such as the number of items included in the scale (e.g. Cortina, 1993). In general, the internal consistency reliability will increase with additional items as long as these items correlate positively with the overall score.

In some cases, a modest positive correlation with the overall score may mask the fact that the items are really measuring a different, but related construct. To check on this possibility, more advanced multivariate techniques such as factor analysis can be used. In general, however, the internal consistency reliability of a measure is quick and easy to compute and does not make demands for additional measures. Therefore, the internal consistency reliability should be reported for any multiple item measure of a construct in a SIS. This requires, of course, that the answers to each item be entered into the as SIS as basic data. This illustrates once again the principle of recording the data at the finest level of detail for maximum utility in the SIS.

Unfortunately, as a group the basic methods for estimating reliability all ignore certain critical issues that surface when using human evaluators for measurement. Therefore, each of the basic methods can provide misleading results due to stable and systematic rater differences. Typically the stable and systematic differences would bias the reliability estimate to be more positive than it really should be.

For evaluators judging team performance, for example, systematic rater differences can affect test-retest reliability. Simple recall of previous Time 1 judgments can contaminate the Time 2 judgments. Even if recall does not occur, systematic harsh or lenient judgment standards among the evaluators can cause a similar biasing of the test-retest reliability estimate. Williams, Holt, & Boehm-Davis, (1997) found that untrained evaluators have strong and systematic individual differences to be harsh or lenient graders. If these differences are stable over time, they would cause a positively biased correlation between Time 1 and Time 2 scores. Clearly, a similar bias would be true for harsh or lenient evaluator standards across alternate forms of measurement of the intended construct. Therefore, alternate form reliability may also be systematically biased by systematic rater differences.

Similarly, internal consistency reliability estimates also ignore some forms of systematic judgment errors made by evaluators. One common systematic error is the halo rating errors. A halo rating error occurs, for example, when an evaluator initially rates a person or crew to have above-average performance. The error occurs if the evaluator continues to rate the person or crew as higher than average on subsequent items regardless of the actual performance. If a person or crew is initially rated negatively, a negative halo effect can bias subsequent judgments to be lower than the actual performance. Systematic harsh or lenient rater differences as discussed above would also create a spuriously high positive relationship among the items. Therefore, to the extent that these and other systematic errors occur across items, the internal consistency estimate of reliability is also inflated. To avoid such rater errors, special approaches were developed for checking rater reliability. Several such methods are compared in the next section.

Methods of estimating reliability for human evaluations

Potential methods for indexing reliability when human evaluators are used for measurement are presented in Table 3.7. The 360-degree evaluation was developed in Industrial/Organizational psychology for performance appraisal of managers. This approach emphasizes the different viewpoints of superiors, peers, subordinates, and the self. Inter-Rater Reliability (IRR) is a set of techniques for improving the coding of crew interaction in aviation. These techniques include several types of systematic feedback and rater training (cf. Williams, Holt, and Boehm-Davis, 1997). Referent Reliability techniques were developed by Johnson & Goldsmith (1998).

All these methods differ from the basic reliability methods in putting additional emphasis on the persons doing the rating as well as the items themselves. Clearly both the items and the persons doing the ratings are important to obtain a highly reliable measurement of complex team or crew performance. Therefore, these methods should be viewed as adding a new dimension or focus to the basic reliability methods presented above. The ultimate value of these methods can be assessed using the basic reliability indexes as the yardstick for acceptable levels of reliability. For example, the judgments of trained raters can be assessed on internal consistency for multiple item scales, or by test-retest reliability for single-item measures.

Table 3.7   Comparison of methods for estimating reliability of human evaluators

Empirical estimates of Reliability	Requirements	Assumed to be important	Ignored
360-degree evaluation	Evaluations by superiors, peers, and subordinates	Different viewpoints for job performance	Time-caused errors, multiple items
Inter-Rater Reliability	Multiple raters evaluating the same sample of objects	Consensus among a group of expert raters	Time-caused errors, accuracy of the consensus judgment
Referent Reliability	Externally-established true scores for objects or cases	Convergence of all raters with a referent standard	Time-caused errors, Initial divergence among raters

360-degree evaluations The 360-degree evaluation was developed to give good developmental feedback to managers rather than to enhance the reliability of assessment (Muchinsky, 1997). Evaluations by the supervisor, peers, and subordinates are generally supposed to give different views of performance that help a manager change. Typically, these evaluations are compared to a self-evaluation, leading to a revised self-evaluation by the manager. Since this revised self-evaluation should be more accurate, the manager can have a more objective basis to change and become more effective.

Taking a measurement viewpoint, however, the multiple evaluation sources offer an improved estimate of performance compared to single-source method. Evaluations by the supervisor, peer, subordinate, and self can be considered multi-person evaluations of performance of the target person. Since the performance information is obtained from different sources, the unique confounds or biases of any particular source will be much less influential in the composite measure.

For example, a common finding in the 360-degree feedback literature is that the self-ratings of managers are more positive than the ratings from the other sources. Conversely, supervisors may be influenced by self-enhancement behaviors or ingratiation behaviors on the part of the manager while peers are not. That is, each viewpoint can have its own systematic bias. If these assessments are combined into one evaluation, the bias of any single evaluator is reduced. If the 360-degree evaluation is used to estimate performance in this way, the average correlation among the ratings of the different sources can be used to estimate overall reliability of the combined evaluation. The use of multiple evaluators is extended in the IRR approach.

Inter-rater reliability Using the IRR approach, each evaluator's ratings of the test cases are compared to the group's judgments by using four indexes. Each index provides information on one aspect or component of reliability. The overall picture is given by the set of index values or an alternative summary measure such as an intraclass correlations that reflect agreement or consistency (McGraw and Wong, 1996). In addition, an estimate of the sensitivity of judgment can be included in this process if the stimuli levels have been accurately determined by SME's.

First, the overall distribution of each evaluator's ratings is compared to the group's distribution of ratings for the same set of items. The *congruency index* measures the extent to which these distributions match. Typical values range from around 0 for chance congruency to 1.0 for a perfect match. For example, a congruency index of .90 means the rater gives the same mix of 1s, 2s, 3s, and 4s as the group. Low congruency suggests that the rater gives a different mix of ratings on the scale compared to the group. The different distribution of ratings is important because this distribution is tied to the possible values of the other indexes like systematic differences and inter-rater correlations. Low congruity will generally signal systematic differences among raters, lack of agreement on specific items, or random guessing.

Second, *systematic differences* of harsher or more lenient grading among the evaluators are tested for the group as a whole and for each individual rater. An analysis of variance of the raters gives an overall indication of whether the differences among the raters in the group are larger than would be expected by chance. For untrained raters in the aviation domain, research using this approach has almost always found significant systematic differences. The individual tests for untrained raters give each person the feedback of whether he or she grades significantly above the group norm or significantly below the group norm.

Third, the *inter-rater correlation* is calculated to see if the raters shift in a consistent manner up and down in their ratings across evaluated items. The index used is the regular correlation coefficient averaged across pairs of raters. The average for the entire group of raters gives an index of the extent to which the raters were consistent among themselves. Separate feedback is provided for each rater on his or her average correlation with the others in the group. This feedback includes a graph of each evaluator's rating profile across items compared to the group's average rating profile. This gives the evaluator feedback on whether he or she consistently moved to higher or lower judgments along with the rest of the group.

Fourth, if the test cases have been externally scored by subject-matter experts, the raters can also be assessed on the *sensitivity* of their evaluations as discussed in an earlier section. Each rater's judgments are analyzed to see if the rater's average judgment was different for stimuli that were known to be higher or lower in value. The degree to which the rater's judgments reflect the real differences is summarized in the omega-squared index. The value of this index and a bar graph of the rater's average judgments for different values of the stimuli are given in the feedback and training session.

For the feedback at the beginning of the rater training session, the congruency, systematic differences, consistency, and sensitivity results are given to individuals. Furthermore, the aggregate results for the set of raters as a whole is reported to the group. This information helps pinpoint either general problems with the group or specific issues for a subset of individuals, as appropriate.

The second part of the rater training session emphasizes a detailed discussion of items that have low group agreement. The index used for measuring agreement on an item is r(wg). The group is presented with the distribution of ratings for each item with agreement below a benchmark value such as .70. Every item with low agreement should be discussed until a common judgment framework is developed that resolves the differences. For detailed evaluations of multiple segments of videotaped aircrew performance, this discussion and resolution can take hours.

In summary, the IRR method compares each rater to the group by using indexes that give the rater information about the congruency, systematic differences, consistency and sensitivity of his or her evaluations. The information from these indexes is then used to improve group consistency and agreement in ratings (Williams, Holt, & Boehm-Davis, 1997).

Referent Rater Reliability   Johnson and Goldsmith (1998) developed referent Rater Reliability (RRR). Their RRR method requires initially establishing a "referent" which represents the true ratings that should be given for each discrete judgment. A group of highly experienced SMEs work as a team to establish the referent by making their own judgments of videotaped crew performance and carefully resolving any discrepancies among their judgments. The accuracy of this standard is critical to establishing the reliability of judgments and training the evaluators to be more reliable.

The referent judgment standard is used to compute two indexes that compare each evaluator's judgments to the standard. Each index estimates one component of reliability in this approach. The first index is the RRR correlation. The RRR correlation is the Pearson correlation of the referent values with the evaluator's judgments. This is conceptually similar to the inter-rater correlation discussed above except that the evaluator's judgments are correlated with an external judgment standard rather than with the other evaluators. Correlating with a true judgment standard rather than other raters makes this index reflect a judge's sensitivity to the different levels of performance represented in the referent standard rather than consistency among other judges. The RRR correlation is also averaged across the evaluators in a group to provide an overall index of this component for the group of evaluators.

The second index is the Mean Absolute Difference (MAD). The MAD index reflects the average deviation of an evaluator's judgments from the referent standard. The index is standardized such that a value of 1 indicates no deviation from the judgment standard, and a value of 0 indicates the maximum possible deviation for all items. As with the RRR correlation, the MAD index is reported for each individual and an average value is reported for the group as a whole.

Johnson and Goldsmith (1998) recommend using both indexes as separate pieces of information for diagnosing reliability issues and rater training. The referent used in RRR training is conceptually an external criterion of what the ratings "should" be. In that sense, the construction of the RRR referent must address issues of validity which are quite similar to the criterion validity issues discussed in the next section on measurement validity.

Measurement validity

Definition of validity

Validity is the extent to which a measure really measures the intended construct (Nunnally, 1978). More specifically, validity is the proportion of variance in a measure that accurately reflects real variation in the targeted construct. Different approaches can be used to estimate the extent to which the reliable variance in a measure validly reflects the intended construct. These approaches are presented in Table 3.8 and discussed below.

Table 3.8 Comparison of basic methods for estimating validity

Approaches to estimating Validity:	Requirements:	Assumed to be important:	Potential confounds, biases, or problems:
Face Validity	Acceptable experts for the domain	The realism of the content of each item	Biases/errors in the judgments of the items
Content Validity	Domain items or topics can be listed	Representative sample of domain items (random)	Biased sampling, context-dependent nature of items
Structural Validity	Specified process or structure for items related to construct	Match of observed relationships to specified structure	Method variance. Structure/Process may not guarantee complete content
Criterion Validity	Acceptable criterion or referent	Match of measured construct to the criterion	Lack of precision or sensitivity in criterion
Predictive and Concurrent Validity	Measures of predicted outcomes	Match of measured construct to the predicted outcome	Lack of precision or sensitivity in predicted outcome
Convergent and Divergent Validity	Specification of constructs related and NOT related to the target construct	Pattern of observed relationships that matches the specified relations	Ambiguous specifications. Imprecise measures of other constructs
Construct and Network Validity	Specification of theoretical network of antecedents and consequences of the target construct	Pattern of observed relationships that matches the specified pattern for all related variables	Ambiguous theory or construct specifications. Imprecise measures of other constructs

Approaches to validity The approaches in Table 3.8 can be roughly grouped into three major families. The first family consists of approaches that compare the measurement items to the desired domain for the construct. Face and content validity methods essentially focus on matching the measurement items to the desired domain, but do so in a different manner.

The second family consists of any approach that examines the interrelationships among the measurement items or sets of items. One approach is represented by structural validity in this table. Structural validity systematically analyzes the pattern of relationships among the items in a measure. This pattern of observed relationships is compared to a theoretically expected pattern to determine the degree of validity.

The third family consists of approaches that compare the result of measurement to external measures. Criterion, predictive, convergent-divergent, construct, and network validity all focus on the extent to which the construct measure has the expected relationship(s) with other measures. This evaluation of the expected relationships ranges from very simple for criterion validity to quite complex for construct and network validity. Each approach is discussed below.

Comparing test items to the domain

Face validity The items used for measuring a construct should be checked for face and content validity (Nunnally 1967). Face validity is the judgment by one or more experts that the items are plausibly measuring the desired construct. Typically, an initial pool of items is evaluated. Items judged as relevant are kept in the final measure while items judged as irrelevant are discarded.

Such judgments by experts are easy and convenient, but there are problems. First, how are the experts objectively defined and selected? Would a different definition or selection of experts resulted in the selection of different items? It is usually impossible to answer these basic questions, and therefore the assessment of face validity must be considered subjective to some extent. In our aviation research, we found that often a more objective item analysis will indicate that items designed by experts to measure a given construct did not, in fact, predict that construct (Hansberger, Holt, and Boehm-Davis, 1999). Face validity is, therefore, easy to establish but only weak evidence for validity.

Content validity Content validity is a more objective method for establishing the fact that a set of items fairly and accurately represents a given domain. However, content validity requires the careful definition of the domain, enumeration of items in the domain, and representative selection of items for the measure. The first step is specifying all possible relevant items for the domain. Unfortunately, the specification of the domain of all possible relevant items for any complex or open-ended domain may be difficult or impossible. Consider, for example, defining all possible relevant items for the job performance of a mid-level manager in a typical firm.

If domain items can be satisfactorily enumerated, content validity can then be demonstrated by showing that the evaluation items are a fair, unbiased, and representative sample of items from this larger domain. Typically this is done by a random selection of items from the domain or a variation of this technique such as stratified random selection that ensures different sub-domains are evenly represented. Unfortunately, the representative selection of items from a domain may not be sufficient to insure a good measure of the construct. Different items may, for example, have unique characteristics that cause variations in response unrelated to the measured construct. Further, items may be too easy or difficult to give useful variability. Thus content validity is still insufficient evidence that a measure will have variability reflecting the intended construct.

Examining interrelationships among the measured items

Structural validity Structural validity is based on finding a unique, expected pattern of interrelationships among the items or sets of items of a measure. Structural validity can be used whenever there is a strong theoretical reason to expect a specific pattern of covariation among the items in a measure. One common method for assessing the pattern of covariation is factor analysis.

The expected pattern of covariation for a single, unitary measure is that all of the items should covary fairly strongly. Conversely, if a construct has several independent components, then the set of items measuring each component should covary, but not necessarily the items measuring different components. Checking for the expected factor structure can be accomplished by the methods of confirmatory factor analysis (see Tabacknik & Fidell, 1996 for a conceptual overview and Waller, 1993 for a review of programs to do this).

For human raters, this idea was adapted to describing the pattern of ratings resulting from a specific rater process by Holt, Meiman, and Seamster (1996). Holt et al were concerned with validity for scenario-based evaluations of aircrew performance. Instructor/evaluator pilots were trained to assess crew performance by first assessing performance on critical technical or management-oriented crew behaviors, then rating the related management or technical skills, and finally rating the overall performance of the Captain and First Officer for that flight segment. As shown in Figure 3.8, the trained sequence of evaluation from specific behaviors to skill evaluations and then overall evaluations of Captain and First Officer set the structure of the expected pattern of relationships.

Depending on the exact pattern of relationships expected for structural validity, the techniques of confirmatory factor analysis, path analysis, or structural equation modeling should be used. Each of these techniques evaluates the extent to which the expected pattern of relationships is found among the items in the measure. Finding this expected pattern of relationships supports the structural validity of the ratings.

Comparing the result of measurement to external variable(s)

Checking the relationship of a measured construct with external variables encompasses a wide variety of empirical techniques ranging from simple correlations to complex structural equation modeling. However, in order to use any of these methods, the relationship of the construct with other variables must be unambiguously specified by the theory. That, in turn, requires that the theory be internally consistent.

Internal consistency An internally consistent theory is one in which the theoretical propositions form a logically compatible set of statements. From a logically consistent set of statements, only one distinct prediction for the relationship between two variables can be derived. For example, the set of theoretical propositions may specify that ability should be positively related to performance.

If, however, a theory is internally inconsistent, one or more propositions in the theory is incompatible with the others. From a logically inconsistent set of statements, logically inconsistent relationships can be derived. From an inconsistent theory you could derive that ability should be positively related to performance using one subset of theoretical statements, but negatively related to performance using a different subset of theoretical statements. More succinctly, the theory can conclude that $A \rightarrow B$ and at the same time conclude that $A \rightarrow$ Not B or the opposite of B. Clearly this type of inconsistent prediction cannot be used to establish validity. Therefore, a prerequisite of empirically establishing the validity of a measure is to have a logically consistent theory.

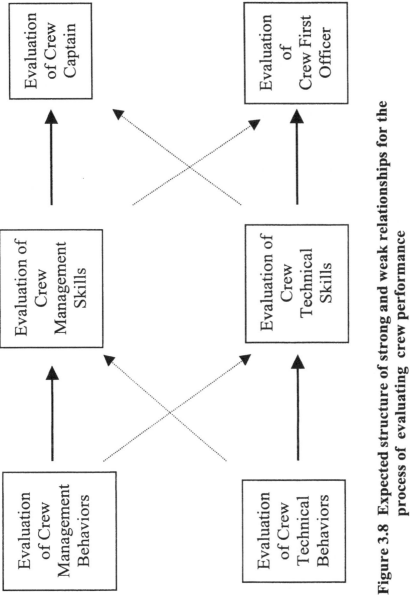

Figure 3.8 Expected structure of strong and weak relationships for the process of evaluating crew performance

In practice, the inconsistency of a theory will seldom be so obvious. Usually a key construct in a theory has a set of postulated antecedents and a set of postulated consequences. Checking the logical consistency of a theory requires checking both the set of links from the antecedents to the construct and from the construct to its predicted effects on the consequences. Logical inconsistency may be found in either the part of the theory leading from the antecedents to the construct or the part of the theory leading from the construct to its predicted effects (Figure 3.9).

In this figure, the central construct is cognitive workload during an abnormal or emergency situation in the cockpit. One antecedent of cognitive workload is the experience of the pilot which has two distinct effects. Pilot experience can increase pilot expertise and thereby decrease workload, or experience can increase the rigidity of habitual thinking that would increase cognitive workload. Thus, pilot experience has two contradictory or inconsistent effects on workload.

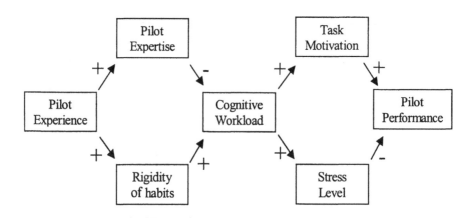

Figure 3.9 Example of a theory with inconsistent predictions

Similarly, the relationship of cognitive workload to pilot performance is also inconsistent. Higher cognitive workload can increase task motivation which will act to increase performance, but it can also increase stress levels which will decrease pilot performance. The difficulty with these contradictory predictions is that any observed relationship between cognitive workload and performance, be it positive, negative, or non-significant, can be explained by some mix of the two processes that have the inconsistent effects.

If inconsistency is found, the theory can sometimes be elaborated to specify the unique conditions under which A → B and the different conditions under which A → Not B. If these conditions are mutually exclusive and can be measured or identified, the theory can still be unambiguously tested. Evidence from these tests under the specified conditions would be acceptable evidence for (or against) validity.

In the above example, the conditions under which one causal path or the other will occur must be unambiguously specified and clearly differentiated. For example, ex-fighter pilots may not experience high stress levels in an emergency situation due to the nature of their previous experience. For these pilots, the link from cognitive workload to performance should be positive because the motivation process will be the main factor affecting performance. If these groups can be clearly distinguished, the distinct predictions for each type of pilot can be unambiguously tested.

Similarly, if moderating conditions for the inconsistency can be identified and clearly measured, the inconsistency may be resolved. In this case, a moderating condition might be the perceived threat level of the emergency or abnormal situation. For high levels of threat, the stress process could dominate the situation and cause a negative relationship between workload and performance. Conversely, for low levels of threat the motivation process could dominate the situation and cause a positive relationship between workload and performance.

The advantage of a SIS for this situation is clear. If the SIS includes a database of pilot background variables such as previous flying experience in the military, total hours as pilot-in-command, and extent of training, the relevant groups of pilots can be clearly distinguished and separated. Using the links to the performance database, the responses of the distinct groups of pilots to simulated emergencies can be analyzed to see if these responses differ in the predicted way. When the theoretical inconsistencies have been resolved, the validity of the theory can be empirically evaluated in several different ways.

Criterion validity Criterion validity requires the presence of a stable and accurate measure of one or more criteria that should be linked to the construct. If a test purports to measure sales ability, for example, a reasonably stable and accurate criterion may be the amount of sales for each person. To be stable, this criterion might have to be aggregated over a time sufficient for random fluctuations to even out as discussed earlier.

Assessing criterion validity is simple when there is only one criterion and the expected relationship between the construct and the criterion is linear. The measure of the construct is correlated with the measure of the criterion. The value of the correlation is the validity coefficient. If, for example, more intelligent shoe sales clerks were expected to sell more shoes, then the intelligence measure could be correlated with adjusted shoe sales. This correlation reflects the degree of connection between the construct and the criterion and must be carefully assessed to evaluate the extent of criterion validity.

To help decide if the criterion validity is acceptable, the scatter plot of values of the construct vs. criterion values should also be inspected. The scatter plot is also valuable for showing possible problems in validation such as restriction in range of either the construct measurement values or the criterion values. The first scatter plot (Figure 3.10) is an example of poor validity. The linear relationship between intelligence and shoe sales is almost flat, and the intelligence of the salesperson only predicts about 1.7% of the variability in the criterion of shoe sales. The R^2 value represents the proportion of the criterion variance accounted for by Intelligence and is converted to a percent of variance by multiplying by 100. The low value of R^2 and the amount of scatter in the plot both indicate poor criterion validity.

Suppose the theory also specified that extraverts should sell more shoes than introverts. The scatter plot below is between the personality trait of extraversion and shoe sales, and it illustrates moderate validity (Figure 3.11). The slope of the line relating extraversion to shoe sales is positive, indicating that extraverts sell more shoes than introverts do. The scatter plot clearly shows that this relationship is not perfect -- about 14.7% of the variability in shoe sales is accounted for by the introversion/extraversion of the salesperson. Although moderate in size, this level of relationship is significantly greater than a chance relationship of $r = 0$. Therefore, the criterion validity of this part of the theory is confirmed. This moderate level of criterion validity is typical for two reasons. First the strength of the observed relationship will be limited by the effects of unreliability of measurement for both constructs and criteria. Secondly, most criteria are jointly affected by multiple causes rather than a single construct. This multi-causality sets a limit on the strength of relationship between any single construct measure and a criterion.

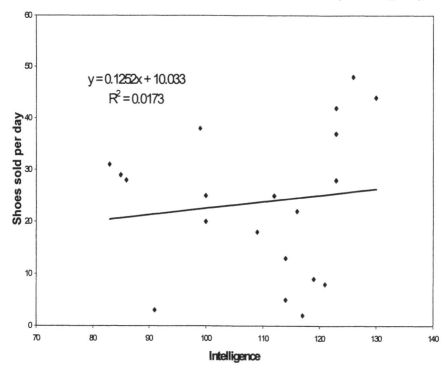

$y = 0.1252x + 10.033$

$R^2 = 0.0173$

Figure 3.10 Poor criterion validity

The value of the validity correlation can be adjusted for unreliability of measurement and for restriction of range. In the first example, the range of intelligence scores was restricted to IQ scores between 80 and 130. The range of IQ in a normal population of 70 to 145 could be used to estimate the correlation between IQ and shoe sales that should occur if clerks with a wider range of IQ were represented in the sample.

Similarly, the measure of introversion and extraversion in the second example may have had a reliability of only .75. The correlation of introversion/extraversion to sales could be adjusted to estimate the correlation between personality and sales that should occur if a perfectly reliable test of personality were used. These adjustments increase the estimated value of the validity coefficient and make it better reflect how well the underlying construct is related to the criterion in the larger population. Typically, these adjustments increase the estimated value of the validity coefficient. These adjustments are routinely made in meta-analysis where the goal is estimating the population value of a relationship between two constructs, each of which are assessed by imperfect measures.

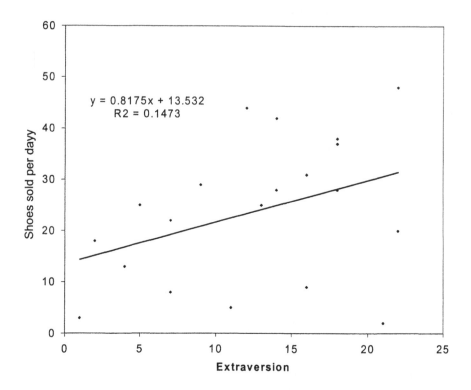

Figure 3.11 Example of moderate criterion validity

However, these adjustments give hypothetical values that do not correspond to the real, demonstrable relationships between a construct and a criterion for any particular sample. For example, although the general population may have an extended range of IQ scores, shoe sales clerks may not include either the extremely high or low segments of the population. Similarly, correcting for attenuation of validity due to the low reliability of a construct's measure relies on the accuracy of the reliability estimate. Since reliability estimates can be biased, the correction for attenuation can have an inaccurate result. Further, since the different reliability estimates discussed earlier could give somewhat different values, the user could end up with several corrected values and not be sure which was "true". Therefore, for routine use in a SIS the safest course is to focus on the basic, uncorrected estimate of the validity correlation and reserve the corrected values for meta-analysis or other cross-sample, population-estimate goals.

One major difficulty with criterion validity is finding an acceptable criterion. The shoe sales example is plausible only because it is a simple job with a plausible single criterion of performance. Even for jobs this simple, finding a good criterion measure can be very difficult. For job performance, for example, many potential criteria such as pay or promotions are crude or insensitive reflections of job performance. In addition, they may also be contaminated by irrelevant factors or confounds such as the effects of ingratiation and favoritism on pay and promotions. More complex jobs may require multiple criteria. Borman & Motowidlo (1993), for example, specified distinct task and contextual performance domains for many organizational jobs.

An aircraft Captain, for example, is managing a team and trying to operate a complex aircraft safely, efficiently, and on-time. The inherent task gives the multiple performance criteria of level of safety, efficiency such as fuel consumption per flight mile, and percent of on-time performance. Note that as in many professions, these task criteria cannot be all maximized at once. In fact, safety may require flight delays that increase fuel consumption and reduce on-time performance. Making up time during a flight might increase the on-time performance at the expense of fuel economy, and so forth. When the teamwork aspects of the Captain's job are included, it is clear that there are many potential performance criteria and the pilot is often juggling conflicting performance demands.

Multiple criteria make the assessment of criterion validity much more complex. The multiple criteria can either be combined into one super-ordinate performance index or analyzed separately. If the former approach is taken, the exact manner of combining the separate performance criteria into one performance index must be carefully analyzed and justified. This may easily become a difficult analysis task in its own right.

If the latter approach is taken, the presence of multiple analyses is problematic. It can easily occur that a measured construct would correlate highly with one criterion but not another. For example, more experienced Captains may have better safety evaluations but a lower percentage of on-time arrivals. In such a case, the evaluation of criterion validity is ambiguous. For this reason, although the evaluation of criterion validity has an appealing simplicity and can result in a single, clear cut index, it is not generally useful and some version of construct or network validity is often used in its place.

Predictive and concurrent validity Predictive and concurrent validity focus on a theoretical connection between a measured construct and another variable. The critical difference between concurrent and predictive validity is the time of measurement of both variables (Figure 3.12). In concurrent validity, the variables are measured at the same time. In predictive validity, the target construct is typically measured first and a predicted effect of the target construct is measured at a later time.

The theory should specify the nature and degree of relationship between the measures based on the theoretical relationship between the underlying constructs as precisely as possible. Since the other variable is typically a measure reflecting another construct rather than an absolute criterion, the issues of sensitivity, reliability, and validity of measurement will also apply to its measurement.

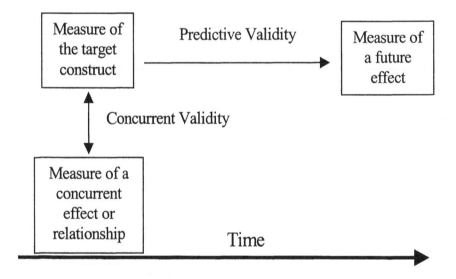

**Figure 3.12 Graphical illustration of predictive and
concurrent validity**

Concurrent validity often uses data from a cross-sectional research method in which a representative sample of a population is measured for both variables. The theoretical link between the measures must be assumed to happen at the time of the measurement or in the past. Concurrent validity can be misleading if there is a time lag involved in the connection of the constructs which is not accounted for in the measurement.

Suppose the theory specifies that salary increases are expected to produce increased job satisfaction and this correlation is to be used to validate a job satisfaction measure. If job satisfaction increases immediately upon receiving the salary increase, then concurrent validity can be used. If, however, job satisfaction would change very slowly after the salary increase, the expected correlation may not be found with a concurrent validity approach. Whenever a theory clearly specifies time lags or the *future* effects of a particular variable or construct, predictive validity methods should be used rather than concurrent validity methods.

Predictive validity often uses data from a longitudinal research method in which a representative sample of the population is first measured for the construct and then later measured for a theoretically-predicted effect of the construct. The timing for measuring the predicted effect must be correct. Some effects will take place quite quickly while other effects will occur more slowly. If the theory is imprecise about the time scale for the effects to occur, then several measures of possible effects at different times might be required.

Both predictive and concurrent validity use the correlation between the measured variables for evidence on validity. Unlike criterion validity, a higher correlation is not always better evidence for validity. If the theory specifies a moderate correlation, for example, finding a very strong relationship can be evidence *against* the validity of the measure for the target construct. Such a stronger-than-predicted relationship may be showing, for example, that the measures are confounded with common measurement variance that is inflating the correlation. Therefore, the observed correlation should match the theoretical value and be neither significantly smaller nor significantly larger than the expected value. When making these comparisons, it may be necessary to obtain better estimates of the validity coefficients by using adjustments for restriction in range in the sample or for the unreliability of measurement, as discussed above.

An example of an implied linear relationship would be the statement that "More intelligent students receive higher current grades and future grades in school." Since the relationship is not specified, a linear relationship can be inferred as the default and tested with the same correlation methods used for criterion validity (Figure 3.13). Since this theoretical statement does not give a qualifier for the expected strength of the linear relationship, the inferred value is that the relationship should be positive and significantly greater than zero.

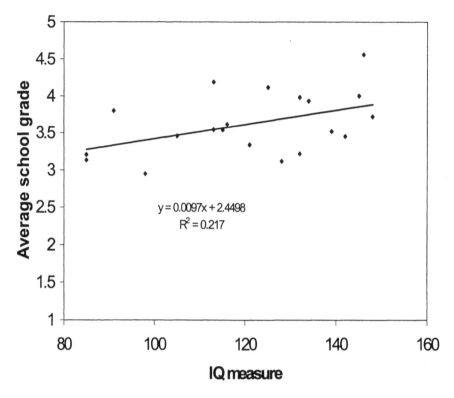

**Figure 3.13   Using correlation to evaluate predictive or concurrent
validity**

The example in the figure confirms this expectation and gives evidence
for the validity of the measurement of intelligence. The slope for the linear
relationship between the IQ measure and the average school grade is
noticeably positive. The relationship is strong enough that about 22% of the
variation in school grades is predicted by differences in the IQ measure.
Since it is plausible that other things besides intelligence affect school
grades, such as motivation, this moderate degree of positive relationship is
reasonable evidence for the validity of the IQ measurement.

Notice that the more imprecise the specification for the relationship is,
the correspondingly weaker is the confirmation evidence for concurrent or
predictive validity. In the above example, *any* positive relationship that was
significantly greater than zero would have confirmed the theoretical
expectation and supported measurement validity. But the probability of
finding some positive relationship between any two randomly selected

variables is about 50 percent, so finding a small positive relationship would not, by itself, be really strong evidence for validity. That is because a slight positive correlation between the IQ measure and average school grades might be there for some reason not related to the theory.

For example, suppose that the IQ measure is a paper-and-pencil measure. In that case, the general reading facility of the students could cause slightly higher scores on the IQ measure. Since general reading facility could also lead to slightly higher average school grades, the general reading facility could cause a slight positive correlation between the IQ measure and school grades. This effect could occur even if the IQ measure did not really index the construct of intelligence as expected.

In general, confirmation of imprecise predictions gives weak evidence for validity while confirmation of precise theoretical predictions gives stronger evidence for validity. Therefore, the more precisely a theory can specify any expected result, the more strongly it can be evaluated. Since many theories emphasize linear relationships, the expected result may be the exact value of a correlation. Finding the precise expected value of the correlation strongly confirms a theory while finding significantly discrepant values of the correlation strongly disconfirms the theory.

Correlations only work for testing theoretical relationships between measured constructs that are linear. A linear relationship is the simplest form of theoretical connection of two continuous constructs, and can be assumed to be the default when the relationship is not clearly specified. Sometimes, however, a relationship that is non-linear is implied or explicitly stated by a theory. Careful reading and interpretation of the statement of verbal theories may be necessary to distinguish which relationships should be linear and which non-linear. Theories expressed mathematically or computationally will typically be quite clear and precise about the specified nature of the relationships. The principle of strong inference also applies to the confirmation of non-linear relationships. That is, the more clearly and precisely the relationship is specified, the stronger the impact of confirmation or disconfirmation on the credibility of the theory.

There are a wide variety of non-linear relationships including exponential, logarithmic, and polynomial forms. These non-linear relationships may also be specified as concurrent relationships or predictive relationships. In either case, once the theory specifies some form of non-linear relationship, the basic principle of assessing validity applies, but methods other than correlations are used to establish the relationships.

The scatter plot describing these relationships will, of course, look different when the predicted relationship is non-linear. For example, the Yerkes-Dodson (1908) law relating arousal to performance specifies that as arousal changes from extremely low to moderate the performance increases, but as arousal continues to increase from moderate to extreme, performance decreases. Intuitively, the increase in arousal from near-sleep to normal conditions helps performance, whereas performance for over-aroused or panicked individuals again decreases. In the aviation domain, over-arousal can be caused by the evaluation process and has been given the name "checkitis" by some pilots. This curvilinear relationship is illustrated in Figure 3.14.

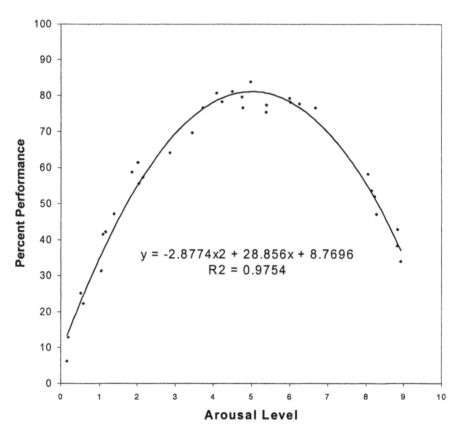

Figure 3.14 Example of a predicted curvilinear relationship

In this example, persons with very low arousal perform poorly. Performance increases as arousal increases up through the moderate levels of arousal (levels 4-6). Increasing arousal above that level leads to declines in performance as predicted by the Yerkes-Dodson law. The equation shows that the prediction includes the linear part of arousal level (29.8 x) and the non-linear quadratic component of arousal level (2.8 x^2). The R^2 for the equation shows that 97% of the variability in performance is predicted by the linear and non-linear relationship with arousal. This degree of confirmation of a non-linear relationship is very strong.

The array of information in the SIS facilitates the evaluation of either concurrent or predictive validity. Any variable measured in the same time-span as the target construct will be represented in the database and is a potential source for evaluating concurrent validity. Predictive validity can also be explored more thoroughly using the connected database information. Since all information in a SIS can be time-stamped, the relationship of a target construct to a wide variety of future measures can be analyzed.

These analyses should be guided by development and extension of the theory surrounding the construct wherever possible. However, if the theory is not sufficiently developed, empirical exploration of predictive validity is also possible. In essence, the empirical exploration focuses on first finding significant concurrent or predictive relationships, and then elaborating the theory to account for the observed relationships. Whenever this approach is used, however, the theory should be constructed so that it explains both the significant relationships with some measures (convergence) and the lack of relationships with other measures (divergence) as discussed below.

Convergent/divergent validity The principle of convergent and discriminant validity is that the theory can specify some variables that ought to correlate positively or negatively with the construct (convergent), and other measures that ought to be completely independent of the construct (divergent) (Campbell & Fiske, 1959). For the convergent part of validity, variables that ought to be related to a construct should converge or correlate with the proposed measure as expected.

For the divergent validity part, variables that ought to be independent or distinct from the construct should diverge or not correlate with the proposed measure. Divergent validity is particularly important if potential confounds could contaminate the measurement process. In that case, measuring the confound and showing that it is NOT related to the measurement of the construct is an important part of establishing the validity of measurement.

One application of convergent/divergent validity is the multi-trait, multi-method matrix (MTMM) proposed by Campbell and Fiske (1959). In the MTMM approach, the major measurement confound is the measurement method. For example, paper-and-pencil measures of attitudes or personality traits may elicit distorted responses from people because they perceive it as a "test" and try to present themselves in the most positive manner rather than answer the questions honestly. For example, Presser & Stinson (1998) found that interviews elicited much higher estimates of religious attendance than self-report measures or time-use measures such as journals.

To check the influence of measurement methods with the MTMM approach requires measuring several constructs that should be divergent (uncorrelated) using different methods of measurement for each construct. The essential idea is that measures of the same construct using different methods ought to converge or agree because they all measure the same construct. Conversely, measures of different, unrelated constructs ought to diverge even if the same type of measurement is used (e.g. paper and pencil test). If this pattern is found, validity is supported because the underlying constructs are influencing results more than the measurement methods. That is, the contribution of the construct to the observed data values is high while the contribution of the method to the data values is low.

To illustrate this method, suppose the theory specified that attitude toward children, attitude toward food, and attitude toward formal education were three independent or divergent attitudes. Further suppose that these attitudes could be measured with a paper-and-pencil (PP) test, a physiological test such as skin conductance (SC), and a biographical data measure based on a person's life history (BD). The expected pattern of results is shown in Table 3.9.

Table 3.9 Example of a Multi-Trait, Multi-Method correlation matrix
  "Hi" indicates high correlations, "Lo" indicates
  low correlations

		Attitude toward Children			Attitude toward Food			Attitude toward Education		
		PP	SC	BD	PP	SC	BD	PP	SC	BD
Attitude Toward Children	PP	1	Hi	Hi	Lo	Lo	Lo	Lo	Lo	Lo
	SC	Hi	1	Hi	Lo	Lo	Lo	Lo	Lo	Lo
	BD	Hi	Hi	1	Lo	Lo	Lo	Lo	Lo	Lo
Attitude Toward Food	PP	Lo	Lo	Lo	1	Hi	Hi	Lo	Lo	Lo
	SC	Lo	Lo	Lo	Hi	1	Hi	Lo	Lo	Lo
	BD	Lo	Lo	Lo	Hi	Hi	1	Lo	Lo	Lo
Attitude Toward Education	PP	Lo	Lo	Lo	Lo	Lo	Lo	1	Hi	Hi
	SC	Lo	Lo	Lo	Lo	Lo	Lo	Hi	1	Hi
	BD	Lo	Lo	Lo	Lo	Lo	Lo	Hi	Hi	1

The expected pattern of correlations for a sample of respondents is shown in this table. The diagonal values of 1.0 simply indicate that the correlation of a measure with itself is the maximum value of the correlation coefficient, 1.0. The three boxed areas of this table indicate the sets of measures that ought to be highly correlated if the constructs are being validly measured. The upper left box shows that the measures of attitudes toward children converge because these measures all intercorrelate highly. Similarly, the middle box shows that the measures of attitudes toward food converge and the lower right box shows that the measures of attitudes toward education converge.

All the cells with "low" correlations show divergent validity. Showing that the measurement methods do not by themselves induce correlations is very important. This means that the measured attitudes mostly reflects the real underlying attitude constructs rather than reflecting the measurement method, which supports measurement validity.

There are clear advantages to using a SIS for clarifying the influence of method variance on the measure of a construct. In general, a SIS should have multiple measures of key constructs, some of which are independent. In this case, the MTMM method can be directly used on the construct measures to establish validity. In aviation research, for example, the knowledge of aircraft systems and functions, FAA regulations and procedures, and CRM principles and procedures might be considered three conceptually distinct and independent domains of knowledge. The multi-trait multi-method approach can be used to establish the validity of measures for these knowledge domains.

For example, different methods of measuring domain knowledge such as an oral exam, a scenario-based flight test (LOE), and a computer-based test (CBT) could be compared in a MTMM. Within each knowledge domain, we would expect these alternate methods should give very similar results. Empirically, the knowledge measures for a single domain using alternate methods should be highly correlated, as in the example (see Figure 3.15). In this example, the heavy black arrows indicate that the Oral, LOE, and CBT measures of knowledge give highly correlated results within each separate knowledge domain. This pattern shows convergence of measurement of the knowledge constructs.

Conversely, the correlation of measures across domains using the same method ought to be noticeably lower. These relationships are indicated by the light dashed arrows in the figure. If good pilots tend to acquire more knowledge in all three domains compared to poor pilots, there may be some positive relationship across the knowledge domains. Although positive, the correlations indicated by the dashed arrows should still be low compared to the within-domain correlations. The pattern of results in this example supports the expected divergence of knowledge in the three different domains. Overall, the MTMM results in this example would also support the validity of the measurements in each of these domains.

Domains of Pilot Expertise

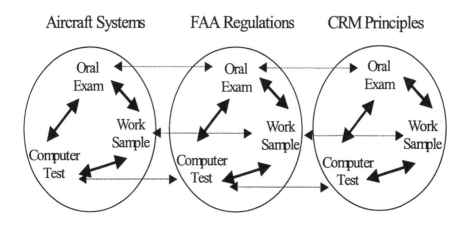

Figure 3.15   Convergent and discriminant validity in aviation evaluations

A second advantage of the SIS is that even if there is initially only one measure of a key construct in the SIS, new methods of measuring a construct can potentially be developed from the information in the connected databases. These new measures can then be used in validation. Both empirical and theoretically guided ways of innovatively combining SIS information may be used to create new measures of key constructs. An example of an empirical method would be the method of criterion groups, and an example of a theoretically guided method would be rational biodata.

Criterion groups   To create a new measure using criterion groups, groups must first be identified that differ on the target construct. Once found, a careful analysis of the responses of the criterion groups across a wide set of items may indicate items that systematically differentiate the groups.

Appropriate criterion groups may be defined by auxiliary measurements or diagnostic information already included in the SIS. Take, for example, the construct of pilot performance in the aviation domain. The domicile captains who supervise all pilots residing near an airline hub could be queried for their judgments about the very best and very worst pilots. these judgments could be used to select the criterion groups of good and poor performing pilots.

Using the SIS information, the good and poor groups of pilots can be tested on a wide variety of measures to find items that significantly separate these two groups. One useful method to do this is discriminant function analysis (DFA), which is discussed in more detail in a subsequent chapter. DFA maximally distinguishes groups using the available items (or scales), and gives classification functions that allow the prediction of other cases into the groups. In this case, the discriminant function analysis would derive classification formulas that distinguish good and poor pilots. Further, the analysis would specify the optimal way of predicting whether other pilots would be likely to perform well or poorly, given their profile of information on the same items. When appropriately combined, these items could become a new scale for measuring pilot performance in the SIS.

Rational biodata   In contrast to empirical methods like discriminant function analysis, the rational biodata approach relies on an explicit theory. The theory is used to select items that measure experience, motivation, attitude, personality, and other constructs (Mumford & Stokes, 1992). For example, an individual's learning motivation could be indexed by the difficulty of the elective classes chosen in secondary education together with the grades achieved in those classes. The connected historical information that is accumulated in a SIS provides a rich base for using this method to measure key constructs, particularly if the item-level responses are recorded as SIS data. If so, all these items are available for constructing theoretically-based measures of key construct.

Once a new measure is constructed, reliability should be assessed. If acceptable, the convergent and divergent validity can be assessed by examining the correlations of the new measure with other validated SIS measures. In this way, the information in the SIS can be used in a bootstrap fashion to create alternative measures of a construct for validation and later use in solving problems.

Construct and network validity   Construct validity uses the theory or theories in a SIS to elaborate a large, comprehensive set of predicted relationships of a target construct with other measures (Nunnally, 1967).

These relationships may include criteria, predicted effects, concurrent relationships. Thus, construct validity may subsume the types of validity discussed above. However, construct validity is more general than a single correlation because an entire set of relationships must be evaluated. Therefore, construct validity is also more difficult to evaluate than other forms of validity.

Network validity has a similar emphasis on pattern of multiple relationships among constructs. Since each construct is associated with one or more measures, the expected outcome is a pattern of relationships among measures. For both network and construct validity, the theoretically-expected network of relationships is empirically compared to the observed pattern of relationships, (Pedhazur & Pedhazur Schmelkin, 1991). This set of theoretical expectations typically includes a set of expected antecedents, concomitants, and consequences for the target construct.

An example for the focal construct of mental ability is given in Figure 3.16. In this figure, aspects of the genetic inheritance of the person as well as the developmental environment are antecedents of mental ability. Conversely, life-span outcomes in the work and economic domains are expected outcomes of mental ability. Analysis of corresponding measures in a SIS would tend to confirm or disconfirm this expected pattern and support or weaken the confidence in the validity of the intelligence test. Methods such as structural equation modeling (Tabachnik & Fidell, 1996) can be used to verify or disconfirm these networks of predicted relationships.

Validity evidence from exploratory analyses

Exploratory data analyses can also give evidence on validity. Exploratory analyses for establishing validity can range from simple searches for relevant correlations to very sophisticated multivariate techniques. Since criterion, concurrent, and predictive validity all involve basic correlations, exploratory data analysis for establishing them consist of examining relevant correlations among SIS measures that could be related to the target construct.

The obtained results from empirical exploration must be carefully checked against chance baselines to ensure that the obtained evidence is not a statistical fluke. The presence of significant correlations, for example, must be carefully judged against how many correlations were analyzed. The more correlations analyzed, the more likely some significant correlations will occur by chance. To compensate for this, the required significance level or size of the correlations can be appropriately increased. Further, these exploratory results can be crosschecked with other samples.

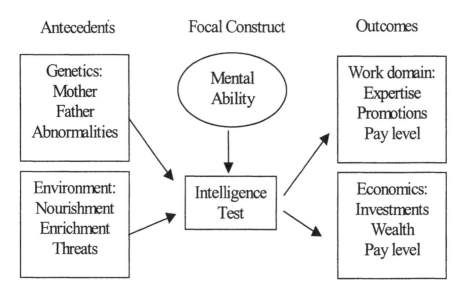

Figure 3.16 Theoretically-expected network of relationships

When an exploratory result has been carefully examined and replicated, then it can stimulate further development of the conceptual framework represented by the theory. The goal is to have any strong, significant results consistently explained and predicted by the theory. The extension of the theory to include new results also makes it easier to evaluate or determine the more comprehensive forms of validity assessment such as convergent/divergent validity or construct/network validity.

For exploratory validity analysis of the more complex forms of validity such as construct/network validity, techniques like exploratory factor analysis or data mining may uncover valuable relationships. The obtained relationships can be compared to theoretical expectations or matched against the best current understanding of the nature and effects of the construct. However, the theoretical connections arising from these analyses are essentially *post-hoc*. *Post-hoc* evidence is always conceptually weaker evidence for validity than the evidence from the *a priori* methods discussed above.

Since one known bias in humans is a tendency to interpret information to support preconceived notions (e.g. Darley & Gross, 1983), the appearance of evidence supporting the theory from exploratory analyses must be taken with a grain of salt. This confirmation bias also affects research scientists (Greenwald, Pratkanis, Leippe, & Baumgardner, 1986). Therefore, two relevant questions are 1) whether the analyst made an equally thorough and vigilant scan for evidence contradicting the validity of the measure, and 2) whether the post-hoc evidence presented could equally well or better be explained by some other cause which does not imply the validity of the measure. For these reasons, exploratory results are generally weak support for validity of a measure.

Once the theory is elaborated, however, the future testing of the relationships is *a priori* and can offer strong validity evidence. The elaboration of an existing theory must, of course, be carefully done to preserve the internal consistency of the theory. Further, the set of predictions of the elaborated theory must be empirically checked against a gamut of relevant results from the SIS to obtain confidence in the modification. Once the theory is consistently elaborated, the validity of the measure can be evaluated by obtaining a new sample of data and performing any of the empirical methods of validity discussed above.

If the original sample of cases in the SIS is sufficiently large, the "new" sample of data can be established as a separate section of the initial sample. The cases for this "hold out" sample must be randomly specified and separated prior to the initial analysis of the base sample. This approach allows intensive analysis and theoretical modification work to be done on the initial base sample without contaminating the independence of the information in the hold-out sample. After the theory is modified, the appropriate validity analyses can immediately be performed on the hold-out sample as if it were a new, independent sample, which is essentially a cross-validation approach. The initial random selection and separation of the cases in the hold-out sample justifies this use and gives more confidence in the results of the validity analyses.

Chapter summary

Scientific data quality requires sensitive, reliable and valid measures. Establishing these qualities takes time and effort. However, these qualities are absolutely essential for having a SIS that contains information as opposed to having a miscellaneous collection of dubious data. The quality of all information analyses of the SIS data depends on these basic qualities. Ignoring these qualities is like ignoring the quality of the foundation of a house and can lead to equally disastrous long-term outcomes.

The gain from establishing sensitivity, reliability, and validity is that the information consumer will know exactly what each measure in the SIS means and how well it indexes the intended construct. The theoretical structure of the SIS gives the precise meaning of each measure and is critical in establishing the quality of the data. Theory is particularly important for establishing the stronger forms of validity such as convergent-discriminant validity or construct validity for a particular measure.

In the initial construction of the SIS, this theoretical structure may be imprecise and incomplete. The evolution of the theoretical framework should proceed hand-in-hand with the validation of the measures in the SIS. This theoretical refinement process is discussed in the next chapters. As the relevant theories are developed, the extensive information in the SIS offers a much broader scope for establishing the scientific quality of the measures. In particular, the variety of measures in the SIS allows stronger, more comprehensive tests of forms of validity such as construct or network validity that depend on the pattern of relationships with other measures.

In turn, the establishment of scientific data quality, particularly validity, can have important consequences for improving the theoretical view and understanding of the information in the SIS. Basic methods for elaborating the theory and evaluating it are covered in the next chapter. Having strongly validated measures and theory allows the information in the SIS to be used to solve a wide variety of practical problems. That process is detailed in a subsequent chapter on solving problems using a SIS and elaborated in chapters on personnel selection, training, and systems analysis.

4 Evaluating Theories

Overview

This chapter covers the basic functions of a theory, advanced issues for theory evaluation, appropriate data for evaluation, and both qualitative and quantitative evaluation methods. Some of the analytical methods discussed in Chapter 3 will also be used in this chapter, in particular, the methods for predictive and construct validity. However, the focus on the use of the methods will be different in this chapter. In Chapter 3, the validity of the theory was assumed to be true so that the empirical methods could be used to evaluate the validity of a measure. In this chapter, the validity of the measure is assumed to be acceptable so that the empirical methods can be used to evaluate the validity of the theory.

Clearly there is a "chicken and egg" problem to establishing measurement validity using a theory on the one hand, and establishing theory validity using a measure on the other hand. As mentioned in Chapter 1, the refinement of the theories and measures in a SIS is a co-evolution process in which the initial form of both theory and measures is relatively simple. From this basis, however, the cumulative effect over time of the scientific development of the theory and measures is to improve the quality of both. A good quality theory has three essential functions that are discussed in the following sections: explanation, prediction, and control.

BASIC THEORY FUNCTIONS

The basic functions of a theory are the explanation, prediction, and control of some focal or target system. That is, a good theory should explain the target system, predict what will happen to the system in the future, and specify how to control or alter specific facets of the system. These three functions can be considered criteria for evaluating a theory. Evaluating competing theories is important for the development and use of the SIS. First, theories with clearly superior levels of explanation, prediction, and control should be chosen for further development and elaboration in the SIS, while inferior theories may be discarded. Second, applications of these theories to problems depend on the accuracy of the theory for adequate solutions (Chapter 5). Inaccurate theories can lead to disastrously wrong problem solutions, so it is important to consider the overall quality of a theory when using it in an SIS.

To be acceptable for evaluation as a scientific theory, a theory must be internally consistent as was discussed in Chapter 3. An inconsistent theory produces inconsistent explanations, conflicting predictions, and ambiguous specifications for changing a system in a desired manner. Inconsistent theories cannot be evaluated using scientific methods and cannot, therefore, be considered scientific theories. Assuming that a theory has been checked for internal consistency as discussed in Chapter 3, the explanatory, predictive, and control aspects of the theory can be evaluated.

Explanation

Plausibility

An adequate theory should plausibly explain how, what, where, when, and why things happen in a target system. This explanation is essentially at the conceptual level rather than the empirical or data level. The principles and processes of the theory must be evaluated as plausible or implausible for the target system.

Typically, this evaluation is subjective. That is, evaluating the plausibility of a theory depends on the particular viewpoint and expertise of the evaluator. This has the undesirable consequence that different evaluators can reach quite different conclusions concerning the plausibility of a theory. Theories involving the causal role of gremlins or fairies may be plausible to persons with New Age beliefs but very implausible to scientists. Conversely, the role of quarks of different colors, flavors, and charm in quantum theory may be quite plausible to theoretical physicists, but quite bizarre to people with New Age beliefs. One way to reduce the subjectivity and variation of this evaluation is to try to find a common ground for evaluation. The common ground may be a conceptual framework coming from the set of theories already existing in the field, or an empirical framework coming from analyses on a set of qualitative or quantitative data, as in a SIS.

Conceptual fit Using this approach, a new theory can be conceptually compared to current theoretical views for a particular domain. Current theoretical views can be abstracted from the major textbooks for a field, technical reports or journals published for each discipline, and books or monographs that reference the relevant domain. To the extent that a new theory uses constructs, principles, relationships and processes that are already proposed for a domain, the theory is conceptually plausible.

The more theoretically unified the domain, the more cogent is this aspect of the evaluation. A domain with a tightly unified and circumscribed set of theories would give a clear picture of the plausibility of a new theoretical viewpoint. On the other hand, a domain with a large and diverse set of theories could give a more ambiguous evaluation of the plausibility of a new theory.

Even in a coherent, circumscribed domain, however, a theory that fits the current viewpoints may turn out to not be empirically supported. Conversely, a theory that does not fit current viewpoints may have convincing empirical support. Therefore, although the basic plausibility and theoretical fit should be used as one selection tool for the theories in a SIS, the empirical validation of the theory must also be considered. Empirical validation may use qualitative methods or quantitative methods. Qualitative methods can increase or decrease the plausibility of a theory and are covered in the next section. Quantitative methods are typically used to confirm or disconfirm specific theoretical predictions and are covered after the prediction function of theories is discussed.

Qualitative methods

Qualitative methods are methods to collect and analyze unstructured information about a system. This unstructured information may come from different sources and take on different forms. For example, this type of information may be detailed descriptions of the people and events in the system, direct interviews or quotes from people in the system, or documents and archival information about the system. A common thread is that qualitative methods try to avoid imposing theoretically specified measurements or manipulations so that the information is an undistorted picture of the real system. Typically this results in unstructured information that must be analyzed with the appropriate qualitative methods. In general, qualitative methods are useful for the initial construction of a theory and for an initial evaluation of the plausibility of a theory, particularly when the theory is undeveloped or poorly defined. Two widely used types of methods for gathering qualitative data are observational methods and interviews.

A variety of observational methods are available including participant observation, direct observation and unobtrusive or trace measures. The focus of all observational methods is to describe the target system as completely and accurately as possible. The different methods all attempt to directly assess a target system, but vary in the degree to which the measurement process may change the system.

Observational methods   Different methods are available to make observations about the functioning of a complex system. Participant observation is having the observer act as an integral part of the target system while gathering the information. The observer may, for example, join a club or group which is the target of investigation. This process raises ethical considerations about the lack of informed consent of the real participants for being studied, but resolving those issues are beyond the scope of this book (for more details, see American Psychological Association, 1992).

The participant observation method is used in training in the aviation domain. Pilots who have just qualified to fly a particular aircraft must fly for a period of time with highly experienced training pilots on regular flights. This is called the Initial Operating Experience (IOE) phase of training. The training pilot must both participate as a normal pilot during the flight and carefully observe the performance of the newly qualified pilot. This observation and training occurs over many flights before the training pilot signs off the new pilot as qualified to fly the aircraft as part of a normal crew.

A participant observer may change the natural system functioning in two distinct ways. First, the participant observer must act as part of the system, and these actions may affect or contaminate the natural processes and outcomes that would otherwise be exhibited by the system. In the IOE example, the highly experienced training pilot may perform in an exemplary fashion that puts less demand and workload on the new pilot compared to a normal line crew.

Second, to the extent that the fact of observation and measurement becomes known among the real participants in the system, they may act or interact in an unnatural manner. For example, the fact that the newly trained pilot knows he or she is being evaluated for final qualification may elicit his or her best behavior. Conversely, the knowledge of evaluation could cause evaluation apprehension in which the new pilot performs more poorly than he or she would otherwise. This latter problem is well known to pilots and some call it "checkitis". Clearly, both participant actions and evaluation awareness can distort measurements of a system in either a positive or negative direction. The alternative method of direct observation offers a different set advantages and disadvantages.

Direct observation has the observer act as an external agent rather than as a participant while gathering the information. In the aviation domain, for example, official observers of commercial aircrews may ride in the cockpit jump seat and directly observe crew interaction and performance. The observer is usually trained to avoid making any direct action in the system. It would be inappropriate for jump seat observers to directly interact with Air Traffic Control during a cockpit observation, for example.

If the observer is well trained and avoids directly interacting with or affecting the system, the biasing of the target system can be minimized. The reactivity of knowing about the observation is still, however, an inescapable part of direct observation and may be an even bigger problem than with participant observation. The evaluation awareness can change the target system and bias the measurements in either positive or negative ways as discussed above.

Variations of direct observation such as using videotapes of the system interactions may decrease the awareness of observation and possible bias. Further, the coding of system interactions may be done in a more standardized environment and under more uniform conditions than would be possible for direct observation. If videotaping is done without the consent of the persons in the system, however, the ethical issues alluded to earlier are relevant. Conversely, if videotaping is done with consent the persons inevitably have some awareness of the evaluation. On the positive side this awareness often seems to fade as the videotaping progresses, possibly due to the decreased salience of the camera relative to other events occurring in the system. The decreased salience and awareness of the videotaping should, in turn, decrease the reactivity of the observation on persons in the target system.

The decrease in reactivity and possible gains in reliability with such a videotaping method must be weighed against the possible loss of some aspects of system information. Videotapes of aircraft crews in simulators, for example, do not have the same field of view and resolution as a direct observer. This limitation may make it impossible to measure fine details of the system process such as the readings on various gauges and instruments when a crew takes particular actions. In general, taping or other methods of recording a system are reducing reactivity while selectively preserving certain aspects of the target system rather than preserving the entire complexity of the original system. A further reduction in reactivity may be achievable if unobtrusive or trace measures of system function can be implemented.

Unobtrusive or trace measures remove both the observer and the awareness of a recording instrument from the setting (Webb, Campbell, Schwartz, Sechrest, & Grove, 1981). This avoids both any direct action of the observer that impacts the system and any reactivity of the knowledge of the observation process.   In the aviation domain, the Flight Operations Quality Assurance (FOQA) program is an example of using trace measures. The data on an aircraft's flight recorder are downloaded after the termination of a flight and used to analyze qualitative aspects of crew performance such as the stability of the approach to landing.   In other domains, there is a wide variety of unobtrusive or trace measures for the constructs and processes relevant to the plausibility of a theory.   For example, the agendas and minutes for committee meetings as well as the contents of email may serve as relevant measures of formal and informal communications in a management system.

All variations of observational methods attempt to study a target system functioning in a normal fashion as directly as possible.   In contrast, qualitative interview methods require one or more persons in the system to function as a respondent or reporter rather than in their normal operational role. The data reflect the respondent's *view* of the system rather than the system itself, and are therefore more indirect measures than the observational methods.

Interview methods   Potential interview methods range from an informal conversational interview to a semi-structured interview or a structured interview with open-ended questions. In a conversational interview, the interviewer interacts with a respondent in a normal conversational manner with the goal of eliciting information about the target system. Other than this goal, there is no fixed agenda of topics or questions. Therefore, the sequence of topics covered can be very idiosyncratic in both content and order.

In the aviation domain, I was talking once with a Standards Captain who supervised newly qualified pilots for their probationary period in line operations. He gave me an example of conversational interviews that he used for this training. He told me he typically elicited a lot of relevant flying information from the new pilot over a beer at the end of the flying day during a layover. The informal conversational setting allowed him to assess and discuss a variety of professional issues in a non-evaluative and non-threatening way, and make unobtrusive suggestions for improvements.

The conversational type of interview may be the least reactive form of interview in that the respondents may not even be aware that they are being interviewed.   This may decrease the effects of such confounds as measurement awareness on the answers but raises ethical considerations about the informed consent of the respondents similar to the participant observation method. Further, the conversational norms for a given situation may restrict what kind of follow-up probes can be asked of the respondent.

A conversational interview also results in less standardized content than either semi-structured or structured interviews.

In a semi-structured interview, the basic list of topics to be covered is specified and the respondent is clearly aware of the interview situation. The interviewer can cover the topics in any order, and is completely free to follow up answers with probing questions. Detailed probing may help clarify the meaning of the original response or uncover more extensive related information.

The semi-structured interview is an attempt to gain some uniformity in the type of information obtained by different interviews without constraining the form, content, and order of the questions. In the aviation domain for example, semi-structured interviews may be used by evaluators assessing the knowledge of aircraft systems. The basic list of topics are the major systems of the aircraft such as hydraulic, electrical, communication, and navigation systems. During the semi-structured interview, the evaluator has to be satisfied that the candidate has adequate knowledge about each aircraft system, but there is no specified order of assessment or list of questions. The flexibility of such assessments must be weighed against the advantages of standardization in an alternative assessment using a structured interview.

In a structured interview the form, content, and order of the essential questions are specified. In qualitative structured interviews, the questions are open-ended and allow a wide diversity of possible responses. The interviewer can ask follow-up questions to clarify responses, but not depart from the overall framework of the interview. Since the questions are precisely specified and given in the order of the interview schedule, the standardization of the administration of the questions is much higher for structured interviews than semi-structured or unstructured interviews. The standardization of the structured interview may restrict its usefulness for exploratory theory development but may enhance its usefulness for theory confirmation or disconfirmation.

An example of a structured interview in the aviation domain is using a computer to administer questions about aircraft systems. Using a computer, the order of questions is precisely specified and flexibility in asking follow-on questions is limited to what has been pre-programmed into the computer. Further, the exact phrasing and administration of each question to each pilot, as well as the allowable forms of response, is exactly the same. Thus, computer-administered interviews represent one of the most structured forms of an interview.

In general, for both observational and interview methods, there is an underlying continuum of the degree of standardization in the measurement situation (Figure 4.1). At the unstandardized end of the continuum, the methods are more purely qualitative and suitable for conceptual exploration, theory construction, and establishing plausibility. At the standardized end of the continuum, the methods merge into quantitative methods and start becoming more suitable for empirical theory evaluation such as the evaluation of prediction. The essential trade-off is that the more standardized versions of these methods correspondingly lose some of their heuristic value for theory exploration and construction, but are more likely to produce systematic, reliable data for evaluating predictions.

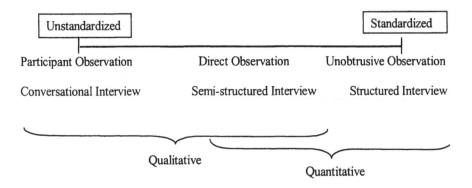

Figure 4.1 Standardization of observation and interview methods

Case study method A case study is the intensive examination of a single system using a wide variety of information-gathering techniques including the observational techniques and interview techniques discussed above. The goal of a case study is to investigate a particular system within a natural context in order to get an idea of both relevant contextual variables as well as the critical system variables. Both qualitative and quantitative methods can be used in conjunction with a case study. However, qualitative methods are more likely in the early development of a theory, while quantitative methods might be used for a case study within the context of a well-developed theory.

The data from a case study will typically focus on one case, or at most a very small sample of cases. The case study should use different ways of assessing the functioning of the system and relevant contextual variables to minimize the effects of method variance or confounding. Due to a sample size of one case, this data is not really suitable for making general tests of theoretical predictions. Assuming the case selected is representative, however, the case study information can help describe and explain the target system. This information may be used to extend the theory about the target system or to add plausibility to a previously developed theoretical account of the system.

The in-depth approach of a case study can be generalized to multiple cases. A small sample of systems may be studied intensively to avoid over-emphasizing the idiosyncratic qualities of any particular one. This approach is sometimes called "panel" research. The sample of systems for a panel may be selected to be representative of a large population or systematically selected by some criterion. In the aviation domain, for example, a criterion for selection might be the occurrence of an accident or other serious event. The National Transportation Safety Board has done extensive case studies of major aviation accidents and compiled the information on that sample (National Transportation Safety Board, 1994). Focusing on systems for which critical events such as accidents have occurred is the hallmark of the "critical incident technique" (Flanagan, 1954).

As the number of systems examined with a case study method increases, however, the finite resources available for gaining complete information will be quickly exceeded. After that point, there is a necessary trade-off between the number of cases studied and the depth and completeness of information about each case. For larger samples of cases, the information will be more limited on each case but the larger sample will allow the use of statistical tools for analyzing and understanding the data. If sufficiently large, a set of case studies can also be used for the quantitative evaluation of theoretical predictions.

One advantage of a SIS is the ability to have extensive information on one or a few cases systematically integrated with less extensive information on a larger sample of cases. This integration allows the comparison of the system(s) for which a case study has been performed with the rest of the sample. The case study results can be compared with the rest of the sample on any qualitative and quantitative measures they have in common.

In the aviation domain, suppose an in-depth case study is performed for an aircraft crew that has a critical incident. Table 4.1 shows the potential set of information about such a crew in the SIS (Crew D in the example). Company policy may require that any crew with a significant incident is interviewed about the incident, assessed for critical life events or fatigue or morale problems, and has their personality assessed by a trained professional.

Table 4.1 Integration of case study information in a SIS

Measures:	Crews: A	B	C	D **2.**	E	F	G	H
Selection Results	X	X	X	X	X	X	X	X
Training Results **3.**	X	X	X	X	X	X	X	X
IOE Results	X	X	X	X	X	X	X	X
Yearly Evaluation	X	X	X	X	X	X	X	X
Significant Incident	No	No	No	Yes	No	No	No	No
Qualitative Interview				X	Additional			
Stressful Life Events			**1.**	X	Qualitative			
Fatigue, morale, etc.				X	Case Study Information			
Personality Factors				X	For Crew D			

The integrated SIS information allows both qualitative and quantitative analysis of Crew D. First, the qualitative information for Crew D can be assessed by itself in the usual manner for case studies (arrow # 1). Second, the qualitative information for Crew D can be compared to Crew D's selection, training, and yearly evaluation record (arrow # 2). This allows the post-incident qualitative information to be compared to pre-incident information for corroboration of hypotheses about Crew D's performance.

Third, Crew D's record can be compared with the other crews on the common measurements (arrow # 3). In the SIS, the entire training and performance record of Crew D can be compared with the population of fleet pilots to determine if Crew D was typical or atypical of crews in the fleet. That is, the question about how Crew D was different or distinct from the other crews can be quantitatively investigated.

Either typical or atypical comparison results can be important information. If the crew is typical of all crews across the training and performance measures (i.e. no noticeable differences), either the routine evaluations are not diagnostic of potential problems or the possibility exists that the other crews in the fleet will be prone to having similar incidents. The former conclusion would require a careful re-inspection of the measures and theory in the SIS. The latter conclusion could have important potential implications for future fleet training.

If the crew is atypical, the precise pattern of differences in training and performance is very important. Combining the differences from other crews may help develop a diagnostic pattern for crews that may have this type of problem in the future. Statistical methods could then be used to detect which other crews may be prone to having similar incidents, which would be immensely important for safety. Thus, it is both practical and important to integrate results from in-depth methods such as case studies with qualitative and quantitative information in the SIS.

Content analysis The result of many types of observations, qualitative interview questions, or case-study results is a free-text type of data. Therefore, the qualitative analysis of text is an important part of qualitative data analysis. The critical variables, processes, and events that are represented in the text should be compared to the current version of the theory or theories of the SIS. This qualitative comparison should increase or decrease the plausibility of the relevant SIS theories.

However, textual information may also be analyzed *quantitatively* to evaluate theoretical predictions. Different methods for analyzing the text can result in quantitative data, ranging from a simple count of key words to more sophisticated analyses of the themes and meaningful structures in the text. Although the sensitivity, reliability and validity of these measures must be established (see Chapter 3), they offer the potential for precisely confirming or disconfirming theoretical predictions.

Prediction

In general, a good theory should accurately predict future or hypothetical states of a system. Hypothetical states of a system are those which could conceivably be created by a particular combination of circumstances, but which are not observed in normal conditions. The criterion of prediction emphasizes the accuracy with which the theoretically predicted results match the obtained or real results at a future time or under the specific conditions.

Prediction must be carefully distinguished from post-diction, which is "predicting" that certain results or states of the system will occur *post hoc* or after the fact. Particularly when dealing with complex or inconsistent theories, it is all too easy to selectively emphasize certain parts of the theory to account for the obtained data, no matter which way the data points. Therefore, the only compelling evidence for prediction is evidence conforming to explicit predictions made prior to any inspection or analysis of the data. Predictions may be more or less explicit, however, and the focus of the predictions may range from an individual case to characteristics of entire distributions of cases. In general, the more explicit the prediction and the more particularistic the focus, the easier it is to disconfirm a theory. For precisely this reason, the confirmation of a theory is stronger when these more explicit predictions are substantiated.

Prediction of an individual case or data point

The strongest prediction that a theory can make is that a specific system will have a specific value on a specific measure. This prediction may, of course, be limited to a specific set of contextual conditions. It is important that the unique predicted value be unlikely to be observed by chance. Specifically, the predicted value should be quite distinct from the average or typical baseline for the system.

To the extent that the predicted value is distinct from the normal expected values for the system, finding that value is strong evidence for the theory. Conversely, finding any other value disconfirms the theory. Disconfirmation can be considered as either "all or nothing" or as a matter of degree. Logically, finding a discrepant value disconfirms the theoretical structure; that is, something in the theory is wrong either in whole or in part. For example, a difference in observed values for the velocity of distant galaxies of just 2.5% above the predicted value has revolutionized the science of cosmology by disconfirming the "steady expansion" model of the universe (Scientific American, 1998).

In fields with more poorly defined theories this type of strong disconfirmation may not be possible. Particularly when a theory is just being constructed for a domain, disconfirming an entire theoretical structure in this strict manner may be discarding correct theoretical components as well as incorrect ones. Therefore, an alternative approach is to consider the degree of disconfirmation. The degree of disconfirmation is stronger to the extent that the measured value is discrepant from the predicted value. From this view, small discrepancies may require minor adjustments of the processes or linkages in the theory while only the major discrepancies require wholesale revision or discarding of the theory.

Weaker versions of point predictions decrease the precision or specificity of the prediction. Weaker versions of prediction could be predictions of the results for a specific case that involve a range of values rather than a specific value, or a prediction for a specific unique value that may occur in a subset of cases rather than a single specific case. These types of predictions are depicted in Table 4.2 for an example of aviation crews performing a critical maneuver.

Table 4.2 Stronger and weaker point predictions from a theory

	Possible values on a scale of maneuver performance:					
Crew	**Violation**	**Unsafe**	**Below Average**	**Average**	**Above Average**	**Excellent**
A	X STRONG					
B						
C	Y?	Y? WEAKER	Y?			
D						Z?
E						Z? WEAKER
F						Z?

The strongest prediction is the precise prediction that Crew A will fly so poorly that they will violate flight standards (cell with "X"). This strong prediction would be disconfirmed by any other performance result for Crew A. A weaker prediction is that Crew C will receive some score below the average category. The range of predicted outcomes for Crew C includes violations, unsafe performance, or merely below average performance (cells with "Y"). This prediction is weaker in that it can only be disconfirmed by Crew C performing average or above. Another weaker prediction is that one of Crews D, E, or F will receive a grade of "Excellent" (cells with "Z").

A very weak prediction (not shown) would be that some of crews A through F would receive grades less than "Average" while other crews would receive a grade higher than "Average". This type of prediction is very weak because normal variability in crew performance could cause both above average and below average performance ratings regardless of whether a theory was correct or incorrect. These predicted patterns differ greatly in how much support they give the theory when they are confirmed, and conversely how much they discredit or refute the theory when they are disconfirmed.

The effect of confirmatory or disconfirmatory results on a theory depends not only on the precision of its predictions but also on the likelihood that the predicted result would occur by chance. The outcome of a point prediction is "all or nothing". The theory wins if the exact value for the exact case is found and loses if it is not. However, the strength of the confirmation or disconfirmation depends on the likelihood of the predicted value. Suppose that violations are very rare events, occurring less than 1% of the time. In this case, for the very precise prediction that Crew A incurs a violation, finding that Crew A does indeed violate a flight standard or regulation gives a strong confirmation of the theory since it would occur by chance less than 1 time out of a 100. Conversely, if violations would occur fairly frequently, this confirmation is correspondingly weaker support for the theory.

The prediction that Crew C has below average performance is inherently a less precise prediction. Therefore, finding that Crew C has a below average grade is weaker confirmation of the theory. This is because one of the three less-than-average grades is more likely to occur for Crew C than is a single specific outcome. Therefore, the relevant baseline comparison for the degree of support of the theory is how unlikely any of the three grades are by chance. In the example above, the likelihood of the three predicted grades must be added together. If these probabilities are relatively small (e.g. 1%, 3%, and 5% respectively), then the observed result gives some confidence in the correctness of the theory because the observed result would happen by chance only 9% of the time. If these probabilities are relatively large (e.g. 5%, 15%, and 25%) then the observed result gives us little confidence in the correctness of the theory because the observed result would happen by chance 50% of the time. Similarly the third prediction, that one of Crews D, E, or F has a grade of "Excellent" is a less precise prediction and empirical support would give weaker confirmation.

Finally, the very weak prediction that some crews are above average and other crews below average is so imprecise that confirmation gives little if any support to the theory. In any large enough sample of crews, by chance we would expect some crews to be graded lower than average while other crews are graded above average. That is, any variability among crews or stemming from the measurement process would create above-average and below-average grades. Since the predicted pattern is almost certain to occur, finding the pattern gives no support to the theory.

Prediction of the mean of a distribution

A theory may also predict some aggregate result. Most commonly, a theory will predict the mean or average over a sample of cases. Alternatively, if a single case is measured repeatedly, the theory may predict the mean or average over a sample of occasions for that case. Other things being equal, this form of prediction is less precise than the prediction of distinct data points discussed above. As in point predictions, however, there are differences in the level of precision for predicting means (Figure 4.2).

As shown in this figure, the most precise level of prediction is the prediction of the exact value of a mean. For any continuous measure with values spread across the continuum of possible values, the exact value of the mean specified by the theory is very unlikely to be observed by chance. Even a very small discrepancy between the specified value and the empirical result will be detected with a sufficiently large sample. Therefore, evaluation of this type of prediction is a very strong test of a theory and the confirmation of this type of prediction is strong support for a theory.

Conversely, predicting the mean is not equal to a certain value is a very weak prediction because by chance the mean will almost certainly be some value other than the specified one. Therefore, the confirmation of this type of prediction gives essentially no support for the theory. For example, Gigerenzer (1993) and others have criticized the method of null hypothesis testing on this basis. This very weak type of prediction will not be covered further.

When a precisely predicted value of a mean is tested, the obtained results should be compared to the predicted mean value while taking into account a chance baseline. One approach to comparing these values is a classical statistical test. One test that can be used to compare a sample mean to a hypothesized value is a single-sample t-test. The t-test compares the difference between the predicted and obtained values of the standard deviation to how much the obtained values would be expected to shift around by chance. The amount of difference that would be expected by chance is called the "standard error" for estimation.

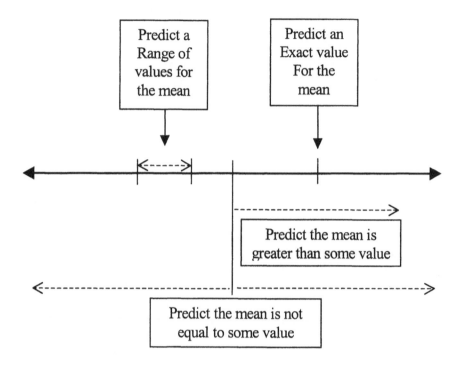

Figure 4.2 Different levels of prediction for the mean of a distribution

The general form of a t-test is:

$$t = \frac{Index \quad value}{Standard \quad Error \quad of \quad estimating \quad Index}$$

For evaluating a mean, the t-test becomes the ratio of the mean divided by the standard error of the mean. Three examples of distributions whose means could be tested with a t-test are given in Figure 4.3. This figure illustrates the hypothetical example of a measure that is scored on a −10 to +10 scale. Suppose the theory predicted an exact mean score of 5.0 for the distribution of scores on this scale. The distributions of scores represented by the Xs, Ys, and Zs in the figure represent possible results of measures on small samples of cases. Each of these distributions can be compared to the predicted value of 5.0 using the t-test.

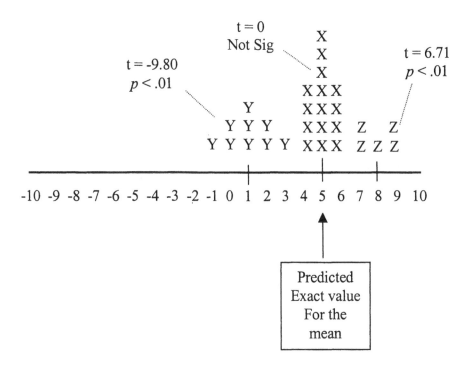

Figure 4.3   Comparing a predicted value for the mean to a distribution

The Y distribution of scores has a mean of 1.0, which appears to be quite discrepant from the predicted mean of 5.0. The t-test compares the degree of discrepancy between the observed and predicted means to the amount of variation in the means that would be expected by chance. The chance variation expected in the mean is estimated from the observed variation of the scores in the sample of Ys and the size of the sample. The larger the observed variation in the sample, the larger the expected variation in the mean. The larger the size of the sample, however, the smaller the expected variation in the mean since large samples give more stable estimates of the mean than small samples.

Disconfirmation of a predicted mean   Despite the fact that the Y sample has only 9 cases, the t-test shows that the discrepancy between the sample mean and the mean of 5.0 predicted by the theory, would occur by chance less than 1 in 100 times. Since this result is unlikely to occur by chance, the discrepancy between the theoretical prediction and the observed result is considered real. That is, the conclusion is that the observed mean is significantly lower than the predicted mean. The theory is, therefore, disconfirmed.

Similarly, the Z distribution of scores has a mean of 8.0 which appears to be higher than the predicted value of 5.0. This sample has fewer scores than the Y sample and an even smaller variation. Nevertheless, the t-test still determines that the observed difference would occur by chance less than 1 in a 100 times. The conclusion from the Z sample would be that the observed mean is significantly higher than the predicted mean. The theory would also be disconfirmed in this case.

For both the Y and Z sample, the disconfirmation of the theoretical prediction of 5.0 has important theoretical implications. The theoretical implications of a disconfirmation partly depend on the degree of predicted – observed discrepancy as discussed earlier for predictions of single data points. For predictions of means, the size of the discrepancy should be considered along with the nature of the scale and scale anchors. The 3-point discrepancy from the predicted value for the Z distribution might be considered moderate for this –10 to +10 scale, but the same 3-point discrepancy would be an enormous discrepancy for a typical 5-point judgment scale such as a Likert (1932) scale. Very large discrepancies might signal the need for discarding a theory or major theoretical revisions, while minor discrepancies call for more minor revisions.

The impact of a single disconfirmation such as this may also depend on the quality of the theory. For domains with loosely coupled or relatively independent theoretical principles, small discrepancies may lead to small or localized theoretical adjustments whereas large discrepancies lead to large-scale revisions. For domains with tightly coupled and precisely integrated theoretical principles, even a small discrepancy may require broad and fundamental theoretical revisions.

Support of a predicted mean The distribution of scores represented by the Xs, tends to corroborate the theoretical prediction of a mean of 5.0. The mean for the X distribution is exactly 5.0, and the t-test finds that this mean is not significantly discrepant from the predicted mean. Finding that the sample mean is not different from the predicted mean does not, in and of itself, confirm that the theory is correct. Rather it is a particular empirical test that the theory has passed rather than failed. The implication of passing this test is that the theory is supported rather than disconfirmed and should be further considered.

First, the theory's predicted value should be compared to a baseline of what would have been expected by chance given the normal scores for that measure. In a SIS, the scores for the entire population can be used to give an idea of what the overall distribution of scores would be, aside from the sample specified by the theory. If the normal baseline for the SIS general scores on this measure is near 0, for example, finding a sample having the predicted value of 5.0 is stronger support for the theory. Contrariwise, if the normal baseline were close to the theory's prediction of 5.0, finding the sample mean of 5.0 has less impact on the credibility of the theory.

Second, even if this sample seems to "fit" the theory and support its predictions, no single piece of evidence can confirm the theory. Rather, the accumulation of a set of instances which confirm the theory and not finding any instances that disconfirm the theory, is the pattern of results which ultimately supports the theory. Therefore, any theory in the SIS should be tested as often and in as many different ways as feasible given the SIS information. This testing should include all theoretical predictions that could add or subtract credibility from a theory. While well developed theories may make exact predictions of a mean value, less developed theories may make more general predictions of a range of possible values for the mean.

Predicting a range of values for the mean   If the theory predicts a *range* of values for the mean rather than a precise value, the observed sample mean can be compared against the end points of the predicted range. Serlin and Lapsley (1993) cover the logic and technical details of these methods. The observed mean must be tested to see if it is significantly lower than the lower end of the predicted range, and tested to see if it is significantly higher than the higher end of the range. To support the theory, the observed mean must NOT be significantly lower than the low limit of the range or significantly higher than the high limit of the range. An example of this kind of testing using the t-test is given in Figure 4.4.

In this example, the theory is predicting a possible range of values of the mean from 4.0 to 6.0. The same three samples of observations (Xs, Ys, and Zs) will serve as the possible observed samples for evaluating this predicted range. Since the prediction is a range of values, each of these samples must be compared to the low limit and the high limit of this range.

The sample distribution represented by Xs has the mean of 5.0 and is within the range specified by the theory. This supports the theory in the sense of not significantly disconfirming its predictions, as discussed above. The support for the theory in this case is weaker than for the precise prediction of a mean of 5.0 discussed above since a value in the specified range is more likely to occur by chance than the single value of 5.0.

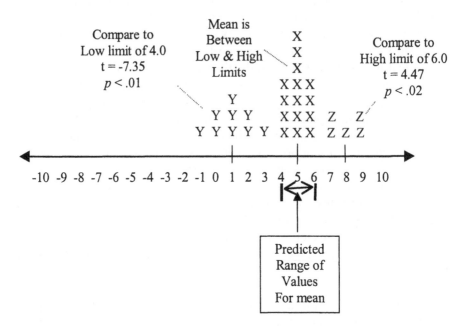

Figure 4.4 Comparing a predicted value for the mean to a distribution

For this reason, as the predicted range of possible values for the mean is extended, the degree of support afforded by finding a mean within the low and high limits is decreased. This is because of the increased likelihood that the observed sample mean will fall within the predicted range simply by chance. If the range of predicted values were to cover the entire middle of the scale from –5 to +5, for example, the observed mean of 5.0 would fall at the upper limit of that range. However, the degree of theoretical support would be much less because many samples would have means in the –5 to +5 range simply by chance. At the extreme, if the range of predicted values were extended to include most of the scale, the mean would almost certainly fall within the predicted range by chance. In such a case, the theory would gain no real support from finding a mean in the extended range.

Returning to the initial example of a predicted range from 4.0 to 6.0, finding a sample such as the distribution of Y values would significantly *disconfirm* the theory. The mean of the Y values of 1.0 is significantly lower than the lower limit of 4.0 for the predicted range. Since this difference would occur by chance less than 1 out of 100 times, the difference is interpreted as real and the theory is disconfirmed. Notice that a theory making reasonably strong predictions can be disconfirmed with a surprisingly small sample of data using methods like the t-test. This is important because the faster this disconfirmation process, the more quickly theories that are good approximations to the target system can be developed and confirmed in the SIS.

Finding a sample such as the distribution of Z values with a mean of 8.0 would seem at first glance to be *prima facie* disconfirmation of the theoretically predicted range of 4.0 to 6.0. The Z distribution has only five cases in it, however. This small sample gives a less stable estimate of where the mean really is and therefore a less clear-cut statistical decision. In this particular example, the mean of 8.0 for the Z sample does not exceed the upper limit of the range enough to be significantly different at the 1-out-of-a-hundred or .01 level. However, the observed sample mean does exceed the upper limit of 6.0 at the 1-out-of-20, or .05 level. More specifically, the observed mean is above the upper limit at a .02 significance level; a difference that would occur by chance about one out of 50 times.

Clearly, the consequences of the empirical evaluation in this case will change depending on the statistical significance level adopted. If the .05 level is adopted, the theory is disconfirmed; if the .01 level is adopted, it is not disconfirmed. This type of ambiguous, borderline result can therefore lead to different courses of action. One way of resolving this ambiguity is to seek out additional relevant cases and perform an empirical test with more statistical power.

In a SIS, the additional cases might be found within the connected databases. In the aviation domain, for example, the extra samples of crews with the theoretically specified qualities could be found in the databases of other fleets for the airline. Adding cases would increase the effective sample size and reduce the expected variability of the sample's mean for the t-test. If the additional cases followed the original distribution of Zs with a mean of 8, the theory would ultimately be disconfirmed. The effect of increased sample size on statistical decisions is further developed in the section on statistical power below.

Predicting the variance of a distribution

Besides the mean, the theory may predict other aspects of the distribution of a variable. The most basic aspect of a distribution besides its mean is the variability or dispersion of the scores. Alliger (1997) strongly emphasized the importance of testing hypotheses about the variability of scores in research. Alliger also provided details on the relevant considerations for the use of tests of variability.

For interval or ratio data, variability is typically summarized by the variance, which is the average squared deviation of scores from the mean. A closely related measure is the standard deviation, which is the square root of the variance. Both indexes reflect variability in the distribution of scores, but the standard deviation is expressed in the same units as the original scores and is therefore easier to interpret.

Since the variance and standard deviation are based on squared differences from the mean, they cannot be negative. The lowest possible predicted value is 0, which represents no variation among the scores at all. The possible positive values of the variance or standard deviation critically depend on the unit of measurement for the variable. The variability of a distribution of peoples' heights measured in millimeters, for example, is much larger than the variability of the same heights measured in inches. Therefore, to accurately test a specific prediction about the variability of a distribution, the unit or scale of measurement should be the one specified by the theory.

Once the predicted value for a standard deviation and the unit of measurement are specified, the test used is a t-test similar to the test for predicted means discussed above. For the particular case of the standard deviation, the obtained value of the standard deviation would be in the numerator. For the denominator, the t-test requires finding the standard error for estimating a standard deviation. This standard error depends, in turn, on how much spread or variation was in the original distribution of scores, and how the distribution was shaped. If the original distribution is shaped "normally" (i.e., a bell-shaped curve), the amount of difference in the standard deviation that would be expected by chance is:

$$\text{Standard Error (Standard Deviation)} = \text{Standard Deviation} / \sqrt{(2 * N)}$$

For example, the distribution of Ys in Figure 4.5 is distributed roughly as a bell-shaped curve. There is a mode at the value 1.0 and the distribution tapers off to either side of 1.0. The standard deviation for the nine Ys in this example is 1.225. The Standard Error for the standard deviation would be 1.225 / $\sqrt{(2*9)}$ or .29. Using this standard error, the obtained standard deviation can be compared to the theoretical prediction. The comparison of the sample to five different values that could be predicted by a theory is illustrated in Table 4.3.

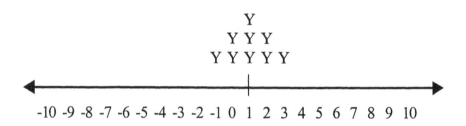

Figure 4.5 Evaluating the variance of a distribution of Y values

Each row of Table 4.3 compares a theoretically predicted standard deviation with the obtained standard deviation for the sample of Ys. The theoretically predicted value of 0.5, for example, is too small. If the population value were really 0.5, the observed standard deviation of 1.22 would have occurred by chance only 4 out of 100 times. That is, the value in the "Probability of the t-test" column for the first row is .04 or 4%. If we are using the.05 level of significance, the predicted standard deviation is significantly too low. Since the standard deviation is too low, we would conclude that the theoretical prediction is disconfirmed. The reasons for getting a lower standard deviation than was expected would have to be explored in modifying the theory.

Conversely, if the theory predicted a standard deviation of either 2.0 or 2.5 (the last two rows of the table), the predicted value would be significantly *larger* than the obtained value. In this case, the theory would also be disconfirmed but the implications for the theory would be different. The reasons for getting a higher standard deviation than was expected would now have to be explored in modifying the theory.

If the theory predicted a standard deviation of 1.0 or 1.5, however, the obtained value of 1.22 could have occurred by chance, and is basically congruent with the theoretical productions. The obtained value of 1.22 lies about midway between 1.0 and 1.5, and is not that far from either value. Further, the sample for estimating the standard deviation is quite small, consisting of only nine Ys. Therefore, within the limits of this small sample of cases, the theoretical prediction would be supported, at least until such a time as a larger sample of Ys could more exactly clarify the precise value of the standard deviation.

Table 4.3 Comparison of five different predicted values for a standard deviation with the obtained value from the distribution of Ys in Figure 4.5

Theory-Predicted S.D	Obtained S.D. in the sample of Ys	Standard Error of the S.D.	t-test (7)	Probability of the t-test	Conclusion (using a .05 criterion)
0.5			-2.51	0.04	S.D. too small. Theory is wrong.
1.0			-0.78	0.46	No significant difference.
1.5	1.22	0.29	0.95	0.37	No significant difference.
2.0			2.69	0.03	S.D. too large. Theory is wrong.
2.5			4.42	0.00	S.D. too large. Theory is wrong.

Wherever it is theoretically possible, a prediction of the variance of a distribution is an important addition to predicting the mean. This is because the variance is independent of the mean, and therefore the additional prediction of the variance is an independent test of the theory.   A theory making both mean and variance predictions can be confirmed in both predictions, disconfirmed in both predictions, or confirmed in one but not confirmed for the other. The possible situations are depicted in Table 4.4.

The confidence in the correctness of the theory is enhanced when predictions for both the mean and variance are confirmed, and conversely the confidence in the theory strongly decreases when predictions for both the mean and variance are disconfirmed. Since the tests for mean and variance are independent, it is quite possible that one will be confirmed and the other not. In such a case, the picture is mixed, and reasons for the disconfirmation should be carefully examined with the other information in the SIS.

Table 4.4 Possible outcomes for a theory predicting both the mean and the variance of a distribution of scores

		Test of the *Mean* predicted by the theory is:	
		Confirmed	Disconfirmed
Test of the *Variance* predicted by the theory is:	Confirmed	Theory is strongly confirmed	Variance is correctly predicted but not mean
	Disconfirmed	Mean is correctly predicted but not variance	Theory is strongly disconfirmed

If the variance is correct but the mean is incorrect, biases in measurement or sampling that would affect the mean of the scores should be considered. The additional information in a SIS will often allow alternative explanations of the results, such as rater bias or sampling bias, to be empirically evaluated. If observers are used to rate performance, for example, the calibration data for the raters should be checked to see if the harsh/lenient rater biases are shifting the mean in the unexpected direction. If a biased sample is suspected, a more representative subsample can be identified, selected, and analyzed to see if the mean for the subsample conforms to the predicted theoretically expected value.

Similarly, if the predicted mean is correct but the predicted variance is not, reasons for the unexpected value of the variance can be explored using the SIS information. It the variance is larger than expected, the reliability of the measure can be checked to see if the low reliability was the source of larger-than-expected variations in scores. In this manner, the evaluation of theoretically predicted means and variances leads to a much more complete overall test of the theory.

Predicting the Entire Shape of a Distribution

The idea of predicting different facets of a distribution can be extended to predicting the entire shape of a distribution of scores for a particular variable and sample in the SIS. The predicted shape of the distribution must cover the range of possible values for the variable. Within this range, however, the predicted shape of the distribution may range from normal to very non-normal.

A normal or bell-shaped distribution is found for many samples of measures such as height, weight, etc. In fact, a normal distribution is found for many sorts of possible measures. This distribution has a single most frequent value or mode. Further, a normal distribution symmetrically tapers off from this mode on both sides, with extreme values becoming more and more infrequent. Besides many basic measures, the normal distribution will generally be found for distributions of mean values, particularly where the mean is averaged over a large sample. This is due to the mathematical properties of taking averages as stated in the Central Limit Theorem. Since a normal distribution occurs so ubiquitously (high baseline probability), predicting and finding this shape of distribution is not very strong evidence for a theory.

Therefore, the distribution-shape predictions of a theory are particularly important when they predict non-normal distributions. The predicted distribution may depart from normal in several different ways. The predicted distribution could be multi-modal; that is, having more than one very frequent or modal value. The predicted distribution could also diverge from normality by being asymmetrically distributed on each side of the mode.

Any theory that successfully predicts the exact non-normal shape of a distribution of scores is correspondingly strongly supported. In the aviation domain, for example, it may be quite reasonable that distributions of crew evaluations on a 5-point scale would be predicted to be asymmetric. This distribution would occur if above standard performance was more commonly observed than below standard performance. This distribution would plausibly occur due to the high levels of expertise and training of the pilot population, which tends to weed out the low performers. Finding the predicted asymmetric (or skewed) distributions of pilot evaluations could, for example, support a theory of pilot expertise and training effects. To be even more strongly tested, however, the theory should specify the exact relative frequencies or percentages of scores for each possible scale value or interval.

The expected distributions of exact values for each key variable can be obtained by different methods. If a measure in the SIS has been used in previous applications or research, the distribution of scores from the previous sample can set the expected distribution for the SIS variable. Alternatively, if a measure is a new measure or does not have any previous empirical work, a panel or focus group of experts can estimate the expected distribution of scores for a particular measure. In the aviation domain, for example, highly experienced evaluators could be asked about their expectations for crew performance on recurring evaluations. Whether the expected distribution comes from precise theoretical predictions, previous data, or expert opinions, the set of expected values can be statistically compared with the obtained distribution of scores for that variable in the SIS.

Using an aviation example, suppose either a theory or a panel of expert instructors gave the expected distribution in Table 4.5 below. This example shows that the instructors expect 1% or less failures, 2% repeats, 10% required debriefs, about 2/3 or 67% Standard Performance, and 20% Above Standard performance. The Chi-square goodness-of-fit test requires that these expected proportions or percentages be converted into predicted frequencies. Suppose a sample of 500 pilots were evaluated. The numbers in the "Predicted Frequencies" column represent the frequencies that would be expected in each evaluation category.

The obtained frequencies differ from the expected pattern, particularly in the relative numbers of Standard and Above Standard performance. Specifically, there are fewer Standard and more than twice as many Above Standard performances as would be expected on the basis of the expected proportions. The Chi-square test results in a value of over 70, which is a large enough value to occur by chance less than 1 in a 100 times. Therefore, the obtained distribution of evaluations is in this case significantly discrepant from the expected proportions of crews, which disconfirms the expectations of the experts.

A theory must be fairly completely specified to predict an entire distribution of scores. Therefore, one goal of theoretical development and elaboration in the SIS would be to achieve theories that can predict the mean, variance, and exact shape of the distribution of values for key variables. Ultimately, the theory should predict values of key variables for specific cases as well. Theories based on comprehensive mathematical or computer models of the underlying system processes may have the completeness and precision to predict distributions. For such theories, obtaining the predicted distribution of scores for a variable is simply a matter of executing the model with appropriate inputs. Subsequent chapters will cover methods for the construction and validation of mathematical-computational theories.

Table 4.5 Comparing predicted frequencies for 500 cases on a 5-point crew performance judgment scale with obtained distributions of scores

Judgement level & label	Performance Description	Predicted Proportion	Predicted Frequency	Obtained Frequency
(1) Fail/ Unsafe	Performance endangers passengers or aircraft	0.01	5	5
(2) Repeat Required	Performance is unsatisfactory but not dangerous.	0.02	10	13
(3) Debrief Required	Performance is satisfactory but has significant weaknesses	0.1	50	37
(4) Standard Perform-ance	Crew complies with company standards for performance in all respects	0.67	335	225
(5) Above Standard perform-ance	Crew exceeds company standards for performance in one or more critical areas	0.2	100	220

Predictions of the mean, variability, and distribution shape are just one set of basic outcomes on which a theory can be evaluated. All of these tests of the theory are focused on one variable or measure. In addition to these single-variable evaluations, a theory will generally specify some relationships among the key variables of a system. These relationships among variables are a second focal point for the empirical evaluation of a theory.

Prediction of the degree of relationship between two variables

Two measured variables in the SIS may be related in different ways. The simplest predicted relationship is that variable Y increases or decreases steadily as variable X changes. If the change in Y is expected to be steady or proportional over the range of possible values for X, the predicted relationship is linear. For a positive linear relationship, Y *increases* as X increases. For example, the idea that increases in task ability or skill should be associated with corresponding increases in job performance is a prediction of a positive linear relationship between ability and performance. This relationship is illustrated in Figure 4.6.

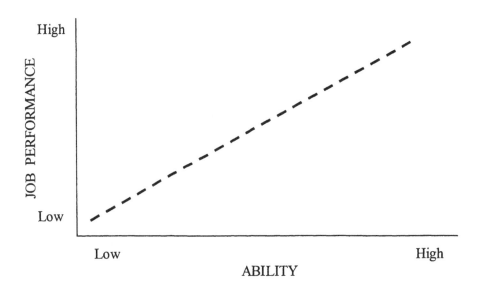

**Figure 4.6 Positive linear relationship between ability and job
performance**

For a negative linear relationship, sometimes called an inverse relationship, Y *decreases* as X increases. When driving an automobile from point A to point B, for example, increasing the average speed or velocity of driving decreases the time required for the trip. Overall, the driving time required for a trip of a given distance should decrease proportionately with the increase in average speed as shown in Figure 4.7.

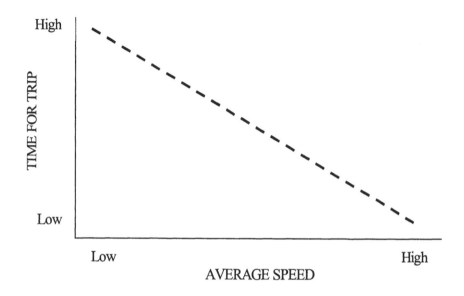

**Figure 4.7 Negative linear relationship between average speed and
trip time**

The estimation of the degree of linear relationship between X and Y is
the focus of regression analysis.   The fundamental idea of regression
analysis is that the relationship between X and Y can be summarized as a
line with a defined slope and intercept. The slope and intercept of the line
are estimated in such a way as to minimize the differences between the line
and the observed data. Specifically, the squared difference between the
observed data and the line is used as a criterion to find the best possible
value of the line's slope and intercept with the Y axis. This criterion is
called the Least Squared Error criterion and is also used for other forms of
statistical estimation.

The result of the regression analysis is the formula $Y' = A + BX$, where
B is the slope and A is the intercept with a Y axis. This formula represents
the line that best describes the linear relationship of X and Y. For positive
relationships, B is positive; for negative relationships, B is negative.   If B
would equal precisely 0, the line would be flat and would represent no
positive or negative linear relationship between X and Y. The value of 0
represents the chance baseline for the estimated relationship between two

variables that have nothing to do with each other or are completely independent.

An example of a 0 relationship is given in Figure 4.8 which illustrates the relationship between the average heights of parents and children for a sample of adopted children. If height at maturity is essentially controlled by genetics, the height of the parents should have no correspondence to the height of their adoptive children. The flat line indicates a zero relationship and describes the situation where the average height of adopted children is unrelated to the height of their adoptive parents.

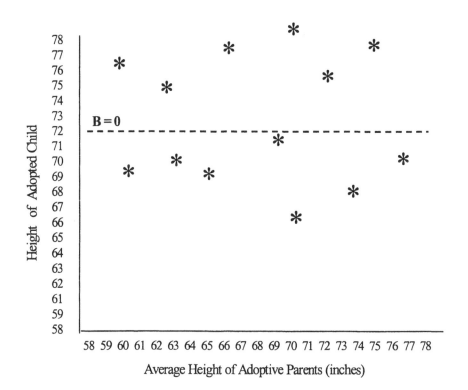

Figure 4.8   Depiction of a zero linear relationship between the average heights of parents and their adopted children

In a natural system, in contrast, most variables will be positively or negatively linked together in some fashion, even if only slightly. For example, the heights of children are positively related to their genetic parents as depicted in Figure 4.9. A truly zero relationship between two measures of a target system will be the exception rather than the rule.

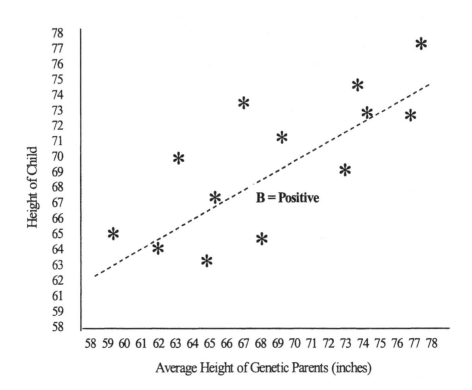

Figure 4.9 **Depiction of a positive linear relationship between the average heights of parents and their genetic children**

Just as for other types of predictions, theoretical predictions of relationships may vary in specificity and in the consequential weak vs. strong support that the evidence can provide for the theory. Less specific theoretical predictions often use the chance baseline for making a prediction. For example, the theory may predict that the linear relationship between X and Y is greater than the 0 or no-relationship baseline. This would correspond to the prediction of a positive relationship in parent and child heights in the above example. For a standard regression analysis, the default test of the slope of the regression line is against the baseline of 0 relationship and would serve as a test of this kind of prediction. Since there will probably be a positive or negative relationship, even if it is very slight, the prediction that the relationship from X to Y is positive has about a 50% chance of being true regardless of the validity of the theory. Therefore, results confirming the direction of just one predicted positive or negative relationship provide relatively weak support for the theory.

A more specific prediction of the exact expected value of a relationship between X and Y allows a much more precise support or disconfirmation of the theory. For evaluating the specific theoretical prediction of the degree of relationship between X and Y, the test will involve the slope of the relationship, B. The observed relationship must be tested against the predicted value of B. The critical test of the theory's prediction is whether the observed positive or negative value of B is different enough from the predicted value to be considered a real difference. If the difference is real at an acceptable level of significance, the theory is disconfirmed. If the expected-obtained difference is non-significant, the theory is supported.

Alternatively, the linear relationship between two variables can be summarized in the correlation index, and this index can be tested directly. The correlation coefficient is essentially similar to a regression relationship in that it indexes the linear relationship between two standardized variables and is a bounded index that has a lower boundary of -1 for perfect negative relationships and an upper boundary of $+1$ for perfect positive relationships. Both of these qualities make it easier to compare correlation coefficients than regression coefficients. The disadvantage of a correlation coefficient as an index of linear relationship is that it is quite dependent on the standard deviation of X and Y for the sample. If the slope of the regression line is not critical information, a linear relationship can be tested equally well using the correlation coefficient.

If the theory specifies a precise expected value for a correlation coefficient, the value obtained from the sample can be compared to the predicted value using a t-test that is similar to the test for the regression slope described above. Confirming the precise value of a correlation offers more support for the theory than confirmation of a general positive or negative relationship. Conversely, disconfirming a precise value offers a stronger disconfirmation of the theory. After the predictions of a theory have been thoroughly tested, other aspects of the theory can be evaluated such as the capability of the theory for altering or controlling the target system. Effective control of the system is important for solving many practical problems.

Control

In general, a good theory should allow for the systematic change or alteration of a target system. Making appropriate changes to the context, parameters, or processes of a system should result in the theoretically expected changes. The criterion of control emphasizes the ability of the theory to specify how to make changes to the system that will result in the expected effects. As with predictions, the expected results will be compared to the obtained results to evaluate the theory's utility for controlling the system.

The difference between predictive accuracy and control utility can be illustrated by a theory of crew teamwork in the aviation domain. If the theory specified that crew teamwork depended on underlying personality traits of pilots that were *not* susceptible to change, the theory could explain and predict teamwork but still not offer much control utility for changing the teamwork of a given crew. Alternatively, if the theory specified that crew teamwork consisted of interaction patterns that were learned and therefore trainable, the theory could offer a wide variety of potential training interventions that would alter crew teamwork in a desired direction. The latter type of theory has the potential for control of the target system, and this potential can be empirically evaluated.

As with prediction, a more precise forecast of expected results from an intervention gives stronger evaluation of the theory's control utility. Confirming precise theoretically expected results gives strong support for the theory, and disconfirming precise results strongly refutes the theory. That is, the more precise and explicit the expected control results, the easier it is to disconfirm the theory. Conversely, the support for the theory is stronger when the more explicit and exact control results are, in fact, found. As for prediction, the focus of the expected results from a control intervention may range from results for specific, individual cases to more general expectations about the mean and variance of a key variable or important relationships among variables.

Obtaining measurements of the target system under the theoretically specified conditions is key for obtaining evidence that will empirically support or disconfirm the control utility of the theory. The conditions may occur naturally as subsets of cases in the SIS or may have to be created artificially. Natural sets of appropriate test cases and designed experiments have distinct pros and cons for evaluating the control utility of a theory.

Natural Sets of Test Cases

A naturally occurring set of test cases that meets the theoretically specified control conditions is sometimes called a "natural experiment". For example, in evaluating the effectiveness of interventions to change the system outcomes, appropriate evidence may be obtainable from the SIS during the normal course of data collection.

In the aviation domain, suppose new, special training is being given to pilots at their yearly recurrent training. Since this training is typically distributed evenly across months for different subsets of pilots, a natural experiment will occur during the middle of the year. At that time, roughly one-half of the pilots will be trained and the rest still awaiting training. If suitable performance measures such as random line checks are available in the SIS, the ability of the training to change performance can be evaluated by comparing the trained and untrained groups. In general, a SIS offers a wide variety of possible test cases for evaluating the expected theoretical results with naturally occurring samples.

The advantage of naturally occurring groups for evaluating a theory is that the test cases are measured in the normal course of events in the normal context. Therefore, there is less opportunity for bias due to a novel measurement or testing situation. Further, if the expected results for different conditions are found, there is a fairly strong presumption that these results will hold true in the actual environment. That is, the anticipated external validity of the result is high.

There are also important disadvantages for a natural experiment. One important disadvantage is that naturally occurring groups have potential systematic confounds that can influence the data. Systematic confounds are systematic differences between the naturally occurring groups other than the ones identified by the theory as relevant.

Such differences are important because they can potentially cause either a false positive or a false negative evaluation of the theory. A false positive error occurs when the theory is false, but confounds create the theoretically expected pattern of results. This may result in a conclusion that the theory is true when it is in fact false.

Conversely, a false negative error occurs when the theory is correct but the confounds counteract the pattern of theoretically expected results. This may result in an apparent disconfirmation of the theory even though it is true. Since avoiding either type of error is important, the problem of systematic confounds is quite important.

If these confounds are known and measurable, these measurements can be added to the SIS. For example, if the target system is the sales performance of shoe salespersons, the known confound of day of the week can be added to the sales information in the SIS. These added measures can be used to formulate an adequate test of the theoretical control expectations. To be adequate, the test must adjust for the influence of the confound or remove its influence from the evaluation data. Two possible methods for dealing with confounds in this situation are the stratification method and covariance adjustment.

In the stratification method, the set of test cases is first stratified by values of the measured confound. This stratification is intended to equalize the specific value of the confound for each subset of test cases. Equalizing the value of the confound will eliminate or greatly reduce its effect on the results for each stratum. The theoretically expected results are then examined within each stratum as well as averaged across strata. This minimizes the possible effects of the confound while evaluating the expected results for the theory.

For testing the effects of clinical treatment in the psychosocial domain, for example, a known confound might be the social class of the client. That is, upper class clients may self-select into clinical treatment more often than lower class clients and also have better treatment outcomes. If the social class of each client is measured and included in a SIS, the clients can be stratified into low, middle, and upper social classes (Figure 4.10). For each stratum, the difference between the treated clients and the waiting-list control group can be compared to see if the treatment has the expected effect on behavior. If the theory is correct, the treated clients should show improved behavior for each social stratum. Further, the difference between treated and untreated client groups can be aggregated across social strata to find if the overall difference is congruent with the theoretically expected difference between treated and untreated clients.

In this figure, the treatment has a large effect for the upper-class stratum, a moderate effect for the middle-class stratum, and no effect for the lower-class clients. These results give not only an idea of the overall effect of the treatment but also an idea of how well it works for each stratum. This information would be valuable for deciding if specific clients would benefit from the treatment and optimizing client outcomes.

Covariance adjustment also uses the measurements of possible confounds, but in a different manner. The covariance analyses essentially use a regression approach to remove the effect of the confound on the key variable(s) of interest. One necessary assumption for the covariance approach is that the relationship between the confound and the key variable is stable across the naturally occurring groups.

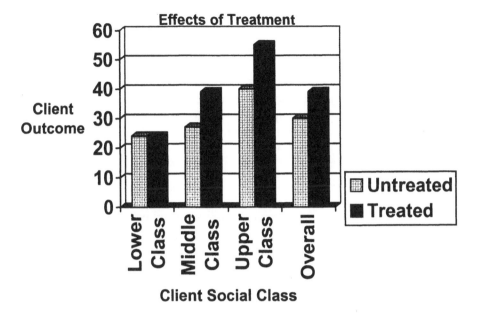

Figure 4.10 Example of stratification

This assumption can be empirically evaluated. If the assumption is plausible, the regression is used to remove the systematic effects of the confound from the key variable. Then the analysis of the theoretically expected results is performed to see if they occur. This "levels the playing field" and the adjusted results can provide an appropriate evaluation of the control utility of the theory. Both the stratification and covariance adjustment approaches, however, require a known, measured confound.

If the confounds are not known or not measurable, neither stratifying nor covariance adjustments can be performed. The data may be inherently biased and result in a false positive or false negative evaluation of the theory. The possible degree of bias may, however, be judged using the completeness and accuracy of the current theory for the target system as a yardstick. If the current theoretical account is a reasonably complete and accurate account of the target system, the scope for potential operation of unknown confounds on the target system is correspondingly smaller. Particularly if the theoretically expected result is a strong one and is confirmed by the data, the likelihood that the result is produced by unknown confounds in this case is relatively low. For this and other reasons, the issue of theoretical completeness is quite important and is covered in detail in a subsequent section.

However, at the onset of development of the SIS, the theoretical base will most likely be quite incomplete. To the extent the theoretical account is incomplete or the validity evidence for the theoretical structure is relatively weak, the potential danger of unknown confounds in natural experiments is correspondingly larger. To help avoid this danger, the evidence from carefully designed and executed experiments are a desirable complement to natural experiments.

Designed experiments

Experiments that are created by design require a careful extrapolation of the theory to particular situations or conditions. Since the situations or conditions will be constructed for the purpose of evaluating the theory, they are not constrained to be natural situations. The important point for evaluating the theory is that the key parameters and processes of the theory will be activated in the specified situation or condition. This activation should be absolutely clear and compelling because the theory will be confirmed or disconfirmed, depending on finding the expected pattern of results for each situation or condition.

How much the designed situation or condition resembles the natural setting for the target system will vary. Conditions used for the experiment may range from almost completely natural to almost completely artificial. Figure 4.11 depicts this range of possibilities. For some evaluations, keeping the situation as natural as possible is desirable. This may be the case particularly when the generalization of the obtained pattern of results to natural situations is important. At the extreme of this approach, the natural situation is used intact except for the systematic manipulation of one or more key features specified by the theory. This approach would be called a field study or field experiment. The difference between these two is that in a field experiment the participants are randomly assigned to different conditions, whereas in a field study the participants are selected into the different conditions naturally rather than being randomly assigned.

An example of a field study in the aviation domain is the evaluation of an Advanced Crew Resource Management (ACRM) training program carried out by Holt, Boehm-Davis, Hansberger, and Beaubien(2000). The ACRM training program consisted of key teamwork procedures that were trained in a one-day session and implemented as Standard Operating Procedure (SOP) for the experimental fleet. Performance of the ACRM-trained fleet was compared to a control fleet of pilots with standard training and SOP. Pilots were assigned to experimental and control fleets by normal airline policies rather than randomly, making this a field study.

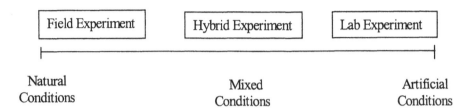

Figure 4.11 Continuum of types of experiments

At the other extreme, the situation may be intentionally designed to be as sterile as possible to avoid contamination with extraneous factors. In this approach, the key parameters and processes of the theory are set into a very artificial situation so that the expected results can occur in as pure and uncontaminated a manner as possible. This design approach may be taken when miscellaneous factors in a more natural situation could influence the theoretically relevant parameters or processes in such a way as to obscure the expected results.

An example of a laboratory experiment related to the aviation domain is Gray and Boehm-Davis's (in press) work on human-computer interaction. They analyzed data from a synthetic laboratory task that mimicked essential parts of the radar and targeting information available to a F-16 pilot in a military situation. Participants completed this task on a computer in a tightly controlled laboratory environment where distracting elements or extraneous events were eliminated. Controlling the environment was important because all responses were timed in milliseconds and this information was the key to constructing detailed models of the participants' perceptions, cognitions, and actions.

Hybrid experiments are situations with mixtures of natural and artificial conditions. This form of experiment may be used when there are some aspects of the situation that should be left natural to enhance generalizability of the results, but other aspects of the situation that must be controlled in a particular manner to create the theoretically relevant condition. In such a case, all the essential theoretically relevant aspects of the situation must be appropriately specified to adequately test the theory, but the remaining aspects of the situation can be left as natural as possible.

In the ACRM evaluation study mentioned above, the theory specified that the effects of improved teamwork would be particularly noticeable in situations that imposed high workload and complex demands on the crew. To adequate test the training and SOP intervention, therefore, a testing situation had to be devised that combined high workload and complex demands on the crews in both fleets. The form of evaluation used for this purpose was a Line Operational Evaluation (LOE) which was part of every pilot's yearly flight test.

The LOE used a naturally scripted flight scenario administered in a complete full-motion simulator of the aircraft. The scenario mimicked a normal flight operation except that significant problems and equipment failures were introduced at specific points. Testing the ACRM trained and untrained fleets in this situation was a hybrid experiment because of the mix of a natural flight scenario with a simulator environment and a set of severe problems or issues for the crew to deal with.

Clearly the balance of a hybrid experiment must be carefully crafted to serve dual purposes. On the one hand, it should adequately create the appropriate conditions for evaluating the theory. On the other hand, it should maintain enough natural or essential similarity to other important situations or contexts to give some confidence in the generalization of the evaluation results to those situations or contexts.

Randomization Randomization can be used in a designed experiment in two distinct ways: random selection of cases, and random assignment of cases to conditions. If cases or persons are randomly selected for a study, they will fairly and accurately represent the intended population. This helps ensure that the results of the study will be true for the population as a whole, which is one aspect of external validity. Random selection of cases is particularly valuable when the expected theoretical result involves the mean, variance, or distribution shape of measures on a set of cases. With a sufficiently large sample, random selection allows an accurate test of the theoretically expected means, variances, or distributions. Without random selection of cases, the representativeness of the mean, variance, etc. to the original population is questionable and the tests of those specific values is not warranted.

The principle of random selection may also be useful in other aspects of the experimental situation. One important aspect of the situation for many experiments is the stimuli that the participant responds to. The stimuli should be an accurate representation of the population of possible stimuli, and one way to ensure this is to take a random sample from the stimulus population. Experiments in recognition of facial expressions, for example, could use a randomly selected set of faces for the judgment task. This is important to avoid the possibility that the accuracy of judgment would depend on the particular faces chosen for the task. In general, the random selection of stimuli will insure that the results of the study generalize to the population of stimuli and minimize the possibility that the results are limited to the stimulus set used in the study.

In contrast to random selection of cases, random assignment of cases does not involve a population. Rather, random assignment involves the fair and equal assignment of whatever cases are available (individuals, crews, etc.) to the different situations or conditions used in the study. Random assignment is desirable because it tends to equalize any potential confounds across the conditions. This helps ensure that the results for the different conditions inside the study can be fairly and precisely compared, which is one aspect of internal validity. If cases cannot be randomly assigned to conditions, the field or laboratory experiment may become a field study or a "quasi-experimental" evaluation situation (Campbell & Stanley, 1959) with decreased internal validity.

If individuals can be randomly assigned to the experimental conditions, the different traits, abilities, or motivation levels among the individuals will tend to be more equal among those conditions. Randomization is particularly important in minimizing the effects of unmeasured or even unknown confounds that might bias the results. In testing a theory of individual job performance, for example, the exact set of abilities and skills (KSAOs) that contribute to the performance may not yet be known. In such a case, random assignment of persons to the different conditions would be critical to attempt to minimize the influence of different individual capabilities on the results.

Random assignment does not, of course, completely cancel or negate the possible effects of these confounds; it merely tends to equalize them. This equalization will be quite good where large samples of cases are randomly assigned to each of the experimental conditions. For small samples of cases (e.g. less than 30 cases per condition), the equalization may be imperfect and inequalities in one or more confounds may still influence the results to some degree.

For extremely small samples of cases (e.g. 10 or less cases per condition), the equalization may be quite imperfect. Particularly if there are numerous possible confounds, the randomization may fail to equalize the samples in the experimental conditions for all possible confounds. The unequal confounds in different treatment conditions could produce spurious confirmatory or disconfirmatory results. For situations in which small samples are inevitable, taking the alternative approach of specifying, measuring, and analyzing the effects of the known confounds is a valuable adjunct to the randomization approach. Identifying the possible confounds to measure will be facilitated by having a theory which is a complete, rather than partial, account of the target system. This completeness issue is one advanced theoretical issue that is discussed in the next section.

ADVANCED THEORETICAL ISSUES

Although all theories should explain, predict, and allow change or control of the target system, there are other desiderata for developing theories in the SIS. The principle of parsimony or simplicity is that a theory should be kept as simple as possible without sacrificing explanation, prediction, or control. According to this principle, if two theories account equally well for the target system and one has fewer parameters and principles, it would be preferable. Other important principles for the construction and evaluation of SIS theories are the principle of completeness and the principles of precision, clarity of meaning, and extensibility of the theoretical representation.

Theoretical completeness

Theoretical completeness is valuable for explaining, predicting, and controlling the entire range of processes and outcomes of the target system. For this reason, it is useful to get an idea of the completeness of the theoretical structures in the SIS. To do this it is necessary to define theoretical completeness and get at least a rough idea how complex a theory must be to have a relatively complete account of the target system.

One approach to defining theoretical completeness is to focus on the key constructs of a target system. These key constructs may be markers for critical processes, key links to outside systems, or key output variables from the system. The completeness of a theoretical account can be defined as the extent to which the processes and parameters of the theory account for the observed variability in these key constructs. This raises the conceptual issue of how much variance in a construct *can* actually be predicted.

Figure 4.12 below gives one way of viewing the variance of a construct. The variation in a key construct may be due to some simple systematic cause, some complex, non-linear cause (e.g. chaotic systems), or inherently and irreducibly random causes (e.g. molecular vibration in a chemical system). Suppose this construct represents a key outcome variable such as the number of widgets produced in a man-machine system. For widget production, the simple systematic causes of different levels of production could be the motivation and skill of the worker, the total time on task, and the quality and efficiency of the production machine. Complex, non-linear influences on production might include the mood or emotional state of the worker, the worker's physiological and cognitive condition, and changes in working conditions or context. The irreducible random error on production may be represented by the inherent variability of the widget production of the machine (e.g. production tolerances), momentary slips or lapses on the part of the worker, and so forth.

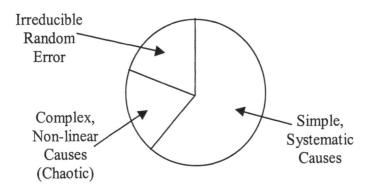

Figure 4.12 Sources of possible variance for a key system construct

A complete theory would account for all systematic (i.e. non-random) variance of the key constructs. This would require specification of all simple systematic and chaotic causes. However, a more modest "good enough" degree of theoretical completeness might be a theory that accounts for all the simple systematic influences on a construct but not the complex non-linear influences. In the example in the figure above, accounting for the simple, systematic influences on production would explain over half of the variance in production. For instance, a theory emphasizing only the role of motivation and skill of the worker, the total time on task, and the quality and efficiency of the production machine could explain over half of the variance in production outcomes, which may be quite sufficient for many applied purposes.

The development of a theoretical account that is sufficient to account for the simple systematic influences on a construct will typically be less complex and require less time and other resources than a complete account that includes all the complex, non-linear influences. Pragmatically, the desired degree of completeness may depend on the ability to use the theoretical account for intended practical applications. For key system variables or processes, however, the research agenda for the SIS might require a complete explanation of all systematic variance, which would in turn require the theoretical development and evaluation of possible complex, non-linear influences on production outcomes in addition to the simple ones.

In practice, however, constructs cannot be directly examined but instead must be measured by some scientific measurement. Therefore, the first step in evaluating theoretical completeness is to establish appropriate measures for the key constructs. Each measured variable will not, of course, have precisely the same variance components as the underlying construct. The measurement process may distort to some extent the relative amount of random error, chaotic influence, and simple systematic influences on the measured variable. This is shown in Figure 4.13 below.

Unreliable measurement, for example, increases the amount of random error in the measure. Reactive measures, however, may sensitize or change the operation of the system and produce chaotic variance in the measure. Finally, measurement that is confounded or contaminated by the measurement of some extraneous construct will reduce the amount of valid systematic variance in the measure.

In practice, the evaluation of completeness of the theory must be based on the measured variables. Therefore, a complete theoretical explanation of the key system variables would imply a theory good enough to account for all the reliable and valid variance in these variables. However, for some applications a satisfactorily complete theoretical explanation might be a theory good enough to account for all the simple, systematic variance in the key measured variables.

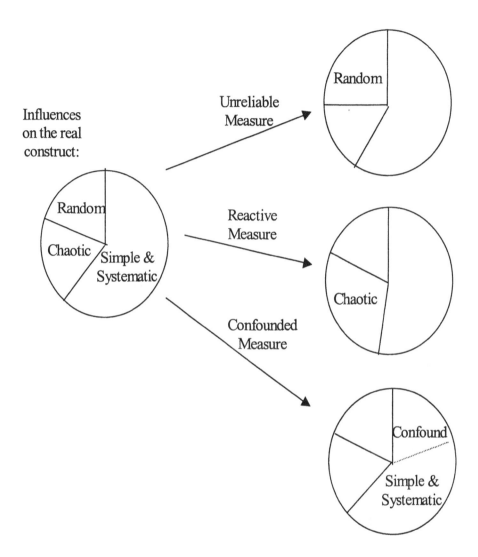

Figure 4.13 Effects of possible measurement problems on the variance of a measure for a key system construct

Theoretical complexity

For any complex, multi-causal target system, constructing a complete theoretical account may require using a variety of theoretical viewpoints. A complete theory is likely to be more complex than a partial or necessary theory that just accounts for some aspects of the target system. The degree of complexity may also depend on the nature of the causal influences underlying variance in the key variables. The relative complexity of the theory depends on how much the variance in each system variable is made up of simple, systematic influences vs. complex, non-linear influences vs. random error. This is illustrated in Table 4.6.

Table 4.6 Theoretical complexity corresponding to variance components

Type of influence	Required theoretical complexity	Resultant predictability
Simple and systematic	Complexity of the theory should equal the complexity of data. (N effects require specifying N causes)	High
Complex non-linear (chaotic)	Complexity of the theory is much less than complexity of data. (N effects require specifying a small number of non-linear causes)	Medium
Random error	Complexity of theory is just the mean and variance of the distribution of random error. (N effects require specifying the basic parameters of an error distribution)	Low

A concrete example of this principle in the aviation domain would be different influences on the key system variable of on-time arrivals for commercial flights. The relative percent of on-time arrivals could be influenced by a variety of factors listed in Table 4.7. The listed causes range from simple and systematic influences on on-time performance such as airline SOP and crew motivation, to chaotic influences such as weather, to essentially random influences such as other aircraft traffic competing for take-off slots at the moment of departure or for landing slots at arrival.

Table 4.7 Influences on the on-time arrival of commercial flights

Type of influence	Example influences for on-time arrivals
Simple and systematic	Airline scheduling of flight intervals and turn-around times. Airline SOP for making up lost time on a delayed flight. Crew motivation and values for on-time performance.
Chaotic	Weather conditions at departure airport. Winds aloft (headwind/tailwind) and turbulence en route. Weather conditions at arrival airport.
Random error	Queue of aircraft waiting for departure. Queue of aircraft arriving at the destination airport. Availability of empty gate for disembarking passengers.

For simple and systematic causes the complexity of the theory must reflect the number of these causes that are found to be significant in the data analysis. If, for example, flight scheduling, SOP, and crew motivation all uniquely and significantly influence on-time performance, then the theory must include these three factors to be complete. In general, if there are N unique simple systematic influences on a system, the theory must specify all N causes to be complete. That is, the theory must be complex enough to include each simple cause as a theoretical component. Clearly the number of simple systematic influences on complex systems could be quite high. However, a high level of complexity of the theory in this case allows for a more complete and accurate explanation of the systematic variance of a measure.

For chaotic causes, the complexity of the theory may be much less than the complexity of the data. The examples in the table are the influences of weather on on-time performance. The generation of different weather outcomes is governed by simple physical processes of temperature, pressure, and moisture changes in the atmosphere. However, the resulting weather patterns are quite complex and difficult to predict for any length of time. Many chaotic systems of this type will show very complex data trends that are caused by fairly simple underlying principles and processes. In such a case, the statement of the theory may only require a small number of basic equations or postulates, but the data that come from the system will be quite complex.

The trade off for the low level of complexity of chaotic theory components is that they will only allow limited prediction of system variance. Concretely speaking, the impact of N measured influences on the target system may be due to a chaotic system based on fewer underlying variables. Estimating or measuring these more basic variables or processes would allow a very efficient theoretical and empirical account of the N effects in the target system. The theoretical predictability will, however, be limited due to the chaotic and non-linear nature of these parts of the system. Chaotic systems may be very sensitive to small changes in inputs or ongoing processes, or have unexpected long-term changes in states or conditions. In the aviation domain the weather, for example, may be somewhat predictable for a flight of short duration or under stable conditions, but much more difficult to predict for flights of several hours or for unstable conditions.

The simplest possible influence on a key variable is random error or some irreducibly random stochastic process. This is illustrated in the above table by the airplane queues waiting for departures and for landings. Since these queues may be inherently unpredictable, the theory becomes a simple specification of the distribution of the mean and variance of error expected due to such influences. For example, the expected variability in on-time performance due to random factors such as the queues of waiting aircraft or the variation in Air Traffic Control commands might be specified as a normal distribution with an average of 10 minutes and a standard deviation of three minutes. The theory in this case has only two distinct values. In general, for random sources of influence the theory can only give a theoretically expected distribution of random error. This simple theoretical statement does not allow any further predictions of variance beyond the distribution of random error.

Therefore, the determination of the extent to which the variance for a measure of a construct is simple and systematic vs. chaotic vs. random has very important theoretical and practical implications. Both the complexity of the theory and the ultimate power of the theory to explain, predict, and control the target system are affected by this determination. In practice, it may be difficult to know before the SIS is constructed how the variance of the key construct breaks down into systematic, chaotic, or random sources.

In the initial construction of the SIS, the determination of all these variance sources may not be a critical issue. The simple systematic variance components will typically be the first ones to be identified, measured, and empirically evaluated in the SIS. However, as the theoretical base of the SIS is developed into a more complete form, a point of diminishing returns may be reached. This is particularly likely once the most important simple and systematic variance components have been theoretically specified. Pragmatically, if the level of theoretical completeness is sufficient for the intended applications, the theoretical development may stop there. If not, further theoretical development will require the addition of other systematic influences, exploring the influence of the complex, non-linear influences on the system, or determining the magnitude and effects of random error on the system. Ultimately, however, the theoretical components underlying all the key variables in the system must be integrated into a complete and coherent theoretical account that can be evaluated scientifically and is pragmatically useful.

When integrating multiple theoretical components for the key variables and processes of the target system, the variety of possible interrelationships is enormous. The theoretical connections may vary qualitatively in different ways such as mediating effects, facilitating or inhibiting conditions, and so forth. Quantitative theoretical connections can also differ in functional form (linear, quadratic, cubic, etc.) as well as complexity of relationship (one to one, one to many, many to one, many to many, etc.). Therefore, the connections among the hypothesized constructs and processes for each component of the complete theory must be precisely, rigorously, and systematically integrated in the SIS theoretical base. If the connections among the theoretical constructs are not carefully specified, the variety of these possible connections may preclude deriving a precise and unambiguous set of predictions for testing the integrated theory. Theoretical integration is, therefore, one important aspect to consider when developing complete SIS theories. The ability to integrate theoretical components precisely will typically differ depending on the form of the theory.

Theories differ in the representation and symbology they use to represent the key variables and processes of the target system. The main forms of theories rely either on using a natural language ("verbal" theories) or on using an artificial representation of some sort ("symbolic" theories). These different forms for representing the conceptual framework of the theory differ not only in integratibility but also in the precision, clarity, and extensibility of basic theoretical components.

Type of theory

Verbal theories

One basic form for representing a theory is *verbal*. Verbal theories are common in social, political, and historical domains of knowledge, among others. Verbal theories have the advantage of using the symbology of human oral communication as a basis. The theoretical concepts therefore have inherent meaning based on the semantic meaning of the words and how these words are arranged in theoretical statements. Further, a verbal theoretical system is extensible by simply adding or augmenting it with additional verbal premises, clarifications, and so forth.

Precision However, verbal theories have several disadvantages. Verbal theories often lack precision, particularly when compared to mathematical or computational theories. A verbal theory will typically result in expected outcomes that are either qualitative or of a fairly simple "more than", "less than" nature.

Clarity Verbal theories may leave the inherent meaning of the key concepts and relationships unclear. This is in part because the semantic meanings of the words underlying the expression of the concept or relationship are themselves unclear, ambiguous, or not quite appropriate for the domain or application. Even such common concepts as intelligence, personality and job performance are difficult to verbally define clearly and concisely. When a theory does verbally define a key technical concept such as "intelligence" using a familiar term, there is always the danger that the common meaning of the term interferes with or obscures the technical meaning. For relationships, stating that "A is related to B" does not specify how, why, where, or to what extent, this relationship occurs, nor does it clarify if this relationship is A causing B, B causing A, or merely a non-causal covariation in A and B. The verbal meaning can always be elaborated with further clarifications and definitions, but this approach increases the risk that the multiple verbal statements will be qualitatively somewhat different. These differences will generate at least theoretical ambiguity of meaning and potentially generate theoretical inconsistencies.

Extensibility Theories are typically extended to cover different conditions or situations, different kinds of people, and so forth. Making sure that these extensions are correct is a critical part of using a theory more broadly. Verbal theories lack a clearly-defined inference procedure that can be used to derive new, legitimate expressions of theoretical principles. Instead, the derivation of new theoretical expectations relies on verbal argument, intuitive plausibility, and other imprecise procedures. Given these imprecise inference procedures, different people can derive different extensions of the theoretical principles. These different extensions can create theoretical ambiguity or inconsistency across different research studies in the domain. This problem impedes finding a clear-cut set of relevant situations and a precise set of expected results that are necessary for evaluating the extended theory.

Integration   Integration across theories requires an extremely clear and precise expression of each theoretical component. The lack of precision and lack of a clearly-defined inference procedure for verbal theories will also hamper theoretical integration. If imprecise verbal connections are used such as "theory X is synergistic with theory Y", the nature of the possible interconnections for a combined theory may be so vague as to preclude empirical valuation.

In the short run, if verbal theories must be integrated in a SIS, the expected set of principles and hypotheses for the combined theory should be evaluated across a set of relevant experts. Only if the experts are consistent in their synthesis of the verbal theories should an integrated version be taken seriously. This intuitive combination procedure may not be necessary if the theories to be integrated are symbolic rather than verbal. In the long run, the development of integrated and complete SIS theories will be facilitated by converting verbal to symbolic theories.

Symbolic theories

Symbolic theories use some form of abstract symbol system to represent theoretical ideas. Symbolic theories can be mathematical, logical, or computational. Although the symbol system used by each one is distinct, the common disadvantage of these forms of expression is that the symbol system is artificial. This makes it initially more difficult for the user to understand and to express the key concepts, relationships, and processes of the theory. However, the artificial systems used in these theoretical forms have several offsetting advantages in precision, clarity, extensibility, and integration.

Precision   Symbolic theories are generally more precise than verbal theories. This precision allows symbolic theories to make more precise empirical predictions than verbal theories. Theories expressed as sets of equations can be solved for precise values of key parameters. Theories expressed as logical systems can use logical derivation to obtain a precise set of logically expected outcomes. Theories expressed as computer programs can be executed and the results inspected for a precise set of expected outcomes. For a situation defined as a set of values for the theoretical variables, each of these symbol systems allows a precise derivation of expected results. These more precise expected results, in turn, and allow a much stronger empirical valuation of the theory. The stronger empirical evaluation, as discussed earlier, facilitates a more rapid confirmation, disconfirmation, or modification of the SIS theoretical structures to form a set that successfully accounts for the target system.

Clarity Symbolic theories also have clear meaning, particularly of the advanced or derived concepts of the theory. Defining constructs symbolically rather than verbally eliminates the ambiguity and possible surplus meaning of verbal definitions. In practice, however, some verbal definition may be required for defining basic concepts or variables in any of these artificial symbol systems. After that point, however, the other variables are symbolically defined from the basic variables. These definitions are explicit, clear, and limited to the precise symbolic relationships.

For example, in Newtonian physics the basic definition of distance and time may require verbal elaboration. Once these are defined, velocity can be symbolically defined as distance traveled per unit of time, and acceleration can be defined as the change in velocity over time. In psychology, a simple example is the definition of intelligence as a standardized result of an IQ test. The IQ test may require verbal definition. Once the test is defined, the derivation of the IQ score is computational. First, the mean and standard deviation of mental performance for each age range must be empirically established. Then the IQ score can be precisely and unambiguously specified by the Z-score that uses that mean and variance, converted to a scale for which the mean is equal to 100 and the standard deviation equal to 15. Thus, the symbolically defined variables are precisely and clearly specified by the theory's symbol system.

Extensibility Symbolic theories are extended by deriving new theoretical statements that must be compatible with the original set of theoretical statements. Unlike verbal theories, the derivation of new theoretical premises or predictions is a formal, step-by-step procedure that can be objectively followed and replicated by different people. Derivation of new axioms in a logical system proceeds by the rules of logic. Derivation of new equations or formalisms in a mathematical theory proceeds by the rules of the appropriate field of mathematics. Derivation of new theoretical predictions for theories expressed as computer programs is typically done by empirically executing the program with the altered input or altering the program code and observing the results. Different persons executing the same program with exactly the same input should obtain the same expected result. In all cases, the extension of the theory to cover new cases, conditions, or contingencies is clear, precise, and objective.

Integration   The precision and clarity of symbolic theories can also aid theoretical integration. Theoretical integration for symbolic theories can take on a number of possible forms depending on the formal expression of the components. For simple, systematic explanations of the key variables that are expressed as one or more equations, the equations can potentially be combined into a larger set that covers all the key variables. Mathematical operations on the combined set of equations can reveal the implications of the integrated theory. Similarly, theoretical components that are expressed as a system of logical statements may be combined in an appropriate logical system if they do not generate inconsistency. The logical implications of the combined system can then be explored in detail. For theoretical components that are expressed as computational programs, the programs can potentially be combined into a combined computational super-model that represents the integrated theory. By executing the combined program, the implications of the integrated computational theory can be explored in detail. In general, theories using the same symbolic form can be precisely and completely integrated.

Integration of symbolic theories using different symbolic forms may be more difficult. Integration will often require the conversion or translation of one of the theories to be integrated. Logical theories might, for example, have to be converted to computational forms for integration with computational theories. Where this conversion is possible without loss of meaning, the theories can still be completely and precisely integrated.

Evaluating verbal and symbolic theories

If the expected results of a theory for a specific condition involve the outcome for a specific case or the mean, variance, or distribution shape of a key variable for a sample of cases, the techniques discussed earlier for evaluating theoretical predictions can be applied. The evaluation of the obtained vs. the expected results for the specified conditions reflects directly on the theory. Obtaining the expected result supports the theory while obtaining discrepant results disconfirms the theory.

Often the precise values of expected outcomes for a given condition are difficult to derive from the theory, particularly if it is in verbal form. However, the theory may still clearly specify that the average outcomes in certain conditions will be better or worse than the outcomes in other conditions. The expected average differences among different conditions can be statistically analyzed to see if the expected results are found.

Most commonly, the theory will specify that the expected outcomes in two or more conditions will show differences in a certain order. In this case, the average results in specified conditions are compared to each other. Alternatively, the theory may specify that certain conditions will produce results that are better or worse than a baseline or "control group". A "control group" is generally a baseline condition in which nothing happens or the theoretically relevant parameters and processes are neutral. In this case, the control group is compared to the conditions for which the theory specifies an increase or decrease in key system variables.

A statistical technique that focuses on evaluating mean differences, such as the analysis of variance (ANOVA), can be used to check on the theory. In its simplest form, ANOVA compares the average level of a measured variable across different conditions. The basic test of this analysis is whether the average levels vary more than we would expect by chance. If the means vary so much that it would occur by chance only 1 out of 100 times, for example, the difference is considered a real difference.

If the expected pattern is complex and has many different groups or treatment conditions, the pattern of theoretically expected results may be supported in some respects but disconfirmed in other respects. Therefore, the evaluation of the theoretical expectations using the SIS data may range from strong overall confirmation, to mixed confirmation and disconfirmation, to strong overall disconfirmation. The theory in the SIS may be preserved, modified, or discarded based on the pattern of confirmation and disconfirmation.

Data for theory evaluation

Data for evaluating a theory should be appropriate to the theory, have acceptable scientific quality, and reflect the target system as broadly and deeply as possible. First, the data for evaluating a theory should qualitatively match the basic, critical facets of the theory (Chapter 1). For example, the data should properly represent the idiographic or nomothetic quality of the theory, and the level at which the theory specifies effects. Clearly a nomothetic measure of intelligence would be inappropriate for a theory that emphasized the unique, different patterns of many different cognitive capacities across individuals. Conversely, a free-response, multi-faceted measure of attitudes toward an object may be an inappropriate measure for a theory that defines attitude as the affective response to an object only.

The level of measurement represented in the data should also qualitatively fit the level of explanation, prediction, and control emphasized by the theory. Although multi-level integrated theories are possible and desirable, in many domains the initial development and refinement of a theory in the SIS will emphasize a certain natural level. In the aviation domain, for example, a theory that predicted crew-level effects should be evaluated with crew-level performance data, and conversely a theory that predicts individual pilot-level effects should be evaluated with individual pilot performance data. Using individual-level performance data to evaluate a crew-level theory (or vice-versa) is generally inappropriate unless there is no alternative. This type of mismatch can cause statistical inference problems as well as conceptual errors.

Scientific data quality The data for evaluating theories should also have acceptable scientific quality as covered in Chapter 3. That is, the sensitivity, reliability, and validity of measurement should be acceptable. Sensitivity, reliability, and certain aspects of validity such as structural validity, predictive validity, and criterion validity, can be evaluated with minimal reliance on the theory. Evaluation of construct or network validity, however, relies very strongly on the content of the theory. Therefore, this evaluation may require a bootstrap procedure for the development and refinement of the theory and its associated measures. Ultimately the cycles of refinement and change should produce a theory and set of corresponding measures that optimally represent the target system.

Data coverage Evaluating more complete theories about the target system will typical require more complete data. To completely represent the target system, the data should be as comprehensive as possible given the limits of measurement opportunities and storage in the computer system for the SIS. This principle includes the measurement of auxiliary data information wherever possible. For example, task performance on any task that can be divided into sub-tasks could include such auxiliary information as the start and end time for each sub-task or information about the order or sequence of sub-tasks. This information would be very important for evaluating a theory of performance at a more detailed level.

Clearly a balance must be struck between recording extremely fine-grained levels of data and the usability of the data in the SIS. The costs of measurement and of data storage and manipulation should be considered in striking an appropriate balance. In the past, however, costs of data storage have tended to rapidly decrease with further developments in data storage technology and software for data manipulation in the computer industry. If these historical trends continue, therefore, the costs of storage or manipulation will become negligible impediments to data collection in the SIS.

However, other irreducible costs, such as the time required for measurement or licensing fees for commercial instruments, may militate against measuring as broad a set of data as possible. Despite the measurement costs, it is important to measure as broadly and as deeply as possible for the data in a SIS. This broad and deep set of data allows maximum flexibility for the potential testing and modification of theories as discussed above, and allows maximum statistical power for the empirical evaluation of the theory or for theoretical applications such as those discussed in succeeding chapters.

The only real limitation on this principle of broad and deep measurement is the extent to which taking additional measurements inherently will bias or interfere with other SIS measurements, or will change the functioning of the target system in some important manner. Interference with other measurements can potentially be identified during the evaluation of the scientific quality of the other measures. If additional measurements decrease the sensitivity, reliability, or validity of measures in the SIS, then there is a real loss of information and use of the additional measures should be re-evaluated.

Equally serious is the potential problem of additional measurements changing the functioning of the target system. Any given measure may be reactive under certain circumstances, but measures vary widely in how reactive they are likely to be in general. In the aviation domain, for example, having a FAA examiner sitting in the jump seat behind the crew during the flight might be rather reactive and alter the likelihood of certain behaviors. Conversely, having the aircraft's flight recorder data electronically transferred to a database after the flight might not influence the conduct of the flight at all. Each additional measure should, therefore, be evaluated on the yardstick of how much the measurement could change target system processes or outcomes. In the early development of the SIS, the judgments of reactivity might have to be subjective.

Later in the development of the SIS, the set of currently validated measures allows the reactivity of other potential measures to be objectively assessed. The possible reactivity of new measures can be empirically evaluated by checking for changes on key measures of the target system. For example, the new measure can be implemented for a representative subsample of SIS cases. The values of key system variables and critical processes in the set of cases with the new measure would be compared with the set of cases that do not involve the new measure. If significant changes are found the new measure may be reactive and should be re-evaluated. If no significant changes are found, the additional measure is non-reactive and should be acceptable for the SIS.

An alternative approach to evaluating reactivity would be to implement the new measurement at a specific point in time. Using this approach, the key system variables and processes before implementation of the new measure would be compared with the results after implementation. Finding no significant or large differences from pre-to post would support the non-reactivity of the new measure. Making these comparisons depends, of course, on the degree of reactive change caused by the measure and the statistical power of the test to detect changes.

Statistical power   In general, statistical power is the ability of a statistical technique to precisely estimate relevant parameters or test theoretical hypotheses (see Cohen, 1977 for a complete coverage of this topic). Other things being equal, statistical power represents a balance between the effect of error in decreasing statistical precision and the effect of the sample size of empirical information (i.e. number of data points) in increasing statistical precision. Estimation and testing are both affected in the same way by error and the number of data points, but the details differ for each application.

When an estimate or test must be made on a sample rather than the entire population, error has the combined effects of taking the sample and making the measurements (called sampling error and measurement error, respectively). This is illustrated in Figure 4.14 for the example of estimating the mean of a population. The first level of error is taking the sample from the population. The sample of cases B, E, J, M, O, U, and W may not perfectly represent the population of cases (represented in this example by the 26 letters of the alphabet).

The second level of error is measuring each case. That is, the true value of the variable for cases B, E, J, M, O, U, and W may not be accurately reflected by the set of measures b, e, j, m, o, u, and w due to imperfect reliability and validity of measurement. Clearly the final estimated value for the mean depends on both of these errors. In a sense, each type of error results from one of the steps in the estimation process.

If the entire population rather than a sample is represented by measures in the SIS, then the estimate will only be subject to measurement error. In the example above, that would mean having all 26 cases in the SIS databases. In the aviation domain, for example, the databases may contain performance measures for all pilots in a fleet. The estimates for the fleet's average performance would be subject only to the effects of measurement error as the entire population is represented in the SIS databases. In other domains were the SIS contains data only for a subsample of target systems, both sampling and measurement error would apply to estimates.

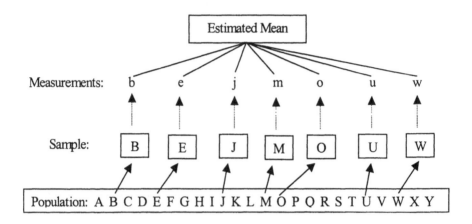

Figure 4.14   Illustration of sampling error and measurement error in estimating a mean

Sampling and measurement error can apply to any kind of estimate or test performed with the data. Estimates would include estimated means and variances of a single variable as well as estimated correlations or relationships between variables. Testing would include tests of hypothesis like t-tests used for tests of differences in means or F-tests for testing differences in variances. Due to this similarity, the determinants of power are similar but not identical for estimation and hypothesis testing.

For estimating parameters such as the mean or variance of a distribution, power depends on the level of error in measurement, the number of data points on which the estimate is based, and the confidence levels set for the measurement. In this case, power represents the degree of certainty that the estimated value of the parameter is correct or nearly correct.

For testing hypotheses, power depends on the level of error in measurement, the number of data points on which the test is based, the size of the effect being tested, and the false positive error rate (e.g. 1 out of 100) used for the statistical test. In this case, power represents the likelihood that an effect of a given size will be correctly detected by the statistical procedure. More specifically, power is the likelihood of concluding that a predicted or expected result is present when it really does exist. This can also be stated as the likelihood of avoiding a false negative conclusion of "no effect" under the conditions that an effect of a certain size is really there.

In both estimation and hypothesis testing, one critical determinant of power is the number of data points on which an estimate or test is based. Sample size and the number of data points for each case (i.e. data density) jointly determine the total number of data points that determine statistical power. That is, power can be increased by increasing the number of cases in the statistical procedure, the number of data points measured for each case, or some combination of these two. Increasing the number of data points is, therefore, one way that estimates can be more precise and that hypothesis testing can avoid both false positive and false negative conclusions. In general, increased power makes the confirmation or disconfirmation of theories in the SIS more clear-cut.

Chapter summary

With a sufficient amount of high-quality, scientifically sound measures, the theories concerning the target system can be strongly evaluated. Three primary evaluation points for a theory are the explanation, prediction, and control of the target system that the theory provides. In addition, theoretical development should strive for a complete picture of the target system, which may require the integration of different theoretical viewpoints. Symbolic theories can be integrated better than verbal theories because they typically have more precision, clarity, and extensibility. The evaluation of the SIS theories requires correspondingly rich SIS data. The scientific evaluation process ensures that the theoretical frameworks that support the information content of the SIS are as good as possible. Having a complete, valid, and accurate theoretical framework sets the stage for effective application of the SIS theories as discussed in following chapters.

5 Solving Problems with a SIS

Overview

The information in a SIS can be used to both identify and solve system problems. In some domains the identification and solution of problems will be the primary functions of the SIS, with the theoretical development and evaluation being more a means to an end than an end in itself. This could be true if the basic SIS goals were issues like accident avoidance or other public-safety issues. This chapter will cover methods that can be used for the identification and solution of problems. There are many different types of problems that could be addressed with a SIS. Considering basic problem issues such as why, when, or what happens in a system helps define problems. Once defined, these problems can be further elaborated and solved using a problem solution cycle.

Problem solution can be thought of as a cyclical process, and the problem solution cycle is illustrated in Figure 5.1 below. The problem solution cycle begins with either well-known problems that have already been identified or with an exploration of the information in the SIS to find problems. The cycle continues through the steps of problem analysis, solution development, implementation, and evaluation. The theory and data of a SIS is important for each of these steps.

Active exploration of the SIS information is necessary to find unknown problems. The problem identification step is aimed at more clearly defining a known problem and the incidence or prevalence of the problem. The problem analysis step is aimed at understanding the exact timing, causes, or conditions of the problem. This elaboration of the origins of the problem sets the stage for using the SIS to develop and evaluate potential solutions. The problem exploration, identification, and analysis steps should result in a more precisely defined problem with a set of associated theoretical and empirical SIS information.

Solution implementation must take into account the real constraints of the system, such as regulatory, legal, or contractual limitations, and still effectively address the problem. The solution implementation stage uses the theory and information of the SIS but also requires a change in the data of the SIS to incorporate new information relevant to systematically evaluating

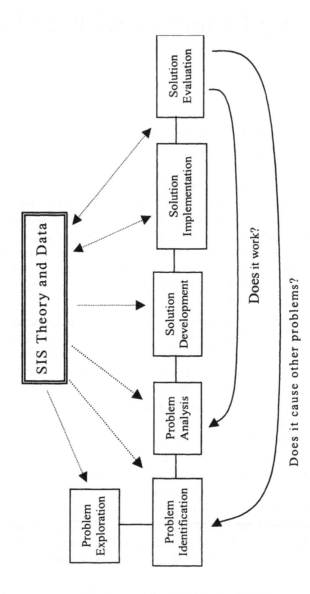

Figure 5.1   The problem solution cycle

the effectiveness of the solution. This two-way relationship is represented in the figure by the bi-directional arrows.

In the short term, the solution evaluation should feed back into the problem solution cycle. First, the information on solution effectiveness (or lack of effectiveness) should feed back to the problem analysis step. Particularly if the solution is ineffective or only partially effective, the problem must be thoroughly re-analyzed to try to understand the reasons for failure of the solution. This reanalysis should be the basis for modifying, augmenting, or replacing the initial solution with a more effective one.

Second, a careful examination of whether the solution is causing any other problems should feed back to the problem identification steps. Since the solution of one problem may cause other problems to occur, the analysis should take that fact into account. In the aviation domain, for example, automation was introduced to solve certain problems, but potentially generated new types of problems (Wiener, 1988).

The evaluation of the solution has both long term and short term effects. In the long term, solution evaluation should ultimately impact the theories in the SIS. Successful solutions increase the credibility of the corresponding theories while unsuccessful solutions decrease theoretical credibility in much the same manner as an experimental test (see Chapter 4). The careful analysis of the effects of a solution should lead to a similar process of refinement and augmentation of the SIS theories as other empirical evidence. The first section of this chapter will cover the questions underlying most basic problems. The next sections cover the use of the SIS during each step of the problem solution cycle.

BASIC PROBLEM QUESTIONS

Critical issues or questions define the basic types of problems. Basic problems concern some state or process in the target system. Problems may be negative things occurring in the target system or positive things that are not occurring in the target system. In the aviation domain, the occurrence of incidents or accidents clearly indicates negative events that are system problems. Positive things that are not occurring could be represented by a lack of crew communication with the passengers concerning flight delays, turbulence or other relevant weather conditions, and so forth. The problem in this case is how to induce crews to do appropriate passenger briefs more frequently.

If the system state or process is undesirable, the problem may be how to detect, avoid, or ameliorate the undesirable state or process. If the issue is a lack of something desirable, the corresponding problem may be how to detect, increase, or facilitate the occurrence of the desirable state or process. Coping with either form of problem requires understanding, predicting, and ultimately altering the occurrence of specific states or processes in the target system.

These functions are similar to the explanation, prediction, and control functions of a theory, but the focus is on a particular problem rather than the theory as a whole. The basic questions underlying the problem are *why* does it happen, *when* does it happen, and *what* can be done about it. Good theory is critical to answering these questions, but the shift in primary focus to the problem rather than the theory is quite important. The elaboration and evaluation of the why, when, and what questions will be centered on problem issues rather than theoretical issues. Answering problem-oriented questions may require that different theories be combined or integrated in a novel fashion, particularly if the problem is complex and multi-faceted.

Why does something happen?

The most basic question about a problem is why the problem occurs, or, if it is the lack of a positive state or process, why the desirable situation does *not* occur. This is essentially the same issue as the explanation criterion for a theory. The explanation of why a problem happens should be plausible in the same manner that a theoretical explanation should be plausible.

In part, the explanation of why a problem occurs will be based on the current set of validated theories in the SIS. In the simplest possible case, the problem will be directly and convincingly derivable from the principles of one theory. In this case, the plausibility of the problem explanation is mainly derived from the plausibility of the theory that accounts for the problem (see Chapter 4).

Alternatively, if the explanation involves an integration or synthesis of several theories in the SIS, the plausibility of the problem explanation will vary with the plausibility of each theory and the nature of the theoretical integration. If the theories are concatenated into an explanatory chain, the plausibility of this combined explanation is no stronger than the weakest link, that is, the theory in the chain with the lowest plausibility.

Explanatory chain

An explanatory chain is illustrated in Figure 5.2. Theory A, B, and C are connected in a causal chain to explain the occurrence of the target problem.

The plausibility of this chain is no better than the least plausible step in the chain of events. For example, Theory A describes the onset or the origin of the problem. If Theory A has low plausibility, the plausibility of the entire account is correspondingly low. The same principle applies to Theory B, which describes a necessary set of intermediate conditions that produce the problem, and to Theory C, which describes the proximal set of necessary conditions to produce the problem.

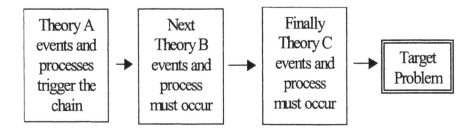

Figure 5.2 Plausibility of a theoretical chain leading to a problem

An example is the occurrence of an automobile collision accident in which car X hits car Y from the rear. Theory C concerning the proximal causes of the accident may involve the relative velocities of both vehicles, the time the driver of car X applied the brakes, and the maximum deceleration possible for a car X. Theory C would be largely based on the physics of objects changing velocity and colliding in the real world.

Theory B concerning the intermediate events in the accident may involve the human factors of the events leading up to maximum braking. The chain of events covered by Theory B would be the time required to recognize the potential accident and start to respond to it, the time required for the driver to transfer a foot from the accelerator pedal to the brake pedal, and the response latency of pressing the brake pedal with maximum effort. Theory B would emphasize a necessary set of events and processes occurring between the more distal causes and the proximal causes of the accident.

Theory A, which concerns the originating cause of the accident, might emphasize personality or situational factors that tend to lead to the accident situation. Personality factors could include hazardous thought patterns on the part of the driver that predispose to being involved in accident situations such as risk-taking, impatience, and perceived invulnerability (Holt, Boehm-

Davis, Fitzgerald, Matyuf, Baughman, & Littman, 1991). Situational factors might include traffic factors such as traffic density and uneven traffic velocities, or vehicle factors such as using a navigational or informational display while driving (Noy, 1997).

The plausibility of the entire explanatory chain depends on each link. If Theory A uses implausible personality, cognitive and situational factors, the explanatory chain is not plausible because the account of starting conditions for the whole set of events is not plausible. Similarly, if Theory B does not convincingly connect the personality, cognitive and situational factors to events just prior to the panic stop, the explanatory chain is not plausible because the connecting link is not plausible. Finally, if Theory C does not portray the physical determinants of the deceleration and collision accurately, the chain is not plausible because the final events and processes leading to the collision are not plausible.

Alternative explanations

However, if each of several alternative theories in the SIS offer sufficient explanation of the problem, the plausibility of the combined explanation may be the plausibility of the most plausible problem explanation. This is illustrated in Figure 5.3. In this figure, Theories, A, B, or C offer alternative explanations for the occurrence of the target problem. The plausibility of the combined account is at least the plausibility of the most plausible explanation. Suppose, for example, that Theory B was quite plausible while Theories A and C were relatively implausible. If either A, B, or C may cause the target problem, the plausibility of the problem explanation is at least the plausibility of Theory B.

For example, suppose the target problem is a high incidence of social aggression. Theory A may focus on situational factors using constructs such as social learning from aggressive models, an aggressive subculture etc (e.g. Bandura,1973). Theory B may focus on the aggressor by emphasizing factors such as goal frustration, dislike of the target person, the emotion of anger, etc (e.g. Berkowitz, 1989). Theory C may emphasize genetic factors that create an human aggressive drive which inevitably produces social aggression (e.g. Lorenz, 1966). The critical point is that the combined explanation would stipulate that aggression stems from personal factors, situational factors, or genetic factors.

These causal factors are basically conceptually independent accounts of social aggression. Therefore, the plausibility of the integrated set of theories for aggression is at least as plausible as the most plausible theory in the set. In practice, the most plausible theory of the set may depend on the particular aggression problem. Some forms of aggression may be plausibly addressed by personal factors, other forms of aggression by situational factors, and still others by genetic predispositions. For aggression observed in a particularly frustrating situation, for example, the focus on personal reactions may be plausible than the social learning and genetic theories. In such a case, the set of theories as a whole still plausibly accounts for the aggression problem.

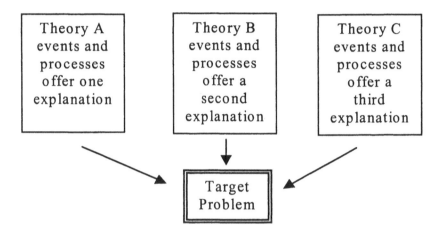

Figure 5.3 Plausibility of alternative theories leading to a problem

The degree of complexity of the theoretical integration also affects the plausibility of the problem explanation. In general, the principle of parsimony suggests that explanations with fewer and simpler principles are preferable to explanations with many or more complex principles. Using this principle, the plausibility of highly complex theoretical integrations is less than simple integrations.

In the worst possible case, the validated theories in the SIS simply do not address the origins of the target problem. Since one goal of the SIS is to build a set of theories that completely explain all important aspects of the target system, this lack of relevance of the SIS theories to the occurrence of the problem is quite serious. In such cases, the problem illuminates a gap in the theoretical coverage that must be remedied by theoretical modifications/extensions or completely new theoretical views coupled with appropriate empirical research. From a theoretical point of view, this is one of the main advantages of problem-oriented research.

When SIS theories do not apply, a careful from-the-ground-up conceptual analysis of the system variables and events related to the occurrence of the problem is necessary to guide the use of the SIS information for problem identification, analysis, and solution. This conceptual analysis will also provide the information needed to appropriately modify, extend, or supplement the current theories in the SIS so that they will explain this target problem. The natural focus for this conceptual analysis is the set of proximal and distal causes of the target problem.

Proximal vs. distal causes of problems

Proximal causes of a target problem are those which immediately and directly lead to the occurrence of the problem. Distal causes of a problem are those which occur further back in time in the causal sequence. In the aviation domain, for example, the proximal cause of a flight problem such as an altitude violation on approach to a landing is a failure to control the aircraft in the appropriate manner. This example is diagrammed in Figure 5.4.

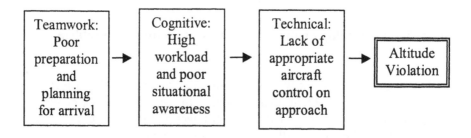

Figure 5.4 Example of a causal chain leading to an aviation problem

In this example, the proximal cause of the problem is the technical issue of appropriately controlling the aircraft's position on the approach glide path. The distal causes, however, may be quite different. The distal cause may be cognitive factors such as the failure of the crew to be aware of the impending violation of altitude limits, or poor situational awareness. The poor situational awareness may in turn be due to more distal causes such as an unusually high workload during the approach to the airport.

This unusually high workload may, however, ultimately be traceable to the lack of foresight and advanced planning for the approach on the part of the crew, which are teamwork issues. In this case, the most distal cause of the altitude violation could be poor crew planning and decision-making on the phase of flight prior to the approach to the airport. This distal cause occurs much earlier in time and is qualitatively quite distinct from the proximal cause of poor technical control of the aircraft. This type of conceptual analysis can guide the use of the SIS for problem identification, empirical analysis, and solution development.

However, problems that consist of the non-occurrence of desirable states or processes in the target system may be far harder to conceptually analyze than problems that consist of the occurrence of negative states or processes. First, it is far more difficult to specify the exact time and conditions of non-occurrence of the desirable state or process. That makes it far more difficult to track back through a causal sequence of proximal and distal causes.

Second, the diversity of the possible approaches to attaining a positive state may be far greater than the possible ways of ending in a negative problem state. A negative problem state often is a series of necessary events such as an "error chain". Failure to attain a positive state, in contrast, may be harder to diagnose because of multiple ways to achieve positive outcomes which point back at different sets of facilitating or causal factors.

The diversity of possible causal paths make a coherent conceptual analysis and plausible explanation of the lack of positive outcomes quite difficult. Particularly for non-occurrence of desirable states and processes, the development of a strong and complete theoretical base in the SIS that will predict optimal states and processes in the target system is critical. Complete and well validated theories provide a solid base upon which to derive the necessary and sufficient conditions for obtaining more optimal outcomes in the target system. This optimization can be done more precisely for theories that are mathematical or computational in nature, which is a major advantage of such theories.

Problem causes vs. problem solutions Identifying the problem cause does not, however, immediately imply the problem solution. For one thing, a given cause may be susceptible to different solutions. Knowing that a disease is caused by a certain bacterium or virus, for example, does not immediately point at a unique solution to the problem of the spread of the disease. Solutions to the spread of such a disease could be as diverse as quarantines, vaccines, antibiotics, or changes in environmental or social conditions fostering the occurrence of the disease. Each of these possible solutions may be effective or ineffective under certain conditions.

Therefore, the conceptual analysis of the causes of the problem is only the first step in the process of solving the problem.

When will something happen?

The question of when or under what conditions a problem will occur is the second basic question about a problem. As before, the problem can be alternatively defined as the occurrence of a negative event or the lack of a positive state or process in the system. This question concerns the same essential issue as the prediction criterion for a theory. The account of when or under what conditions the problem occurs should be as precise and accurate as possible. To answer this question, the theories and data in the SIS must address the timing and specific conditions of critical problems. If this information is not available, the theory or data in the SIS may have to be refined or augmented to be able to specifically address when problem states or processes occur.

For an aviation example, suppose the problem of poor automation use by crews is the focal problem. Both SIS theories and data must be reviewed for relevance to automation use problems. The current data in the SIS may not have automation-related measurements of performance. Lack of appropriate data precludes the use of empirical methods for problem analysis.

Even if appropriate data are lacking, relevant theoretical analysis may still be possible. That is, if the current set of empirically-supported theories in the SIS may cover automation tasks and issues sufficiently well to predict the timing and conditions of automation use problems. This extension of using well-confirmed theories is one important long-range advantage of the development of the theoretical base of the SIS. Although extensions of SIS theories to solve new problems should be checked by empirical methods, solution by theoretical extension is much quicker and consumes less resources than initiating a new research and development program aimed at solving each new problem. However, if neither data nor theory cover automation use problems, both aspects of the SIS should be extended to cover the focal problem.

Extending SIS data to appropriately address the problem depends on whether the timing or conditions of occurrence is the critical issue. If the critical issue is the timing of the problem, the relevant SIS data would have to be time-stamped precisely enough to identify the time of problem occurrence. For problems in automation use, for example, an overall evaluation at the end of the flight that listed automation problems would not be useful for precisely identifying the time of problem onset. Evaluations of automation use completed after each phase of flight would offer somewhat better information on when automation use problems were occurring, but still be imprecise. A better form of data would be a time-stamped evaluation of automation use problems recorded as a trace measure by the flight simulator or made by an evaluator noting the time and nature of the problem as soon as it is observed.

If the critical issue is the onset conditions for the problem, the relevant SIS data would have to describe the conditions in sufficiently rich detail for the pattern of relevant conditions to be determined. In the aviation domain, for example, the National Transportation Safety Board (NTSB) investigations of fatal accident delve into the conditions of the crew, the aircraft, and the flight in extensive detail (NTSB, 1994). The focus of these investigations is to find the pattern of conditions that tends to lead to aviation accidents. One NTSB study of fatal accidents found a pattern of crews having accidents during high workload conditions on the first flight of the duty day when they had not previously flown together as a crew. On a much broader scale, the information in a SIS can be similarly analyzed for patterns of conditions associated with the key problems.

What can we do about it?

The question of what can be done about a problem is the third basic question. If the problem is the occurrence of a negative event, the control of this event may involve prevention, detection, mitigation, or some combination of all three. This question of what can be done about a problem is essentially the same as the control criterion for a theory. If a theory in the SIS is directly applicable, it should point to manipulations of the context, parameters, or processes of the target system that will reduce the likelihood or effects of a negative event. Different methods may be necessary for reducing the likelihood or mitigating the effects of a problem.

For example, in the aviation domain, Helmreich and Merritt (1998) have proposed that error avoidance, error trapping, and error mitigation are three fundamental ways of addressing the problem of flight errors. Error avoidance implies error prevention, that is, eliminating the occurrence of the error in the first place. Error trapping implies the immediate detection and solution of an error once it has occurred. Error mitigation is reducing the negative results of errors after they have occurred. Each form of controlling flight errors may require different interventions. Common interventions for controlling undesirable states or processes would focus on changing the system context, system components, or system processes.

Changing system context

Changing some aspect of the system context is one method of solving problems. System context includes any aspect of the situation and conditions in which the target system operates. In the aviation domain, for example, the system context might include the corporate culture or climate of the organization, the staffing policies for assigning crew members to flights, the corporate rules for rest intervals between flights, and so forth.

Changing context to avoid automation use errors would, of course, depend on the precise cause of those problems. Suppose a theory specified that a major causal factor for these problems was pilot fatigue due to inadequate rest intervals between flights. In such a case, a change in the corporate context which would require adequate rest intervals between flights should solve the problem.

Changing system components

Changing some aspect of the system itself is a second method of solving problems. Typical systems will include physical components and human components, either of which may be a focus for change. Physical components include all physical aspects of the situation such as tools, the person-machine interface, lighting, computer hardware and software, etc. Human components of the system include everything related to human performance at different levels of system such as the person, team, and department levels. The relevant factors may include experience, training, knowledge, skills, abilities, and so forth.

Human factors research has identified many basic principles of changing systems for better human use. Both the physical and cognitive demands of the system can be reduced by appropriate changes. In the automation use example, the relevant theory could be a human factors theory that specifies that the problems are due to an unnatural, difficult-to-use automation interface. Human factors principles could then be used to appropriately redesign the interface, and this intervention should solve the problem.

Alternatively, the focus of change can be in the human aspects of the system. Theories in the field of Industrial/Organizational (I/O) psychology typically focus on the human aspect of system behavior at an individual, group, or organizational level. These theories can be applied to change workplace behavior. Suppose the automation use problems were due to inaccurate knowledge of the system. In this case training to correct the knowledge deficits should solve the problem.

Changing system processes

Changing system processes is a third major approach to solving problems. Of course, changing system context or components may work by indirectly changing system processes, but some interventions are directly focused on changing system processes. These processes may involve processes executed by humans, processes executed by computers or other machines, and processes that jointly involve humans and machines.

Targeting human processes for change to solve automation use problems would be appropriate if a relevant theory from I/O psychology specified lack of crew communication and situation awareness as critical causal factors. The change in process might involve instituting a required procedure in which the pilot working with the automation would communicate the initiation and completion of each significant task to the other pilot in the cockpit. This change in team communication processes should increase the mutual awareness of the state of the automation on the part of both pilots and reduce or eliminate the automation use problems.

Targeting machine processes for change would be appropriate if the relevant human factors theory specified a lack of information from the automation system to the pilots as a critical causal factor. This might occur if the automation had sufficient autonomy to change its mode of operation under certain flight conditions, but was not clearly informing the crew about such mode changes. The appropriate change in process in this case might be a change in the processes of the flight computer such that the shifts in the automation modes would be annunciated by both visual and auditory messages to the crew.

Table 5.1 gives examples of the three basic questions as they apply to each facet of a target system in the aviation domain. The focal point of the analysis is potential performance problems for routine flights. This table offers examples of some of the types of problems that would be the focus of SIS applications for aircrew performance in the aviation domain.

The basic why, when, and what to do questions concerning an initial problem will motivate the use of a SIS in the problem solution cycle. However, the initial problem may not be the real problem. That is why the first steps of the problem solution cycle should always involve problem exploration, identification, and analysis to ensure that the problem is correctly defined and delineated before developing solutions. Using SIS information in these steps may result in more precisely defining a problem, extending the definition and scope of a problem, or replacing the initial problem with a more critical, important problem for solution.

Table 5.1 Aviation domain examples of why, when, and what questions for problems with different system facets

| | | Problem Question | | |
		Why?	**When?**	**What to do?**
Focal Facet of Target System	**System Context**	Why is on-time performance worse for shuttle operations?	When (what conditions) do crews have problems with on-time performance?	What can be done to alter the shuttle operational context so that on-time performance improves?
	Physical System	Why is aircraft type A harder to fly than aircraft type B?	When (what conditions) do crews have problems flying aircraft A?	What can be done to change aircraft A and make it easier to fly?
	Human System	Why do fleet A's pilots perform better than fleet B's pilots?	When (what conditions) do the crews of Fleet A perform better?	What training can be given to Fleet B pilots to improve performance?
	System Process	Why is checklist performance slower than desirable?	When (what conditions) is checklist performance unusually slow?	What procedural changes would aid speedier checklist performance?

STEPS IN THE PROBLEM SOLUTION CYCLE

Problem exploration

Practical use of a SIS starts with problem exploration. Problem exploration is the conceptual and empirical exploration of the SIS for potential problems. This exploration is aimed at finding currently unknown or undocumented problems. Finding unknown problems is particularly important when the prevailing consensus is that there are no significant problems in the target system. The complacency induced by such a consensus can lead to ignoring tell-tale signs or symptoms of system problems. Janis (1982), for example, describes the state of "group think" in which a group ignores negative input, often with disastrous results. To break this type of mind set, the active, intense, and dedicated search for problems should be an important use of the information in the SIS. The problem-search function can, for example, be defined as a legitimate task for a corporate safety, efficiency, or quality assurance department using the SIS. This search for problems can use either the conceptual or empirical aspects of the SIS.

Conceptual problem exploration

The conceptual exploration of potential problems in the target system uses the theoretical part of the SIS. This exploration will become easier and more thorough as the SIS theories evolve into better accounts of the target system and are more thoroughly integrated into a coherent whole. Initially, conceptual problem exploration may have to depend on less precise and relatively unintegrated theoretical perspectives, but the development of the SIS should lead to more precise and integrated theories over time.

Less precise and more loosely integrated forms of theories make conceptual problem exploration more difficult and dependent on the knowledge and assumptions of the explorer. This is particularly the case for loosely-knit verbal theories. In such a situation, the conceptual exploration will require a qualitative elaboration or extension of the SIS theories. This qualitative process will often require some outside knowledge or domain expertise. Suppose, for example, that an aviation domain SIS identified the corporate goals and procedures for operating the aircraft as critical parameters for crew performance. In this case, the set of corporate goals and procedures may be examined for potential goal conflicts. One corporate goal might be "Safety" while other corporate goals could be "Efficiency of Operation" and "On-time Performance".

The goal of on-time performance may require crews to hurry their normal pace of task activities to meet a flight deadline. If the SIS theory specified a speed/accuracy trade-off in the quality of performance, the increased speed of performance would dictate a decreased accuracy of performance. Decreased accuracy of performance on critical flight tasks could lead to a less safe outcome. Basically, the goals of on-time performance and safety could conflict and adversely affect crew performance.

Real world constraints could also create a basic conflict in this set of corporate goals. Fuel use is more efficient, for example, when fewer engines are used or the aircraft is flown at slower speeds. Therefore, the goal of "Efficiency of Operation" could induce fuel-conserving procedures that would interfere with either Safety or On-time Performance. For example, taxiing from the gate using a single engine could save fuel but put the extra cognitive workload of starting the other engines on the crew during the taxi to the runway, thereby reducing situation awareness and safety. Similarly, reducing power to save fuel during cruise may increase efficiency of operation at the expense of on-time performance at the end of the flight. Conceptual problem exploration should include these "what if" scenarios.

More precise and integrated forms of theories such as sets of mathematical functions or computer programs allow for a great deal of quantitative as well as qualitative "what if" exploration that may identify potential problems. In this process, unusual conditions or situations can be examined by specifying the appropriate values of parameters and processes in the theory. Execution of the program or solution of the equations yields a picture of performance of the target system under the specified conditions. In contrast to loosely-knit verbal theories, obtaining this picture of target system performance is relatively objective and independent of the biases or viewpoints of the person exploring the problem.

Where quantitative results can be obtained, the "what if" picture of target system performance can be compared to current system performance using appropriate benchmarks. If the "what if" performance is noticeably lower than current performance, the conceptual exploration may have identified conditions for producing a potential problem. If the "what if" performance is noticeably higher than current performance, the conceptual exploration may have identified a potential solution or more optimal state for the system. This type of conceptual exploration uses precise empirical methods but is fundamentally conceptual since the theories and key input values are conceptually derived. The conceptually derived empirical data used in this process is sometimes called "Monte Carlo" data or a Monte Carlo simulation to distinguish it from real measures and values in the SIS databases. Of course, the real measures and data in a SIS can also be used to empirically explore for problems.

Empirical problem exploration

The empirical exploration of potential problems in the target system uses the measures and data in the SIS. In the initial development of the SIS, the empirical exploration of problems may be hampered by poor quality or incomplete data. As the SIS is developed over time, however, this exploration will become more effective as the SIS data is refined into a more sensitive, reliable, valid, and complete set of measures concerning the target system. Empirical problem exploration can involve either qualitative or quantitative data analysis.

Qualitative data analysis could involve many forms of SIS information such as written documents, trace measures of the operations of target systems, and so forth. For example, in the aviation domain, the free text notes or comments made by evaluators during flight tests can be qualitatively analyzed for possible crew problems. Similarly, trace recordings of relevant parameters from simulator training sessions could be qualitatively examined for critical incidents that indicate potential crew problems (Flanagan, 1954). One instance of such critical incidents is the occurrence of aircraft tail strikes due to over rotation on takeoff.

The analysis of the SIS information may be augmented with the analysis of associated informational databases. In the aviation domain, the associated informational databases could be the Aviation Safety Reporting System (ASRS). The first-hand narratives of potential problem situations in these databases can be analyzed for themes or principles underlying potential problem incidents. Combing through this type of textual information for indications of possible problems will often be laborious and time-consuming. For more efficient exploration, a preliminary scan of the quantitative data in the SIS can be used to guide a more in-depth qualitative analysis of selected cases or incidents.

Quantitative data analysis for problem exploration in a SIS may use empirical techniques that range from simple to complex. Simple procedures will typically involve a single variable (univariate) and simple transformations or analyses with that variable. Complex procedures will typically involve multiple variables (multivariate) and more complicated transformations or analyses of these variables.

Outlier analysis One simple analytic procedure is identifying outliers of key indicator variables for the target system. Outliers represent statistically rare or unusual cases. Typically, outliers are extremely high or low values on a variable. To find outliers, the value for each case is compared to an overall distribution of values. If this comparison indicates that the value is sufficiently rare or unusual, the case may be evaluated further to see if the extreme value represents a problem.

For an aviation example, the time required to complete a required checklist during flight may represent a key variable. Suppose crews were evaluated on this variable and the distribution of times in Figure 5.5 was found. This figure is the frequency distribution of how many crews required a certain amount of time to complete the checklist (e.g. less than 20 seconds, 21-25 seconds, 26-30 seconds, etc.). The data in this figure show that overall crews take an average of about 60 seconds to perform this checklist, with a standard deviation of 10 seconds. However, crews W and X were much faster than would be expected from the overall distribution of times, and crew Y was much slower. Statistically, these outliers would be evaluated to see how likely they were to occur by chance, and outliers that were sufficiently unlikely (say a likelihood of less than 1 out of 1,000), would be selected for further qualitative and quantitative evaluation.

One simple empirical method of identifying univariate outliers is to compute a z-score for each case where z = (Case score – average score) / Standard Deviation of scores. The z-score method works well when the distribution of scores approximates the normal, bell-shaped curve. The distribution of times in this example has a unimodal, bell-shaped curve that approximates the normal distribution. In this example, crews W, X, and Y have a z-score so extreme that it would occur by chance less than 1 in 1,000 times. Such rare cases can be found by using the mean and standard deviation of values from the database and establishing empirical cut-off values of $z > +3.0$ or $z < -3.0$. Crews W and X have z-scores less than the cutoff of -3.0, and crew Y has a z-score greater than 3.0, so these three cases would be selected for further evaluation.

The evaluation explores whether the significantly lower or higher checklist times represent problems. The further evaluation is critical since these extremes may or may not represent real problems. Take the significantly lower times for crews W and X, for example. Crews W and X may be too hasty and slipshod in performing the checklist, or alternatively these crews might be much more efficient and timely in performing the checklist. Analyzing the qualitative data (e.g. instructors comments) and quantitative data (e.g. associated performance ratings) for these crews should clarify if the performance is problematic.

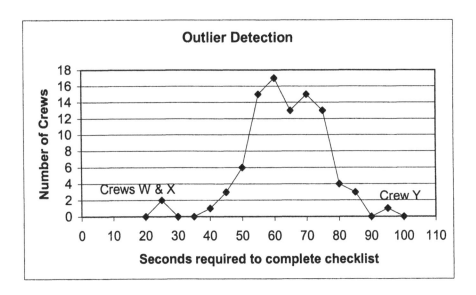

Figure 5.5 Example of detection of outliers for timed performance

Conversely, crew Y's extremely slow performance may or may not indicate a problem. If crew Y is slow because they have difficulty executing the checklist tasks, their slow performance may indicate a true problem. Alternatively, if crew Y is slow because they are more thorough and double-check their checklist actions, their slow performance may not indicate a problem at all but rather superior performance.

Multivariate outliers The concept of a univariate outlier can be usefully extended to the concept of a multivariate outlier. A multivariate outlier is a case which has a rare or unusual pattern of scores across a set of variables. Essentially, this method looks for cases with profiles of scores that are unexpected or atypical given the average scores in the sample and the interrelationships among the set of variables. The analysis for multivariate outliers involves the calculation of the Mahalanobis distance for extreme cases based on a set of key values. The Mahalanobis distance basically combines the divergence or distance of one case from a profile of sample means for a set of measures. The Mahalanobis distance index combines the differences across all the measures while taking into account the interrelationships among the measures.

This distance index essentially summarizes how unusual or atypical a case is for the entire set of measures. This index can be tested to see how likely the pattern of differences for each case would occur by chance. This probability can be used like the probability of the z-score in the univariate example above to screen out the cases that have extremely unlikely or rare values. These cases can be further examined to see if the unusual profile of scores for each one represents a problem or not.

The importance of considering multivariate outliers is that a case may appear to have acceptable, within-range values on all variables when they are looked at separately, but still be a very atypical or rare case when the profile of scores is looked at as a unit or set. This is because the nature of the relationships among the measures might make the entire set of scores in the profile unlikely even though each separate score was possible.

An example of this is given in the Figure 5.6 for a situation where speed and accuracy of performing a series of items in a checklist is examined for a sample of pilots. Suppose that in general the faster the checklist is performed (lower time) the more errors of omission, commission, or change of order are made by a pilot. Conversely, taking more time generally leads to fewer errors.

Given this relationship, it is possible for a pilot to have speed scores and accuracy scores that are quite unusual in combination although each score separately is not that rare. Pilot A, for example, has an average of 1.5 errors while executing the checklist and an average time of 45 seconds. Neither the time nor the errors is significantly different from the means for the sample ($z = -1.50$ for time, $z = -1.39$ for errors). Together, however, Pilot A's profile is quite unusual in that this pilot is able to complete the checklist more quickly than the average for the sample while producing fewer errors than is typical for the sample. This is an example of a profile that could be identified as a multivariate outlier, but would not be a problem.

Pilot B, on the contrary, might represent a problem. Pilot B's overall profile of speed and accuracy is also quite unusual given the speed/accuracy trade-off of the entire sample. Pilot B takes longer than most pilots ($z = +1.50$ for time) but still incurs more errors than is typical for the sample ($z = +1.24$ for errors). Although neither the z scores for time or accuracy are significantly unusual by themselves, the combined profile of scores is significantly rare or atypical given the Mahalanobis distance index. Since this profile combines slower and more error-prone execution of the checklist, it could well represent a problem.

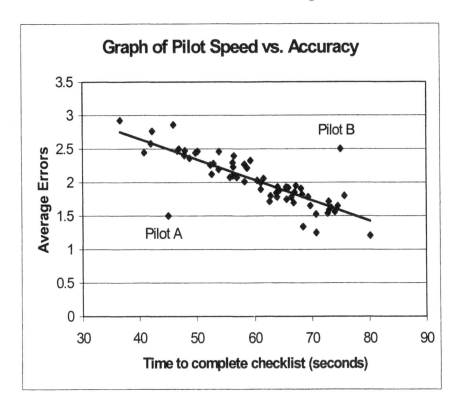

Figure 5.6 Multivariate outliers for speed and accuracy of performance

Ascertaining if a pattern such as Pilot B's profile is really a problem, as opposed to an unfortunate fluke occurrence, can be aided by other quantitative methods that combine or synthesize profile information. Different methods can be used for this purpose, but two quantitative methods that may make valuable contributions are cluster analysis and multidimensional scaling. These methods use the concept of a multivariate distance and can be used for assessing patterns of information across multiple cases in problem exploration.

Cluster analysis Cluster analysis links together cases with similar profiles of scores into groups or clusters of cases. This linking is based on how related each case is to the others. Relatedness is indexed by the "proximity" or nearness of cases to each other. That is, the set of scores for a case is compared to a set of scores for another case or a set of mean scores for a group or cluster. The case is linked to another case or to a cluster with which it is most similar. This process is repeated until all cases are joined to clusters.

Although all linking methods use distances or proximities for linking, the details of different linking methods are different for different versions of cluster analysis. Typical options for linking a new case to a cluster are nearest neighbor, centroid, and farthest neighbor options. These options differ on how the critical distance between a new case and the cases already in a cluster is defined.

In Figure 5.7, the new case A could be joined in the cluster B, C, D, or in the cluster E, F, G. The nearest neighbor to A is D in the first cluster and E in the second cluster. The nearest neighbor option would join A to the first cluster as A is closer to D than to E. The centroid represents the average of the cases in a cluster and is point C for the first cluster and point F for the second cluster. The centroid option would join A to the second cluster as A is closer to F than to C. The farthest neighbor option would use point B in the first cluster and point G in the second cluster. The farthest neighbor option would also join A to the second cluster as A is closer to G than to B. As the farthest neighbor option is emphasizing the similarity of a new case to the most *dissimilar* case in the group, it generally makes more sense to use either the nearest neighbor or the centroid as the basis for joining new cases to clusters.

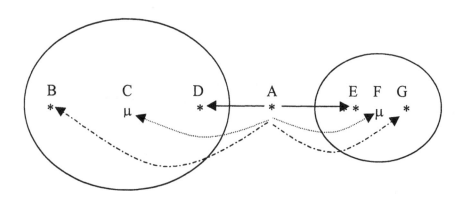

Figure 5.7 Clustering a new case into one of two clusters using nearest neighbor, centroid, and farthest neighbor options

Cluster analysis can be quite useful in problem exploration. Clusters can help create a typology of cases that aids in problem definition. For example, clusters based on performance measures can indicate if there are distinct *types* of high or low performance. Distinct types of low performance profiles, in particular, may indicate qualitatively different types of problems that require different types of solutions. Cluster analysis makes very few assumptions about the data beyond the fact that different scores on the variables can be sensibly combined into some type of distance, proximity or similarity information about the cases that can be used to group them. Since the information is synthesized across variables, the variables should be on a common measurement scale or be standardized to avoid the possibility of a measure with highly variable scores dominating the solution. In comparison, multidimensional scaling can also be used for problem exploration but makes more assumptions about the measures.

Multidimensional scaling Although multidimensional scaling has a similar starting point in using a set of distances among cases or objects, the goal of the analysis is different. Rather than establishing clusters or groups of cases, the goal of the analysis is to establish a set of underlying dimensions that parsimoniously account for the observed distances among the cases. As with cluster analysis this method can be used to find the dimensions underlying either cases or a general set of objects. The distances among the cases or objects are typically called dissimilarities for multidimensional scaling, but conceptually the idea of similarity is the same as proximity or nearness and the idea of dissimilarity is the same as distance.

The critical issue for using multidimensional scaling for problem exploration is whether it makes sense to search for a small underlying set of dimensions (typically 2-4) to concisely describe the differences in the profiles of scores. Establishing a common set of dimensions assumes that all cases can be sensibly placed somewhere on each dimension. That assumption would NOT be appropriate for situations where the cases are qualitatively distinct and not quantitatively comparable. If, for example, apples, oranges and other types of fruits were truly incomparable, using multidimensional scaling to find the underlying dimensions of these fruits would not be appropriate.

For many other domains, of course, the idea of common underlying dimensions is reasonable. If the analysis is successful, these dimensions would offer a parsimonious way to describe the differences among cases. In the aviation domain, for example, the set of performance differences among crews could condense into the basic dimensions of speed of performance (fast vs. slow performance) and quality of performance (good vs. poor performance). This condensation would then guide problem exploration by focusing it on the two major issues of performance speed and performance quality. In particular, finding two dimensions would suggest that slow performance and poor quality of performance are two distinct problems that might require distinct types of solutions.

The net result from the conceptual and empirical exploration for potential problems must be a defined problem or potential problem. This problem can then be added to the list of known problems which is used in the next steps of problem identification and analysis. A precise and concise definition of problems found by problem exploration is necessary for a preliminary analysis of relevant aspects of the problem such as the incidence and prevalence of the problem. Good definition is also required for further problem analyses that ultimately lead to an appropriate problem solution.

Problem identification

Problem identification starts where problem exploration ends—with a list of known or potential problems. These problems must have an initial definition that is sufficiently clear for the problem to be explicitly connected to information in the SIS. The definition of a problem may be changed, of course, as the analysis of information in the SIS indicates necessary revisions. The problem identification step will typically involve basic analysis on the incidence and prevalence of the problem, time trends, and so forth. All of these analyses focus on the problem itself. Further elaboration of how the problem is related to other SIS information is carried out in the problem analysis step.

A preliminary step that may be useful or necessary in problem identification is to check the list of problems for redundancy, errors, or gaps. The initial set of problems should be conceptually organized in a coherent way. Sometimes this organization may be a taxonomy of errors or problems in the target system. These taxonomies can be simple and yet have powerful implications for further problem analysis and solutions. An example of a simple, powerful error taxonomy is Reason's taxonomy of slips, lapses, and mistakes as qualitatively distinct types of errors (Reason, 1991). A good taxonomy should have mutually exclusive and exhaustive categories for classifying cases or relevant observations. These categories should be conceptually connected to the theories underlying the SIS.

Alternatively, the problems can be directly tied to the best current theories about the parameters and processes of the target system. If the set of theories in the SIS offers a consistent and coherent picture of the target system, tying the problems to the conceptual elements of these theories is one way to organize them. In the initial development of the SIS, the theories are apt to be somewhat piecemeal and unintegrated, and in this case the taxonomic approach may be a more feasible first step. As the SIS theories evolve over time, important and persistent problems should be tied to the theories in the SIS wherever possible.

Prevalence of problems

Once the potential problems have been clearly defined and distinguished, the data in the SIS can be used. The first issue is the prevalence of the problem in one or more populations. Prevalence is defined as the relative proportion, frequency, or percentage of cases in a defined population that have the problem. To analyze prevalence, marker variables in the SIS that indicate the problem must be selected and the values which constitute problems specified.

Once the variables and values are specified, estimating the prevalence of the problem for a given population may be as simple as selecting the appropriate sample from the databases and counting the relative frequency of occurrence of the specified values for the critical variables. Prevalence information is extremely important both to understanding a problem and to designing appropriate and adequate solutions for the problem.

If the percentage of cases having the problem is low, the solution may be directed at changing the occurrence of those particular cases. If, on the other hand, the percentage of cases having the problem is high, the solution may require systematic changes in broad aspects of the context for the system. In the aviation domain, for example, if only a few crews are having trouble with on-time flight performance, the solution may involve training those crews. Conversely, if almost all crews in a given fleet were having problems with on-time flight performance, the solution may possible involve changes to corporate policies, flight time tables, operating guidelines for the aircraft, or other contextual factors.

Incidence of problems

Conceptually, the incidence of problems is the rate at which the problems develop in the population. This is independent of prevalence because high or low prevalence may be associated with either high or low incidence. Table 5.2 gives examples of high and low incidence associated with high or low prevalence using the epidemiology of diseases as the focal domain.

Table 5.2 Illustrating the independence of prevalence and incidence using common diseases as examples

		Prevalence of the problem in the current population is:	
		High	**Low**
Incidence of the problem in a new set of cases is:	**High**	Problems occur easily and remain for a period of time: e.g. Herpes virus	Problems occur easily but are removed quickly from population: e.g. Flu virus
	Low	Problems occur slowly but remain in the population: e.g. Tuberculosis	Problems occur slowly and do not remain in the population: e.g. Mad Cow disease

Herpes simplex virus infections, for example, have a high rate at which uninfected people can catch the disease, and the disease is durable in people who are infected. The flu virus also gives a high rate of infection to uninfected persons, but typically runs a one to two week course after which the person is not contagious, which effectively removes the case from the population. Tuberculosis is not as infectious as the flu, but is much more durable in people who are infected. Finally, Mad Cow disease spreads only slowly and indirectly through consumption of contaminated products. The severity of the disease tends to eliminate the infected cases from the population in relatively short order, so prevalence is low.

One method to examine the incidence of problems is to take a representative (e.g. random) sample of the population and follow it over time. The rate of increase of problems in the problem-free part of the sample will correspond to the incidence rate. This rate will typically be expressed as the proportion of new problem cases expected in a certain time frame (e.g. one year).

Other analyses could address the prevalence of problems in the population over time. One approach is to analyze trends. Trend analysis can be used to track the relative frequency or severity of problems in an appropriate sample (or the population) over suitable time intervals.

Trend analysis

Trend analysis is essentially searching for systematic effects of time on either the frequency or severity of defined problems in a sample or population. There are at least three distinct questions for using time-based analyses in problem identification.

- Are there any significant (non-chance) changes over time?
- Are there systematic trends (e.g. linear or curvilinear changes) over time?
- Is the current state of the target system different from baseline periods?

At the most general level, the analysis should answer the question of whether problems are significantly changing over time or not. If nothing is changing over time and the status quo is acceptable, no further analysis is necessary. If the frequency or severity of a problem is changing over time, the trends for that change can be analyzed in more detail.

If a problem is indexed by a specific variable, changes in the problem over time can be tested by an analysis of variance (ANOVA). In the aviation domain, an example would be flight delays for a fleet of aircraft across months. On-time arrival data would be checked for differences over months of the year. If the monthly differences were non-chance (significant at a specified level), the differences in average values across time can be further analyzed for long-term and short-term trends. For example, the ANOVA might detect that average arrival times were significantly later in the winter weather months of January, February, and March, possibly due to delays caused by the extra time required for de-icing the aircraft, or other aspects of winter operations.

If the problem is defined by a set of variables rather than one variable, a similar analysis can be performed using the entire set of variables. An example of a problem defined by multiple measures in the aviation domain is an unstabilized approach. A good approach is jointly defined by altitude and airspeed as the aircraft approaches the runway. A stabilized approach requires smooth, joint decreases in both airspeed and altitude. An unstabilized approach may deviate from a stabilized approach by either an altitude that is too high or too low, or by airspeed that is too fast or too slow. The focus of the trend analysis might be to analyze unstabilized approaches at a specific airport or runway context.

For this type of situation, the analysis is similar except that a multivariate form of ANOVA called multivariate analysis of variance (MANOVA) is used. The MANOVA would check if the total set of items significantly varies across time intervals, while taking into account the interdependence among the variables in the set. This type of analysis could be used to check if the final approach altitudes and airspeeds were significantly different across months. Any deviations over time in either approach airspeeds, altitudes, or the combination of these two variables would be detected by this analysis.

Types of trends If there are significant changes over time, the next level is to find any systematic long-range trends in this problem. The long range trends may be steady increases or steady decreases that would indicate a linear trend, or various cyclic or non-linear patterns of increases and decreases over time. Figure 5.8 gives examples of extremely strong trends in scores over months. Typically the focal measure would be a score representing the percent correct or percent satisfactory performance on a key system variable.

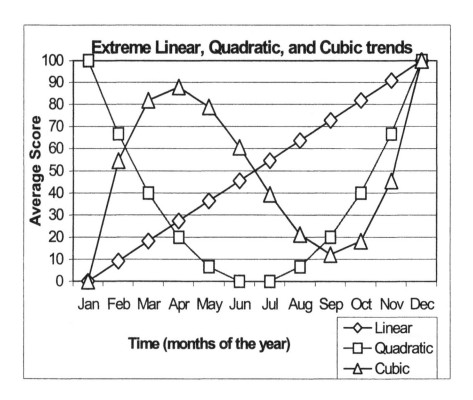

Figure 5.8 Examples of extreme long-term trends

The three lines in this figure show three very distinct trends over time. The straight line connecting the black diamonds shows a steady, gradual increase in performance from 0% in January to 100% in December. This type of trend is a linear trend and corresponds to a linear relationship between month and average score. The implication of this trend is a steady improvement on the average scores over time. Notice that there is no change in direction of this trend, it just keeps going steadily upward. The other lines in this figure illustrate two very different types of non-linear trends. Non-linear trends can be identified by the fact that the trend changes in direction somewhere during the total time interval.

The curved line connecting the white squares shows a decrease from January to June followed by an increase from July to December. This type of non-linear trend is a quadratic trend and is sometimes called a U-shaped trend due to its shape. The implication of this trend is a sharp decline in scores from January to June followed by a sharp increase in scores from July to December. Notice that there is a major change in direction of this curve—the trend changes from a downward trend to an upward trend in the middle of the summer.

The line connecting the gray triangles shows a different type of non-linear trend. This line shows an initial increase in the spring followed by a summer slump and a resurgence of scores in the fall. There are two noticeable changes in direction or inflection points on this curve, which identifies it as a cubic trend. These long-range trends set the context for the final aspect of trend analysis, which is the analysis focused on the current time interval.

Current interval The focal question for this level of analysis is whether the current state (e.g., this week or this month) is significantly different from the immediately preceding periods. That is, is the problem currently increasing or decreasing in frequency or severity compared to a reasonable baseline? If significant long-range trends have already been found, a variation of this question is whether the current state is greater or less than what we would expect from the long range trend. To test this variation, the long-range trend is extrapolated to the current time interval and the observed results of the current state are compared to the extrapolation.

To accomplish an analysis of immediate trends, the first step is to determine the window of time that defines the current state. Naturally, the appropriate window of time greatly depends on the type of target system in the SIS domain. For a performance SIS in the aviation domain, for example, the time window for the current state could reasonably be the current month's performance data. Faster-changing systems with more fine-grained data in the SIS may use a correspondingly smaller time window of weeks or days. In contrast, the performance record for mutual funds and other long-term investment systems is typically measured in years.

The second step in the analysis is to establish the appropriate time window for a relevant comparison baseline. Different domains could have different appropriate baselines depending on their natural timeframe. For the aviation domain, the appropriate comparison for the current month's performance data might be the preceding year's performance data. However, within any particular domain there may be more than one appropriate baseline comparison that could give pertinent information. For example, is this month's performance better or worse than last month? last quarter? last year?. Choosing multiple baselines gives more comparison information, but also creates the possibility that different time comparisons will give different results. Figure 5.9 illustrates this possibility.

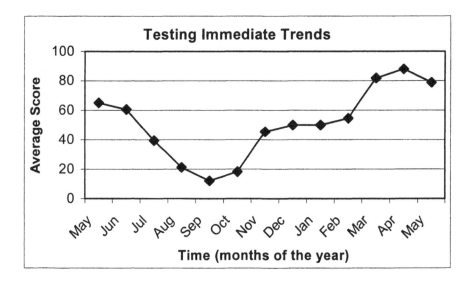

Figure 5.9 Comparatively testing the current May's performance to April, the last year, and last May

In this figure, the current month's average performance is represented by the average score of about 80 for May (far right of the graph). This average score is worse than last month (April) but better than the average for the year. The current month's performance is also better than the performance in May a year ago. The results of these comparisons help identify the problem as a short-term decrease in performance apparently due to something happening this month. The exact nature of what is happening this month to decrease performance is the appropriate focal point for problem analysis, which is the next step in the problem cycle.

Problem analysis

Problem analysis consists of elaborating the knowledge about the problem, its possible causes, and its context. This involves examining other aspects of the system that are related to the problem either by the theories in the SIS or by the empirical data in the SIS. If the SIS theories are well-developed and validated, they can efficiently guide the empirical exploration of issues related to the problem. If the SIS theories are not yet well-developed, the analysis of connected issues can be done empirically but should still be guided by focal questions. There are a variety of possible focal questions that can guide empirical analyses, but common ones may be "Why do some cases have this problem and other cases don't?", "What other variables predict the occurrence of this problem?" , and "What other variables are related to this problem and how are they related?" Each of these questions has corresponding empirical methods that can be used for problem analysis using SIS data.

Cross-classification analysis

One of the first steps in problem analysis is to localize the occurrence of the problem. These analyses should attempt to localize the problem by characteristic time, situation, or personnel. That is, the analyses should find when, where, and with whom the problem is occurring. One of the simplest methods for localizing a problem is to cross-classify the occurrence of the problem by time, situation, or personnel.

The collateral information about each case in the SIS is invaluable for performing a wide variety of possible cross-classification analyses. The retrievable information from the databases should include time measures, situation variables, and information about personnel. Using this information, the identified problem cases can be cross-classified on each of these facets.

When possible, of course, the facet selected for cross-classification should be theoretically relevant to the occurrence of the problem. In general, using available theory to select relevant variables for cross-classification will be an efficient approach to finding diagnostic patterns. Given the plethora of available information in a large-scale SIS, analyzing all possible cross-classifications is not practically feasible and runs the risk that any pattern found may represent a non-reproducible statistical fluke.

Since the analysis must be targeted at some subset of variables, the theory offers the best approach to find candidates for relevant variables.

The clearest possible outcome from the cross-classification analysis would be a strong association of the problem with one facet but not the others. For example, the problem may occur only for certain personnel, but be relatively evenly spread across times or situations. This example is illustrated for the aviation domain in Table 5.3. In this table, the problem of late-arriving flights is cross-classified by crew and by month of year.

Table 5.3 Percentage of late-arriving flights by crew and by month

	Jan	Feb	Mar	Apr	May	Jun	Jul	Aug	Sep	Oct	Nov	Dec	Aver.
Crew A	1.6	2.7	0.3	0.9	0.3	0.6	0.6	3.0	0.5	0.4	1.2	2.1	1.2
Crew B	1.2	2.7	2.7	1.3	0.0	0.2	2.3	1.8	2.2	1.4	2.2	0.8	1.6
Crew C	**5.6**	**6.2**	**9.1**	**9.7**	**8.1**	**7.6**	**6.1**	**5.6**	**6.1**	**6.3**	**5.8**	**5.5**	**6.8**
Crew D	**5.2**	**9.6**	**8.9**	**8.4**	**8.2**	**7.8**	**8.0**	**8.3**	**5.4**	**6.9**	**7.8**	**8.5**	**7.8**
Crew E	0.4	1.5	1.8	1.7	1.9	1.9	2.5	2.3	1.3	2.6	1.6	0.8	1.7
Crew F	1.1	2.3	1.3	0.1	0.4	1.5	1.1	2.2	2.6	1.6	1.1	0.9	1.4

Each cell of the table contains the percentage of late-arriving flights for particular crew and month. In this case, late arrivals seem to be associated with crews C and D (rows with bold print), but not crews A, B, E or F. Crews C and D have around 7% of their flights arriving late whilst the other crews all have less than 2% of their flights arriving late. The percent of late-arriving flights is not associated with seasons; it varies only slightly over the months of the year, ranging from 2.5% to just over 4%. The strong covariation of the problem with certain crews but not months suggests a personnel cause for the problem rather than a time cause (Kelley, 1972).

This pattern could be further analyzed by doing a more fine-grained, second level of cross-classification analyses. These "drill-down" analyses further explore the contribution of other variables to the occurrence of the problem. The breakdown variables can represent different potential causes of the problem. In this example, the set of problem cases for the high-problem subset of personnel like crews C and D could be further cross-classified with relevant personnel variables such as age, experience, or training records. The results from the drill-down analyses of personnel variables could further elaborate the possible causes of the problem.

An example of a drill-down analysis exploring situational factors is given in Table 5.4. In Table 5.4, the late arrivals of Crews C and D are cross-classified by time of day and by direction of flight. As before, each cell represents a percent of the flights made by Crews C and D that were late.

Table 5.4 Drill-down analysis of late arrivals for crews C and D

Direction of Flight					
	N. to S.	S. to N.	W. to E.	E. to W.	Aver.
Morning	5.7	5.1	2.9	15.3	7.3
Afternoon	5.0	5.7	2.1	15.9	7.2
Evening	5.9	5.6	2.8	15.9	7.6
Average:	5.6	5.5	2.6	15.7	7.3

This drill down analysis helps to pick apart the role that time of day or direction of flight might play in their pattern of late arrivals. Time of day does not seem to have a noticeable effect on the rate of late arrivals as the rate is reasonably constant for morning, afternoon, and evening flights. This lack of difference helps rule out factors such as crew fatigue that would plausibly vary across time of duty day.

The direction of flight, however, does seem to be a relevant factor in the tendency of Crews C and D to have late arrivals. In this example, these crews had a particular problem with being late on East to West flights. Their on-time performance for West to East flights is essentially normal for the overall sample of crews, and the results for North to South and South to North flights are intermediate. This pattern of results could occur, for instance, if crews C and D were prioritizing fuel efficiency over on-time performance and reducing power to save fuel on their flights. The prevailing West to East winds would help them be on time for the West to East set of flights, but impede their making the destination on time for the East to West flights.

The pattern of results from a cross-classification may not be as clear as in the above examples. When the pattern is unclear, statistical methods can help determine if the pattern of variations could occur by chance or not.

Statistical analyses can also be used to check on the relative strength of pattern even when the pattern of results seems clear-cut. Both the knowledge that a pattern is not attributable to chance and the strength of the pattern help the interpretation process.

Statistical methods that can be used to analyze the patterns in a cross-classification analysis are chi-square and log-linear model analyses. Both types of analyses check the frequency of events in each cell against the likelihood that the frequencies would occur by chance. The chi-square analysis is simpler to execute and useful for simple cross-classifications, while the log-linear analysis is more complex to execute but also offers more power for analyzing complex cross-classifications. An example of a chi-square analysis is given in Table 5.5.

Table 5.5 Observed and expected frequencies for the late arrivals of crews C and D

	Direction of Flight			
	N. to S.	S. to N.	W. to E.	E. to W.
Observed Frequency	8	8	4	24
Expected Frequency	11	11	11	11

This table is based on hypothetical results for these two crews during the year. The total number of flights on which they had late arrivals was 44 in this example. If these 44 late-arrival flights were distributed evenly across the different flight directions, we would expect 11 late flights in each direction. The observed frequencies of late flights are, however, noticeably different from this even chance baseline, with far more late flights than we would expect in the East to West direction and far fewer in the West to East direction. The chi-square test of the observed vs. expected frequencies confirms that this pattern of differences is very unlikely to occur by chance (i.e., would occur less than 1 out of 100 times by chance). Therefore, the statistical analysis supports the conclusion that there is something about the East-West direction of flight that is contributing to late arrivals for these two crews.

Log-linear analysis proceeds in a somewhat analogous fashion to the chi-square analysis. The occurrence of some event, such as the late arrivals in this example, is analyzed for differences due to one or more facets of classification. However, the log-linear analysis can simultaneously evaluate the impact of multiple facets of the cross-classification on the occurrence of the problem. For this example, the rate of late arrivals can be analyzed for potential impact of the direction of flight, the time of day, and any interaction of direction of flight by time of day. Thus, having the capability for analyzing the separate and joint impact of different cross-classification facets on the occurrence of problems is an advantage of log-linear model analyses.

If no theoretical or conceptual structure exists to guide the analysis, the pattern of problems can still be analyzed by purely empirical methods. These methods are sometimes described as "data mining" methods in the data warehousing literature (e.g. Barquin, & Edelstein, 1997). Different data mining methods are appropriate for different kinds of problems and types of data. Since these methods are atheoretical, they typically rely on a statistical criterion for performing the analysis.

For example, a logic tree analysis tries to find a sequence of cross-classifications on different variables which will maximally predict the occurrence of the problem. In a way this is analogous to a chi-square analysis of a cross-classification analysis, but the analysis in this case is guided solely by the accuracy of prediction at each step and only variables that significantly contribute to problem prediction are added to the logic tree. That is, the computer is selecting the cross-classification variables on a statistical basis rather than the data analyst selecting the cross-classification variables based on theoretical relevance.

Clearly these approaches can reach very different results. Since the data mining analyses are statistically guided, the results of these analysis will strongly depend on the exact set of cases analyzed. The results of purely empirical methods should, of course, still be evaluated in light of the theories in the SIS. Where connections can be made between the data mining results and relevant theory, appropriate theoretical changes should be made. To ensure the data mining results were not due to statistical flukes, these theoretical modifications should be both cross validated on a new sample and checked with converging analyses.

From a theoretical point of view, using purely empirical methods is inefficient and unprofitable. The likelihood of efficient and coherent theoretical progress coming from the sole use of atheoretical methods is quite low, so preference should be given to theory-guided methods. When theory is unavailable or still being developed, however, the purely empirical methods have a legitimate role in SIS analyses. Once the pattern of occurrence for a problem has been determined by either theoretically guided or purely empirical methods, other analytical methods can be used to further describe, explain, and predict the problem.

Discriminant function analysis

One approach to further analyze problems in the SIS is to compare cases where problems occur with cases having no problems. This can lead to a better understanding of the differences that may cause the problems. One method for comparing different sets of cases is Discriminant Function Analysis (DFA). The goal of DFA is to find a linear combination of variables that will maximally differentiate or discriminate known groups. At a minimum these groups may represent categories such as a "set of cases with problems" and a "set of cases without problems".

If problems can be taxonomically classified, different groups of cases would represent each type of problem. Suppose, for example, in the aviation domain that we could isolate different sets of crews having operational problems like poor on-time performance, technical flying problems like unstable approaches, and teamwork problems like poor communication. Using an additional "control" group of crews with no problems, the DFA could explore what combination of information in the SIS would differentiate these four groups. The ways that these groups can be distinguished would potentially help not only in understanding and predicting the problems, but also in designing appropriate solutions.

Differences on quantitative variables can also be used to establish groups for a DFA. If only one quantitative variable is involved, appropriate cut-points should be established that separate cases into qualitatively different groups. Cutting scores on a continuous variable into groups is justified when the distribution seems to fall into natural clumps or clusters of cases that might indicate qualitative differences. The apparent clustering can be double-checked with cluster analysis to be sure that it is reasonable before proceeding with the DFA. Conversely an even, normal distribution with no natural groups might be better analyzed with the regression method (discussed below).

Multiple quantitative variables can also be used to establish groups for a DFA analysis. The quantitative variables of height and weight, for example, could be split at the median and used to define four groups of persons: above median height and weight (big), below medium height and weight (small), below median height but above median weight (thick), and above median height but below median weight (thin). Cases from these four groups could be further analyzed with other information in the SIS to determine if these groups were describable or predictable from other factors.

If there are more than two different groups, discriminant function analysis will calculate different ways of distinguishing the groups. Each way of distinguishing the groups corresponds to a separate discriminant function. Each discriminant function is defined by a unique linear combination of the variables and differentiates the groups in an independent way.

Suppose, for example, in the aviation domain that a fleet of pilots were drawn from four very different backgrounds: military fighter pilot, military bomber pilot, civilian flight instructor, and Captains in commuter operations. Due to past experience, these groups of Captains might differ in how they organize and run their cockpits. The hypothetical example in Figure 5.10 concerns the teamwork and communication patterns of these Captains.

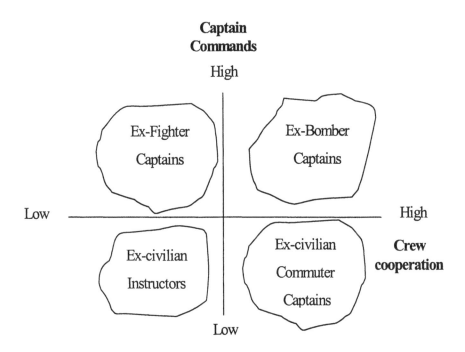

Figure 5.10 Example of two discriminant functions distinguishing three pilot groups

In Figure 5.10, two discriminant functions distinguish the four groups of Captains. The discriminant function on the abscissa of the graph is a linear function which emphasizes items measuring crew cooperation. On this discriminant function, the ex-bomber Captains and ex-civilian commuter Captains are high while ex-fighter Captains and ex-civilian instructors are moderate to low. This would be a natural result of the bomber and commuter Captains working with a crew during their military aviation career and transferring those team practices to the civil aviation domain. Both fighter pilots and civilian aviation instructors would lack this extensive crew coordination background and score lower on these items.

The second discriminant function is independent of the first and emphasizes items having to do with the Captain issuing commands during cockpit communication. This discriminant function is the ordinate in the graph and separates the pilot groups in a different way. On this function, both of the military groups are relatively high in Captain's commands, in contrast to the civilian pilots who are relatively low. Both of the military aviator groups may have become used to a command-oriented style of communication during their military careers and tend to carry this style over into their civilian cockpits. Both groups of civilian pilots, in contrast, would have been interacting in a different context and might consequently have avoided giving commands unless necessary.

This example illustrates that a set of groups may be usefully distinguished in more than one way. In fact, a set of k groups could conceivably be distinguished in k-1 different ways if there were sufficient cases in each group and sufficient items for making the distinctions. The DFA analysis is designed to find the ways that maximally discriminate or separate the groups first, and only then the weaker discriminating functions. Each function is tested for significance. This test helps evaluate how many functions are "real" or discriminate sufficiently well to consider further in problem analysis.

Discriminant function analysis also gives an empirical method for classifying either the old cases or new cases into the specified groups. This "classification analysis" can be used on the old cases to check how well the discriminant functions really separate out the members of the different groups. Classification analysis can also be used on new cases to predict which group these cases most resemble. If one or more of the groups represents a problem, then being able to classify cases into the problem group is particularly important. In the aviation domain, for example, this predicted classification could be used to direct more training or testing resources to pilots who are classified into a problem group.

Multiple regression analysis

If the problem can be measured in terms of its severity, then a natural problem exploration technique is finding which other factors in the SIS can predict the occurrence of the problem. In this approach, the measure of problem severity is considered the criterion. Given this criterion, the object is to find how to combine a set of predictors or explanatory variables that will maximally predict or explain the variation in problem severity.

Multiple regression analysis finds a set of weights for the set of predictors that combines them and maximizes the prediction of the criterion. Several statistical tests give useful results. First, the overall test of the significance of prediction can be used to decide if the prediction is really above a chance baseline. Second, statistical tests of the unique weight for each predictor can be used to decide if that predictor is uniquely predicting the problem above a chance baseline. More sophisticated tests can be made of the difference in the weights to see if certain variables are significantly more predictive of the problem than others (see Cohen and Cohen, 1983 for details).

Corresponding to the significance tests of each predictor are measures of the relative strength of the results. The simplest measure of the strength of prediction is R-squared, or the square of the multiple correlation coefficient. This value indicates the amount of variance in the criterion that can be explained or predicted by the linear combination of predictors. This is an important bottom-line indicator of how well current SIS variables can predict the problem.

Similarly, the strength of prediction of each predictor variable is indexed by the semi-partial correlation coefficient (sr). It indexes the unique linear relationship between a predictor and a criterion. When squared, this value reflects the amount of variance in the criterion that is predicted by each predictor. Predictors with a high percent of variance accounted for are statistically more important than those which predict less.

A variation of this approach called logistic regression can be used when the criterion is a dichotomous or nominal scale, such as pass versus fail outcomes in training. Logistic regression also results in a set of weights for the predictors that reflects their relative importance.

The end result of the problem analysis step should be a better theoretical understanding of the problem based on empirical results. This better understanding should include a more precise knowledge of when, where, and to whom the problem is occurring as well as information on how problem and non-problem cases differ and how to best predict problems from the collateral information in the SIS. This understanding is critical to developing, implementing, and evaluating problem solutions.

Solution development

Efficiently developing effective solutions hinges on the scope, grain, and validity of the applicable theories in the SIS. If SIS theories are appropriately modified or elaborated based on the information gained during the problem analysis step, they should more effectively address the problem. A theory-based solution to a problem partly depends on the characteristics of the SIS theories as discussed in previous chapters. The theory should have broad enough scope to cover the problem's essential states or processes. The granularity of the theory should be sufficiently fine to account for when, where, and to whom the problem occurs. The validity evidence for the theory should give confidence that it can be correctly and effectively applied to the problem. If the relevant theories are mathematical or computational, the exact predicted effects of a solution can be evaluated.

However, if the information from problem exploration, identification, and analysis is not integrated into the SIS theories, the development of the solution must depend to some extent on human intuition and supposition. The human judgment steps in solution development are a large potential source of error. This potential error can be minimized by using systematic solution development methods.

Solution development methods

Nominal Group Technique One method for developing solutions is using a Nominal Group Technique (NGT) where multiple experts first develop solutions independently, and then discuss the set of possible solutions as a group (Van De Ven & Delbecq, 1971) This method is illustrated in Figure 5.11. In this method, the first stage is solution generation. Each expert individually considers the problem and produces his or her list of possible solutions. The objective of the solution generation stage is the maximum range of possible solutions. Having the experts solve the problem individually avoids the potential negative effects of the group setting on creative thinking.

The second stage is the solution-evaluation stage, which is performed in a group setting. The group uses the combined lists generated by all the experts to avoid overlooking anyone's unique solution. The group's initial focus is evaluating the solutions, but this focus can shift to synthesizing or modifying the proposed solutions if none of the solutions on the list seems optimal. The final output from the NGT group stage is a proposed optimal solution along with the reasons that the solution was evaluated as best. These reasons should point to the expected effects of the solution that should be incorporated into the evaluation plan (discussed below).

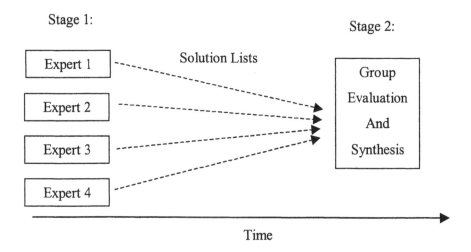

Figure 5.11 Generating and evaluating problem solutions by Nominal Group Technique

Delphi method An alternative technique is a Delphi method (Dalkey, 1969) in which proposed solutions developed by a committee or set of experts are brought together and synthesized by a central person or small sub group (Figure 5.12). The synthesized solution or small set of solutions plus relevant information about the apparent pros and cons of each solution is then redistributed for further evaluation by the large set of experts. This process is continued over cycles of proposing and synthesizing solutions until an optimal solution is developed that the majority of experts can agree on. In this method the large set of experts does not need to actually meet unless the problem is not solvable over several cycles of the method.

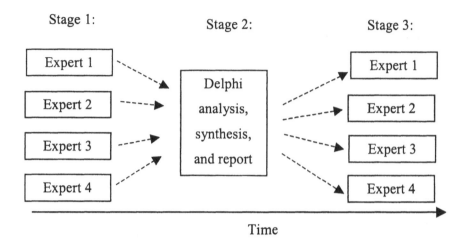

Figure 5.12 **Generating and evaluating problem solutions by
Delphi Method (1 cycle)**

The common element of these methods is that solution development is
not dependent on one person. To allow solution development to depend
solely on one person is to risk a highly idiosyncratic solution that will
inadequately address the problem. Using a group is not an absolute
guarantee of a better solution as group process is also subject to a known set
of biases and errors. Care must be taken, for example, to avoid negative
processes such as "Groupthink" (Janis, 1982). Therefore, the combination
of a set of domain experts with a development process that minimizes the
influence of either individual or group errors is the best way to develop
solutions when the SIS theories are not precise guides to solutions.

Theory-based solutions

If a good theory is available (i.e. it covers the problem and accounts for the
collateral information found about the problem in the problem analysis
step), using this theory to initially develop solutions is more efficient than
relying on domain experts. Theory-based solutions can, of course, be
checked by domain experts to ensure against errors caused by either the
misapplication or misuse of theoretical principles. This type of final check
on potential solutions is an efficient use of the domain experts' time and
helps prevent catastrophic errors in theory application. Catastrophic

application errors might occur when the SIS is initially being developed and the theoretical coverage is sparse rather than complete. Particularly in the early stage of SIS development, therefore, checking theory-based solutions with domain experts is advisable.

Even if a good theory is available, however, applying the theory to the problem will still require some level of human contribution. This contribution can be very minimal where the theory is both complete and precise (e.g. a set of equations or a computer program). In this case, the human contribution consists of solving the equations or running the program until the conditions are found that resolve the problem. While some judgment may be required to do this efficiently or to make very large qualitative changes in the theory, in many situations the development of the solution can be largely automated and require minimal human time or effort. For well-specified quantitative theories, essentially automatic methods using hill-climbing or other forms of parameter adjustment can be used to find optimal solutions.

Other automatic methods can be used to find optimal configurations for complex processes that can be expressed as computer programs. In the genetic algorithms approach, the computer program or algorithm expressing all significant aspects of the processes in the target system is considered as a genetic code like DNA. Following this analogy, the algorithm's component processes are sliced up into small pieces analogous to genes and recombined into many different possible DNA syntheses. Each synthesis is automatically evaluated and preference given to better versions of the algorithm in the next splitting and reproduction cycle. By a process mimicking mutation and natural selection, this approach may evolve a specification for processes in the target system that produces a much more optimal result.

The human contribution is more intensive when mathematical or computational forms of the theory are not available. Lacking such theory, a person must explore for satisfactory solutions by systematically changing the key system parameters or processes specified by the theory. This may involve setting up "What if" scenarios (e.g. what if we increase training on topic A or procedure B?). Each scenario should be evaluated as precisely and thoroughly as possible for problem solution and for avoiding undesirable side effects on other aspects of the system.

Human contributions are also necessary when distinct theories in the SIS must be combined to build a more complete account of the problem that can be used to develop a better solution. This forced synthesis of theoretical viewpoints is a major advantage of problem-focused research using the SIS. Each theory emphasizes a certain view of a certain domain of information concerning the target system. Particularly in the early development of the SIS, however, the domain for each theory will be a partial subset of the potential information about the system, as illustrated in Figure 5.13.

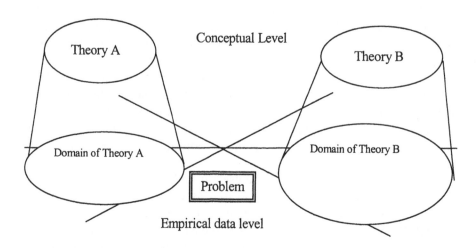

Figure 5.13 Example of a problem requiring extension of two theories

In this figure, the problem and the domains of Theory A and Theory B are at the level of measured variables in the SIS. The theories are at the conceptual level and involve the principles and processes that allow explanation, prediction and control of the information concerning the target system. In a sense, these theories act as a lens through which certain aspects of the empirical information can be sensibly viewed and interpreted. As these theories currently stand, however, neither accounts for the problem very well.

To account for the problem sufficiently well to design an effective solution, the domains of theories A and B must be extended and possibly integrated. Since the problem is near both the domains covered by Theory A and Theory B, both theories might be relevant to the problem and its solution. The integration of Theory A and Theory B should offer not only a more complete account of the problem but also a combined scope of operation in the SIS which is much more extensive than either theory separately. Theoretical integration and synthesis is particularly important when the solution to a problem *must* be optimal.

Not all problems, however, may be serious enough to merit an all-out search for an optimal solution. The solution development for minor or non-critical problems may be aimed at finding a "sufficiently good" solution rather than finding a best or optimal solution. Finding a sufficiently good solution is sometimes called "satisficing" (Simon, 1983), and is practical when the cost of developing and elaborating solutions is high relative to the incremental benefit derived from a better solution. For system-critical problems, however, the potential benefits of better solutions over a long time period will generally outweigh the cost of developing and evaluating multiple solutions in the short-term. In these cases, the solution development process should be allocated sufficient time and other resources to develop the best possible solution to the problem. When a satisfactory or optimal solution is available, the next step is implementing it.

Solution implementation

Evaluation measures

The implementation of a solution involves some systematic change to the context or key parameters and processes of the target system. Systematic evaluation of the solution is clearly necessary to evaluate if it worked as planned. In order to evaluate the solution, the appropriate data must be gathered in the SIS. In some cases, the set of measures already in the SIS will contain a complete and efficient set of measures to evaluate the solution. In many cases, however, the set of measures may be either incomplete, indirect, or inefficient for evaluating the effectiveness of the solution. The data in the SIS should be appropriately augmented with new or revised measures and an analysis plan designed for evaluating the solution.

New or revised evaluation data can be kept apart from the rest of the data in the SIS or integrated into it. If the evaluation consists of a one-time study to evaluate a particular solution, the data may be kept apart from the rest of the SIS. The data should still, however, be linkable to the appropriate SIS databases. This can be done, for example, by using appropriate keys or identification numbers. If this is not done, the other information in the SIS cannot be brought to bear on the ultimate evaluation of the solution. This would result in an evaluation that is informationally isolated from the other information in the SIS. Since the effects of the solution cannot be cross-checked with the other SIS information, the evaluation will be correspondingly less complete and weaker.

If the problem and the effects of the implemented solution are important enough to be evaluated on an ongoing basis, the evaluation data should definitely be integrated into the SIS. The major cost of this integration is the time for database augmentation or redesign. The major benefit of this integration is the easy cross-analysis of the evaluation information with the other SIS information. The measures used for solution evaluation must be checked for sensitivity, reliability, and validity like any other scientific measures. The evaluation of construct or network validity will involve the other SIS information as a necessary part of the evaluation process. Therefore, in most cases the benefit of a scientifically acceptable evaluation measure will outweigh the cost of database redesign.

Establishing the sensitivity, reliability, and validity of the evaluation measures is just the first essential step in the analysis plan for a problem solution. If the evaluation measures are basically faulty, there cannot be an effective evaluation of the solution. The subsequent steps in the evaluation of the problem solution depend on the evaluation design.

Evaluation design

Many evaluation designs are possible (Campbell & Stanley, 1966, Boehm-Davis, Holt, & Beaubien, 2001). One of the simplest designs is a pre-post evaluation. This design relies on establishing a pre-solution baseline and then comparing the post-solution measurements to the pre-solution baseline. The important point for implementation is that such a design requires an adequate time of measurement before the solution is implemented in order to establish a stable baseline. If this requirement is not recognized or if the solution has to be implemented immediately and no appropriate pre-solution measures exist in the SIS, it may not be possible to scientifically evaluate the effects of a solution.

Alternatives to the time-line evaluation design are designs that use different groups and compare the results across groups. The solution can be implemented in one group, for example, and the results compared to the measurements from a group without the solution intervention. This is essentially an evaluation experiment. If natural groups are used, this approach is subject to the problems discussed under natural experiments in a previous chapter. If cases can be randomly assigned to solution and no solution groups, the strength of the evaluation is much stronger as it is less likely to have systematic confounds between the groups. In any case, selecting the evaluation design should be followed by a careful specification of the expected effects of the problem solution.

Expected evaluation results

The expected results from a problem solution should be specified as completely and precisely as possible. Completeness and precision of the expected results is necessary for the strongest possible statistical evaluation of the solution in the final step of the problem solution cycle. At least two aspects of specifying the expected results are critical: First, the specification of exactly where and when in the target system the solution has an effect, and second, the specification of exactly how this effect will show up in the SIS measures.

Where Specifying where in the target system the solution will have an effect helps set the scope and level of granularity for appropriate measures of the effect. If the target system is a multi-level system, such as the pilot-crew-fleet system in the aviation domain, this specification should designate the exact level for the expected results. Depending on the specified level, different measures may be required such as crew-level versus individual pilot-level measures of performance. Some higher-level effects, such as expected changes in fleet performance, may be defined as a simple aggregate of lower-level measures such as the crew performance measures. This type of aggregation requires, however, the tacit assumption that the data are suitable for combination and that the combination correctly represents the real upper-level result.

If the target system is a complex mix of states, processes, and parameters, the exact aspect or component of the target system that will be changed by the solution should be specified. Different measures will be required, for example, to measure an outcome variable such as on-time performance versus a process variable such as the quality of crew interaction. The available SIS measures should be evaluated to see if they offer the appropriate scope, granularity, and level for measuring the effects of the solution.

Similarly, if the focus of change is the personnel of the target system, this specification should designate what aspect of the persons will be affected. Many different aspects of persons such as their attitudes, beliefs, values, motivations, behaviors, or even emotions, could be affected by a problem solution. The content of the SIS personnel measures must be appropriate for the expected effects.

When Specifying exactly when the solution will have an effect in the target system helps set the necessary timing of measures of the effect. Some solutions may take a period of time to become effective, particularly if they require change of well-established processes in the system or in human habits. In the aviation domain, for example, Holt, Boehm-Davis, & Hansberger, (1998) found strong effects of Advanced Crew Resource Management (ACRM) training only after a year of formal implementation and practice had elapsed. Other solutions may have a brief and transient expected effect. The timing of the SIS measures must be appropriate to capture the expected result when it should occur. To play it safe, of course, the implementation of measures for a solution may cover a wider range of times than the precisely anticipated time for the effect. In this way, the evaluation is safeguarded against the possibility that the effect may occur slightly earlier or later than expected and not show up in the evaluation measures.

How   The specification of exactly how the anticipated effects of the solution will show up in the SIS measures is the final key aspect of specifying the expected results. Exactly how the solution will affect the target system requires a detailed understanding of how the target system works. If the SIS is well-developed, this understanding is codified in the SIS theories about the system, and these theories can be used as the basis for specifying how the solution should affect the system. If the SIS is in its initial stages of development, the specification of exactly how the solution should affect the target system will have to rely more on expert opinion than validated theories. This specification can be derived, for example, as part of the development of proposed solutions using the Nominal Group Technique or Delphi method discussed above. Often a solution will affect a system by changing system processes as well as outcomes, and the effects of such changes must be carefully considered for evaluation.

Change of processes

Solutions that change the processes in the target system can have a wide variety of expected effects. If a process is changed, the effects of that change may depend on the values of situational variables or particular states of the system. Either the occurrence of the process itself must be directly measured, or the effects of these possible changes must be carefully tracked and specified so that the a precise pattern of changes in outcomes can be specified.

In the aviation domain, for example, Crew Resource Management (CRM) typically involves the processes of crew communication and coordination in the cockpit. Solutions changing these processes can have a wide variety of outcomes. Measuring the processes themselves may be more feasible in videotaped evaluation sessions using a simulator than on a typical line flight. Nevertheless, the anticipated outcomes of the CRM changes on certain aspects of typical line flights such as on-time performance or fuel consumption should be predicted by carefully tracking the effects of the process changes on flight outcome variables. If a change in process must be evaluated by a change in outcomes, the change in outcomes should be carefully assessed as this becomes the focal point of the evaluation step.

Change of outcomes

Univariate changes When specifying exactly how the solution affects outcomes for the target system, it is important to consider the full range of possible effects. The full range of effects includes both direct and indirect effects. Direct effects are the expected changes in frequencies, average scores, variability, or distribution shape for a univariate measure as well as the expected multivariate changes discussed in the next section.

The complete specification of the full range of possible effects should be summarized in an analysis plan that will guide the statistical evaluation of the solution in the final step of the problem solution cycle. This helps the evaluation be more systematic and complete. The analysis plan offers a strong evaluation of a solution to the extent it is complete and explicit about the exact expected effects of the solution.

Commonly, a solution should decrease the frequency of the problem or some measure of the degree of severity of the problem. This expectation translates into an expected difference in means on the appropriate measure of frequency or severity. Using the pre-post design discussed above, the mean frequency or severity of the problem after the solution is implemented (post) should be significantly lower than the mean for the baseline measurement before the solution is implemented (pre). Similarly, for the different groups design discussed above, the mean frequency or severity of the problem should be less for the group for which the solution is implemented than for the control or comparison group for which no solution is implemented.

It is important, however, to carefully consider possible effects on variability as well as expected differences in means (Alliger, 1997). The effects of a solution could be to either increase or decrease the variability of a measure. In the aviation domain, for example, consider a grading scale for crew performance that would be "Unsatisfactory, Satisfactory, Standard, Above Standard, Excellent". Suppose a problem is observed with non-standard performance patterns of crews and a problem solution is developed that emphasizes performing precisely according to company standards. This solution may have an expected effect of reducing the variability of performance as illustrated in Figure 5.14.

In this figure, the expected variability of performance should decrease after implementation of the solution. That is, if the solution is successful, all crews will be performing exactly according to company standards. Notice that a solution emphasizing a strict adherence to a standard could have the unintended byproduct of reducing the occurrence of above standard performance as well as the desired effect of reducing below standard performance. Looking for this kind of unintended byproduct will be a part of the final evaluation step discussed below.

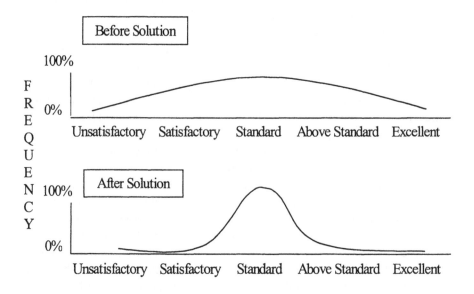

Figure 5.14 Example of a problem solution that decreases variability in a measure

Contrariwise, an expected effect of a solution could be to increase variability on certain measures. Suppose, for example, that part of a solution to a high frequency of late-arriving flights is to change company policy to allow the Captain of the flight more discretion in sacrificing fuel efficiency to achieve an on-time arrival. In the normal course of events, some flights could still be flown with efficient fuel settings and arrive on time, while other flights would require the more inefficient high-thrust settings to achieve on-time arrivals. Therefore, an expected effect of this solution would be to increase the variability in fuel consumption across flights.

Besides the mean and variability of a measure, the exact shape of the distribution of scores on some measure may also be specified to change as a result of a problem solution. The expected results may, for example, involve a specific sub-range of scores in the distribution. In the aviation domain, the expected effect of a safety innovation may be to prevent catastrophically bad outcomes on a very small percentage of training evaluations. In such a case, the expected results may only involve the extreme tail of a distribution of scores, with the mean and variance being relatively unaffected.

Multivariate changes Although expected changes in means, variances, or the distribution of a single variable will be the most common form of expected change, the expected effect of a solution may also involve joint effects on two or more variables. That is, the expected results of a solution may be to change the relationship between measures or to change a unique pattern of values among a set of variables. Both predicting and evaluating these multivariate changes is facilitated by the broad variety of validated measurements in the SIS.

A problem solution may be expected to change the strength of relationship between two SIS variables. In the aviation domain, for example, the degree to which the Captain communicates with the First Officer may be positively correlated to the degree to which the Captain is an extravert versus an introvert. If the problem solution involves the tight scripting of crew communication patterns into Standard Operating Procedure, the Captain-First Officer communication may become far less dependent on the Captain's personality. This decreased dependency on personality factors could possibly be expected with an intervention that proceduralized interaction such as Advanced Crew Resource Management. If the intervention is successful, the correlation between the Captain's extraversion and cockpit communication should significantly decrease from the baseline value before ACRM.

Extending this idea, a problem solution could affect the relationships among a set of variables. For example, the ability of a set of predictors to account for a criterion could be predicted to systematically increase or decrease after the implementation of the solution. This would involve checking for changes in the R-squared from the multiple regression analyses. Similarly, the exact pattern of which predictors will most strongly predict a criterion could be expected to shift as a result of implementing the solution. This would involve checking for significant shifts in the pattern of the regression weights for the regression analyses.

Alternatively, the anticipated effect of a problem solution could be to either prevent or create a certain pattern of values in a set of measures. The pattern could be a sequential pattern of certain values over time. In the aviation domain, a problem solution could be aimed at preventing an "error chain" of poor crew decisions and actions, as in Figure 5.15. The error chain is essentially the concatenation of several errors over time that leads to a catastrophic result. If the crew decisions and actions are faithfully represented in SIS measures, the expected result of the solution could be to prevent one or more steps in the sequential pattern of the error chain, thus preventing the catastrophic outcome.

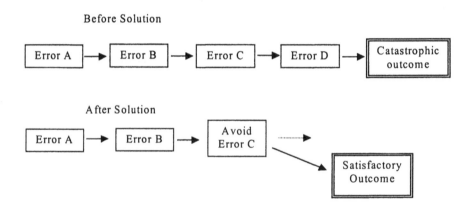

Figure 5.15 Illustration of the effects of a solution on an error chain

Alternatively, the expected result of a solution could be a unique pattern of values in a set of variables measured simultaneously. A problem solution aimed at increasing the quality of landings should result in a better set or pattern of values across several measures. For example, desirable combinations of values for approach angle, airspeed, position over the runway numbers, and height above runway should occur more frequently after the solution is implemented.

It is very important that the entire set of expected results from a problem solution be laid out as early as possible and definitely prior to the implementation of the solution. First, this expected set of results specifies exactly what must be measured in the SIS to evaluate the solution. These measures must be implemented sufficiently early to allow for statistical evaluation of the solution in the next step. Second, this thorough exploration of expected results may uncover one or more counter-productive expected results. If these counter-productive effects are identified early enough, the problem solution can be appropriately revised to avoid or mitigate these negative byproducts of the solution. Third, the specificity and completeness of the expected results makes the inference that the solution either works or does not work in the solution evaluation stage that much stronger.

Solution evaluation

The evaluation of a solution should use the information in the SIS to answer two fundamental questions: "Does the solution actually solve the problem?", and "Does the solution produce other problems?". To answer the first question, the set of relevant measures plus the set of expected results should be combined into appropriate analyses. Answering the second question involves a more *post hoc*, exploratory use of the collateral information in the SIS.

Analysis plan

The analysis plan should specify the data collection, analysis method, and a complete set of expected effects for every important facet of the solution. The heart of the analysis plan consists of the details of the expected direct and indirect effects of the solution. These details should be coherently organized to facilitate use in the data analysis. Although the analysis plan can be quite detailed and elaborate, it can also be as simple as a detailed table that specifies all the required information. An example is Table 5.6.

This table contains a brief example of the two expected effects of a new training program for airline crews. In this example, the program was designed to increase the quality of crew communication and coordination and to ultimately result in more on-time flight arrivals while conserving fuel. The crew interaction is measured by CRM items on line checks given to crews on a routine basis. The quality of communication and coordination is expected to be significantly higher after the solution is implemented compared to before. The on-time rate is measured by the extent of late arrivals at the gate for all flights for the fleet. The frequency and extent of late arrivals is expected to significantly decrease after the implementation of the solution. Similarly, fuel consumption is expected to decrease. Note that disconfirming results are also listed in this plan.

Table 5.6 Illustration of the essential elements of an analysis plan

	Data Collection			Data Analysis		
Solution Focus	**Sample**	**Time**	**Measures and Setting**	**test**	**Expected Result**	**Disconfir-ming Result**
Better CRM	Crews	Pre & Post	CRM items on line checks	t-test	post > pre	No difference, post < pre
Better on-time rate	Flight legs	Pre & Post	Gate arrival times (minutes late)	t-test	post < pre	No difference, post > pre
Lower fuel consump tion	Flight legs	Pre & Post	Fuel per leg	t-test	post < pre	No difference, post > pre

Once the analysis plan is specified, the actual data collection and analysis can proceed as planned. Usually, the data collection will either proceed until there is a sufficient sample size to ensure adequate statistical power for the analyses or until there is a natural stopping point caused by limited resources such as time or the availability of a sample. Occasionally, analyses can be designed so that the data collection proceeds only far enough to either confirm or disconfirm the expected result. There are variations of some of the basic statistical techniques that allow for testing the data on an ongoing basis for the expected effects.

These techniques (e.g. sequential analysis of variance) are not routine and are not included in typical data analysis packages, but their use should be considered when data collection is quite expensive or potential results are time-critical. In a safety domain, for example, the evaluation of whether a training program prevents catastrophic outcomes may be done in this manner so that if the training is found to be effective, it can be extended to all parts of the population as soon as possible.

Exploring for side effects

In addition to the basic evaluation of the solution, the analysis should search for the occurrence of unintended negative byproducts of the solution. If the evaluation data has been integrated in the SIS, this exploration is greatly facilitated by presence of the collateral information in the SIS. This exploration is *post hoc* and care must be taken in considering the number of *post hoc* analyses performed when interpreting the real significance of an empirical result. That is, in a rich set of data such as that available in the SIS, performing many different analyses is bound to lead to a result of one kind or another. Therefore, any obtained post hoc results should be examined for theoretical fit, sensibility, and so forth.

If negative byproducts are found and verified, the initial implementation of the problem solution may be altered or refined to avoid the side effects. If any of the side effects is important enough to constitute a problem in its own right, it may require another iteration of the problem solution cycle. In general, however, successful solutions will generate far fewer and less severe problems than they resolve.

Chapter summary

This chapter has reviewed the basic types of problems and a problem solution cycle. The problem solution cycle emphasizes the identification, exploration, and analysis of problems prior to the development, implementation, and evaluation of solutions. The information in the SIS is critical for a complete and thorough execution of these steps. Typically, the implementation and evaluation of solutions will require augmentation and enrichment of the data and measurements in the SIS. This potentially increases both the relevance and completeness of the SIS measures. Finally, the confirmation or disconfirmation of solution effectiveness will help substantiate or discredit the theory upon which the solution is based. The empirical extension of the databases plus the theoretical development will greatly enhance the quality of available information about the target system.

6 Structural Modeling Methods

Overview

The previous two chapters covered the basic methods for evaluating and modifying simple theories and for identifying and solving simple problems. The information in the SIS also facilitates the use of much more advanced and sophisticated methods for evaluating theories and solving problems. Some advanced methods focus on stable structures in the target system. Basic examples of these methods are covered in this chapter. Other advanced methods focus on the dynamic, changing aspects of the target system. Those methods are covered in the Chapters 7 and 8.

Models for assessing structure are particularly appropriate for theories that emphasize the static, underlying components of the target system. Typically, these components are assumed to be stable and to be organized in some type of structure. Understanding these critical components and how they are structured becomes the focus point for understanding the system. The tacit assumption is that the target system has negligible dynamic processes, at least for a given period of time. When gathering appropriate data in the SIS, the expected time scale of stability is important. When choosing a method of analysis to determine the structure, the available data and the qualitative nature of the expected structure must be considered.

Time scales for stable structures will differ tremendously depending on the exact content and time scale of the theory. This is shown in Table 6.1 below. On the one hand, theories concerning subatomic particles emphasize the existence of very short-lived structures. On the other hand, theories of cosmology emphasize structures that endure over eons. Theories relevant to human systems may have time scales ranging from milliseconds to decades, depending on the exact theoretical focus. What is important is that the structure upon which the theory focuses is stable for the time scale addressed by the theory, and can be assessed by appropriate measures in the SIS.

Some theories may assume that the stable structure is the basic true state of the target system. This would be the case, for example, for McRae and Costa's (1990) view of the stability of personality. They assume that in adulthood the structure of personality traits or dimensions is essentially stable. For theories like this that do not have a dynamic component, the types of analyses covered in this chapter are particularly appropriate. Since stability is assumed, the measurements can be conducted at any convenient time and the resulting data entered into the SIS.

Table 6.1 Time scale for stable structures in different theories

Theoretical Domain	Structure	Time scale for stability
Particle Physics	subatomic	femtoseconds
Computation	computer states	nanoseconds
Neuropsychology	neuronal structure or neural network	milliseconds to hours
Human Learning	interrelated cognitions or knowledge structure	hours to weeks
Crew resource management	team structure	days to months
Human development	cognitive/social structure	months to years
Adult personality	personality trait structure	decades
Civilization	social structure	decades to centuries
Cosmology	universe	eons

Other theories may assume the target system is inherently dynamic but there are long periods of steady states where the system achieves an absorbing state, a stable equilibrium, or some other form of stability. This stability may become permanent or may be only temporary. For example, balance theories of cognition (e.g. Cartwright & Harary, 1956) describe the dynamic process of adjusting a cognitive structure to achieve balance. After adjusting for a non-congruent cognitive input, the cognitive structure stays stable until another discordant cognitive input occurs. For the periods of stability, the methods covered in this chapter are also appropriate. To assess structure, data has to be gathered during these stable periods and included in the SIS. Once suitable data are available, the appropriate method of analysis must be determined.

The analytic methods discussed in this chapter will all focus on resolving an underlying stable structure, but they will approach this goal in different ways depending on the qualitative nature of the structure. There are several qualitatively distinct ways to define a stable structure, and each may require a different type of analysis. Three common definitions of structure are covered in Table 6.2. These definitions have correspondingly different types of appropriate analyses and results.

The first definition of structure emphasizes a stable structural relationship among measured variables in a system. This type of stability can involve the stability of relationships such as a pattern of correlations among measured variables. Usually the stable pattern is taken to indicate underlying common constructs called factors that summarize the variation or covariation among the measured variables. For more detailed coverage of the factor analysis approach see Harman (1976) or Mulaik (1972). When factors summarize covariation among a set of variables, they may represent latent or underlying common constructs among that set. This structure can also involve the stable relationships among a set of factors. Factors may interrelate in a variety of different ways that can be explored by structural equation modeling.

Table 6.2 Different forms of system structure with associated analyses

Definition		Analyses	
Form of structure	**Examples**	**Type of analyses**	**Type of results**
Stable covariation or relationship among variables	Personality structure, Basic Ability structure	Factor analysis, Structural Equation Modeling	Latent variables & linear relationships among them
Stable distances or dissimilarity among a set of elements	Semantic structure, sociometric structure	Cluster analysis, Multi-dimensional scaling	Dimensions for describing the space, positions & distances among elements
Network of structural connections among elements	Knowledge structure, mental models	Pathfinder	Basic network of elements and connections & qualities of the network

The second definition of structure emphasizes a stable structure of elements in a space defined by dimensions. For more detailed coverage of the multidimensional scaling approach see Romney, Shepard & Nerlove

(1972). In this view, the structure is the spatial structure in which these elements are arranged. One example is the arrangement of atoms in a crystal lattice. Another example is the arrangement of cognitive elements in a knowledge structure. For example, animals may be represented in a knowledge structure that has the underlying dimensions of size, ferocity, and degree of domesticity.

Commonly, the structure of the elements is represented by the set of their respective distances in an n-dimensional Euclidean space. For example, the normal 3-dimensional Euclidean space is typically represented by the x, y, and z dimensions. In some cases, such as a Cartesian coordinate system, these dimensions are arbitrary. If the dimensions are arbitrary, they can be rotated or even converted to another representation, such as polar coordinates, without any loss of meaning. Sometimes, however, these dimensions have a real meaning for the theory of the target system. If the dimensions represent meaningful facets or constructs of the theory, they cannot be arbitrarily rotated or converted to another representation.

An example of meaningful dimensions comes from the domain of air traffic control. Air traffic controllers must deal with the positions of a set of aircraft in their assigned sectors. One important structure is the positions of the set of aircraft in the 3-dimensional space in that sector. In this case the x, y, and z dimensions represent longitude, latitude, and altitude above mean sea level. These dimensions can completely describe the structure of distances among a set of aircraft in the sector at a given instant. In this case, the x, y, and z dimensions each has a qualitatively different essential meaning. This set of distances could not be converted to another representation without losing the match to the cognitive structures of the controllers.

The third definition of structure emphasizes the set of connections among a set of elements. From this point of view, structures can be represented by a graph where the elements are nodes and the connections are links or arcs in the graph. Qualitatively distinct kinds of links can be represented by appropriately labeling the connecting arcs. The entire structure can be depicted and analyzed using graph theory and associated methods such as Pathfinder. For more detail on the Pathfinder method, see Schvaneveldt, Dearholt, & Durso (1988).

Typically, the initial use of all structure-determining methods is to explore what kind of structure exists in a domain. These exploratory solutions should be checked on a separate sample by either cross-validation or other confirmatory methods. The basic functions of data exploration, cross-validation, and solution confirmation will be discussed for each form of structural analysis covered in the sections below.

STABLE COVARIATION OR RELATIONSHIP AMONG VARIABLES

One approach to finding structure is to examine the relationships among variables that index distinct aspects of the target system. To the extent that these measured variables are related, the most parsimonious assumption is that a single underlying construct is producing the observed relationships. This construct may represent a common theme among the measures, connected functions in the target system that are reflected by the measures, or a common set of underlying processes that produce the set of measures. In any case, finding the construct and understanding its role in the target system will allow better explanation, prediction, and control of the target system.

Two common methods of examining the covariation among measured variables are factor analysis (FA) and structural equation modeling (SEM). The methods are similar in that covariation among measures helps define a latent variable or factor which is assumed to produce the measured scores. The methods are different in that SEM allows for a more varied structure of linear relationships among the latent variables than does FA and distinguishes between variables exogenous or external to the system and variables endogenous or internal to the system. Conversely, FA focuses on only one set of variables and is correspondingly simpler. Either method can be appropriately used for data exploration, cross-validation, and solution confirmation.

Factor analysis

Factor Analysis starts with a table or matrix that summarizes the covariation or correlation among the variables. In a typical correlation matrix, the diagonal elements would have the value 1.0 that reflects the total variance of the variable. In FA, however, each 1.0 is replaced by a "communality" which is an estimate of how much variance the variable *shares* with the other variables. This is important because the factors are based on the shared or common variance among the set of measures under consideration. The communality estimates range from a minimum value of 0 (no shared variance) to a maximum value of 1.0 (100% shared variance).

For an example from the aviation domain, consider a flight evaluation in which both technical and teamwork (CRM) aspects of crew performance were assessed for each segment of a real or simulated flight: take off, climb, cruise, descent, and landing. If these ten measures were available for a large sample of crews, each pair of measures could be correlated and the results summarized in a table such as Table 6.3. In Table 6.3, these measures are numbered 1 through 10 and the correlation between each pair of measures would be entered in each cell.

Different patterns of covariation among these measures would indicate a different structure of underlying factors. The pattern of correlations shown in Table 6.3 supports two distinct factors, each indicated by a cluster of high correlations. The pattern of high correlations for take off and climb items indicates an "airport departure" performance factor. Similarly, the high correlations among the items for the descent and landing phases of flight indicate an "airport arrival" performance factor. Both technical and CRM items contribute to the departure performance factor and the arrival performance factor.

Table 6.3   Example of a pattern of correlations indicating separate departure and arrival performance factors

	Measure									
	Take Off Tech 1	Take Off CRM 2	Climb Tech 3	Climb CRM 4	Cruise Tech 5	Cruise CRM 6	Descent Tech 7	Descent CRM 8	Landing Tech 9	Landing CRM 10
1	C1	Hi	Hi	Hi						
2	Hi	C2	Hi	Hi						
3	Hi	Hi	C3	Hi						
4	Hi	Hi	Hi	C4						
5					C5					
6						C6				
7							C7	Hi	Hi	Hi
8							Hi	C8	Hi	Hi
9							Hi	Hi	C9	Hi
10							Hi	Hi	Hi	C10

Communality estimates in Table 6.3 are indicated by C1 to C10 for the ten measures, and these values would be entered in the diagonal of the matrix. In this case, the communalities of measures 1 to 4 and 7 to 10 would be quite high as these items are part of the two basic factors. That is, each of these items shares a significant amount of variance with the other items in its factor. Items 5 and 6, on the other hand, would have low communalities as they do not form part of any factor and would not have much shared variance.

For a contrasting example, suppose that there was a basic factor of technical skill at flying the aircraft that was consistent across all phases of flight. In that case the technical measures should correlate highly with each other but not necessarily with the CRM measures. This pattern is shown in Table 6.4. Note that this is a very different pattern of correlations than the pattern indicating departure and arrival factors shown above. For this example, the communalities would be high for measures 1, 3, 5, 7, and 9 as these items share common variance in the underlying technical performance factor. Conversely, the communalities of the measures 2, 4, 6, 8, and 10, which do not share a common underlying factor, would be low.

Table 6.4   Example of a pattern of correlations indicating a technical performance factor

	Measure									
	Take Off Tech 1	Take Off CRM 2	Climb Tech 3	Climb CRM 4	Cruise Tech 5	Cruise CRM 6	Descent Tech 7	Descent CRM 8	Landing Tech 9	Landing CRM 10
1	C1		Hi		Hi		Hi		Hi	
2		C2								
3	Hi		C3		Hi		Hi		Hi	
4				C4						
5	Hi		Hi		C5		Hi		Hi	
6						C6				
7	Hi		Hi		Hi		C7		Hi	
8								C8		
9	Hi		Hi		Hi		Hi		C9	
10										C10

Factor analysis is necessary because the patterns of correlations indicating the basic underlying factors are often not clear-cut. As the complexity of the data increases, the direct inspection of a matrix of correlations to find patterns that indicate underlying factors is generally not feasible. Instead, the factor analysis process mathematically condenses the pattern of covariation in a matrix into underlying dimensions. Whenever there is a set of potentially related measures, this process can be used to explore the pattern of covariation among the measures and reduce this pattern to a simpler set of fundamental dimensions or constructs. This reduction preserves the essential information in the original set of measures in a more condensed and usable form.

Data exploration

There are several basic steps in using factor analysis to explore the underlying structure among a set of variables. These steps include selecting the measures and cases for the analysis, preparing the correlation matrix, extracting an initial set of factors, rotating the factors for a more meaningful solution, estimating factor scores, and validating the factor scores.

Step 1: Selecting measures and cases The very first step is to select an appropriate set of measures and an appropriate set of cases. Particularly in a SIS, a wide variety of measures may be available, and an appropriate subset should be selected with some kind of theoretical, rational, or pragmatic justification. The temptation to throw an ill-assorted set of measures into a factor analysis and hope the process will sort it all out, must be resisted. The selected set of measures should have some potential common themes or elements which might be underlying factors. Specifying up front the expected number and nature of factors as clearly as possible will help prompt a careful selection of measures and provide a valuable crosscheck on the results from succeeding steps in the analysis. For example, Hansberger, Holt & Boehm-Davis (1999) specified communication, workload, and situation awareness factors for instructor evaluations of pilots at an airline. Their factor analysis confirmed the presence of communication and workload factors, but indicated a planning factor rather than the expected situation awareness factor.

One part of selecting measures is to determine the appropriate level of measurement. This requires determining the appropriate level of measurement both for items and for cases. The different possible levels of measurement are illustrated in Table 6.5 for an aviation domain example of evaluating flying performance or proficiency. Selecting the appropriate level of measurement for the items involves a trade-off of the scope and granularity of the desired information. There are pros and cons for selecting each level.

For example, selecting the item level of measurement gives a fine-grained, detailed basis for finding a suitable set of factors. Selecting the item level also avoids making any assumptions that the items cluster or hang together in certain ways, which would be made by using the sub scale or scale level of measurement. However, if the item level is selected for several measures with large numbers of items, this will quickly result in a very large number of items in the correlation matrix. Such a large correlation matrix requires, in turn, a large sample of cases to precisely and accurately estimate all the correlations. The effective number of cases that are available depends on the appropriate level of measurement for cases (discussed below), and may not be sufficient.

Table 6.5 Examples of different levels of measurement for items and cases

		Level of Measurement for Items		
		Item	Sub scale	Scale
Level of measure-ment for cases	Pilot	Individual score on each item	Individual Tech and CRM scores	Individual Overall Performance score
	Crew	Crew score or average score	Crew Tech and CRM scores	Crew Overall Performance score
	Fleet	Fleet average for each item	Fleet average for Tech and CRM scores	Fleet average for Overall Performance score

In contrast, selecting the sub-scale or scale level of measurement greatly reduces the size of the correlation matrix. Using the scale level allows the inclusion of more scales in the factor analysis, which increases the potential scope of the analysis. However, the fine-grained item-level information is not available when the scale level is chosen, which precludes finding factors that may represent small subsets of items across several different measures. Essentially, the scale score is *presumed* to be a correct reflection of a true, unitary score for the measure.

Making the decision on the appropriate level of measurement for items can be aided by the theoretical and empirical information available in a SIS. If a measure has a strong theoretical basis as a unitary measure and has previous empirical results such as reliability and validity analyses that support the unitary definition of the measure, then using the scale level of measurement is justifiable. If a measure does not have an explicit theoretical basis in the SIS nor any empirical support for being a unitary measure, then using the item level of measurement may be wiser as using either the sub scale or scale level would be making unsupportable assumptions. Using the item level of measurement will allow the coherence and structure of the relationships among the items to be determined by the factor analysis.

As the SIS develops over time, both the theoretical and empirical support for the measures will become clearer. In the long term, therefore, analyses using scale scores will be justifiable even if initially many analyses must be targeted at the item level. Even if the item level is initially chosen, variants of factor analysis, such as hierarchical factor analysis, can be used to give some empirical indication of appropriate sub scales and scales for a set of items.

Once the level of measurement has been specified, the critical issue is obtaining a representative set of cases for the factor analysis. It is just as important to define the appropriate level of measurement for cases and select an appropriate set of cases as it is to choose an appropriate level of measurement for items. Particularly if lower levels of measurement for cases are specified, such as the individual or crew level in Table 6.5, a SIS may have a large number of potential cases. In this situation it is important to select a representative sample for the desired population. A representative sample is important for factor analysis since the sample helps determine exactly how the measures are interrelated, which is the basis of the factor structure. That is, very different samples may give different factor structures. Therefore, it is important that the sample represents the target population as accurately as possible.

Fortunately, a properly connected SIS should connect the case information with a large amount of auxiliary information on each case. This auxiliary information can be queried from the databases and used to help select a good sample. To ensure the representativeness of the sample, the mean and variance on the auxiliary measures for the sample of cases in the SIS should be compared wherever possible to known values for the target population. A representative sample should have means and variances that match the established means and variances in the target population for each of the auxiliary measures.

Using an aviation example, suppose the selected level of cases was the aircraft crew and that auxiliary information was available on the amount of time each crew had been flying together, the average age of the crew members, and the average time the crew members had been flying for a particular fleet. Clearly the sample of crews selected for analysis should have an average time flying together, average age, and average time flying for the fleet that matches the known mean values of these variables for the fleet.

Any systematic differences on these auxiliary variables indicate an unrepresentative sample that could alter the pattern of correlations and the obtained factor structure. For example, if the sample were composed of crews who had not flown together very long, the factor structure might not show a well-developed teamwork factor that would be true of more experienced crews. For similar reasons, the average age and average time flying for the fleet of the sample must match the population of crews for the fleet. This checking of representativeness on the auxiliary measures can be combined with a careful, empirical checking of the data for missing or out-of-bounds values prior to the factor analysis.

Step 2: Preparing the Correlation matrix After the level of measurement has been specified and the cases selected to represent the population, the measures can be correlated using any standard statistical program. The basic data of means and standard deviations as well as the correlation matrix and relevant multivariate data should be examined to cull out any inappropriate cases or measures.

Inappropriate measures The means for each measure should be in a range that would be reasonably expected for that population and sample of cases. In the aviation domain, for example, items measuring pilot performance on a five-point scale should reasonably have means above the midpoint of 3.0. This would be expected due to the intensive training and accumulated expertise of the professional pilot population. Conversely, average ratings for untrained student pilots might be expected to be below the midpoint for many items. Measures with means that do not make sense for the sample should be considered for deletion.

The variances for each measure should be greater than zero. A typical problem with some types of evaluations or measures is that there is little or no variation in the resulting values (Figure 6.1). If a measure has little or no variance, it cannot covary or correlate well with the other measures in the factor analysis due to its limited variance. Therefore, measures with little or no variance should be deleted from the analysis. If suitable substitute measures from the SIS can be identified, then the problem measure can be replaced with its substitute.

Similarly, a distribution for a measure that is extremely skewed or otherwise distorted from the normal bell-shaped distribution will limit its possible correlation with other measures. These distortions will often occur when the mean is near its upper limit of possible values (called a "ceiling effect") or its lower limit of possible values (called a "floor effect"). These distortions will limit the relationship of a measure to the others in the analysis, particularly if the other measures are also distorted but in a different way. Although these distortions may limit the size of possible correlations, they do not absolutely prevent a measure from having a sufficient relationship to play a role in a factor. Therefore, measures showing these distortions should be considered for deletion to the extent the distortion is severe. The decision to delete such a measure may also hinge on the availability of a suitable alternative replacement measure in the SIS.

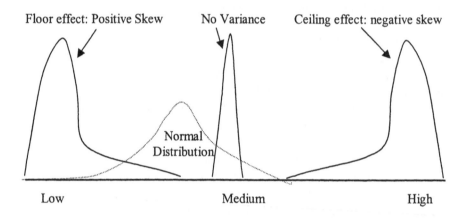

Figure 6.1 Examples of measures with poor distributions of scores

The correlation matrix should also be carefully examined for potential problems. First off, the correlation matrix should have correlations of sufficient magnitude to justify a factor analysis of underlying common dimensions. If all the correlations are small and non-significant, for example, underlying common dimensions probably are not present because if common factors are present at least some of the measures should be strongly related. Even if the entire set of measures shows strong relations, however, a particular measure may fail to be strongly related to any of the others. Such a measure is unlikely to play a role in the underlying factors and can be deleted or, if possible, a suitable alternative measure be substituted.

Conversely, a problem may also be indicated by a measure being too strongly related to another. Although possible, this will occur more rarely than the problem of too weak a relationship discussed above. If two scales are perfectly correlated, for example, they are essentially duplicate measures of the same thing. In such a case, one measure can be deleted without losing any information about the measured construct.

Raw correlations may not show strong relationships between measures due to the impact of unreliability of measurement, which tends to decrease correlations. If the reliabilities of each measure are known from the preliminary scientific data quality screening in the SIS, they can be used to correct the raw correlation for the effects of unreliability. If the corrected correlation is at or near 1.0, the measures can be considered duplicates. Since each measure is a slightly unreliable measure of the same construct, the most reasonable action is to take the average of the two measures and use that new, more reliable composite in the factor analysis.

Inappropriate cases   The distribution for each measure and the joint distribution of all measures should be evaluated for outliers.   Extreme outliers can strongly influence the correlations in either a positive or negative direction, and thus strongly influence the results of the factor analysis.   Typically a cut-off point of + or − 3 Standard Deviations is used for identifying possible outliers on each measure.   Similarly, the Mahalanobis distance is used to identify multivariate outliers that are significantly different from the sample on the entire set of measures. These outlier identification methods are heuristic methods rather than hard-and-fast rules, however, so each case should be individually considered before deletion.

In particular, cross-checking the values of the measures with the auxiliary information in the SIS may help clarify whether an extreme score is real or a fluke. If the extreme score is supported by a consistent pattern of auxiliary evidence and seems to be a legitimate score, its value may be "trimmed" rather than discarding the case. Trimming the value for a case is simply substituting the nearest acceptable value for the extreme value. The nearest acceptable value can be the highest acceptable value among the set of cases or the value set by the + or − 3 Standard Deviations heuristic. After the data have been examined and appropriately cleaned, the correlation matrix is ready for extracting factors.

Step 3: extracting an initial set of factors   There are several different methods for extracting an initial set of factors. Each method has slightly different assumptions about what constitutes the "best" or optimal solution and how to derive the solution. One of the simplest methods is "principal axis" factoring, which uses an iterative process to estimate the underlying factors that account for the observed pattern of covariation. This method is the default in some statistical packages.   During this process, the communalities or shared variance estimates for each measure are re-estimated on each step.

The iterative estimates of factors and communalities stop when the solution is stable and is no longer improving by even a small amount. This stable, "best" solution is a set of orthogonal or independent factors that are arranged in order of strength (Figure 6.2). That is, the first factor accounts for most of the covariation pattern in the original correlation matrix, the second factor is orthogonal to the first but accounts for the most covariation possible with that constraint, the third factor is orthogonal to both of the first two factors, and so on.

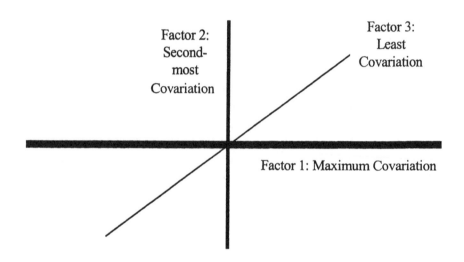

Figure 6.2 Initial extraction of three factors in order of descriptive strength

The initial solution is an efficient condensation of the pattern of covariation in the original correlation matrix into a smaller set of basic dimensions or factors. However, this solution is typically not easily interpretable because the first factor will have the majority of items related to it while the succeeding factors have relatively few. The first factor will be difficult to interpret because of its complexity, and conversely the remaining factors may not have sufficient relevant items to clarify their inherent meaning. To increase the interpretability of these dimensions, the factors can be rotated to a more meaningful structure where items are more evenly related to different factors.

Step 4: rotating the factors for a more meaningful solution The idea of rotating a factor solution for a more meaningful interpretation is justified by the fact that the initial factor extraction actually defines a small subspace that accounts for the covariation among the measures. This subspace can be defined by different sets of x, y, and z axes and still account for the same amount of covariation. If the new x, y, and z axes are rotated from the initial extraction but still define the same subspace, they are a functionally equivalent set of underlying factors.

This is illustrated in Figure 6.3. The rotated factors are essentially a new x, y, z coordinate system that can serve as an alternative to the dimensions that were initially extracted. This new set of dimensions is functionally equivalent to the original set since any unique x, y, z point defined on the original set can be precisely defined as a corresponding unique x, y, z point on the new dimensions. The critical issue is whether the rotated factors are more meaningful. The interpretation of the meaning of each factor requires interpreting a pattern of factor loadings.

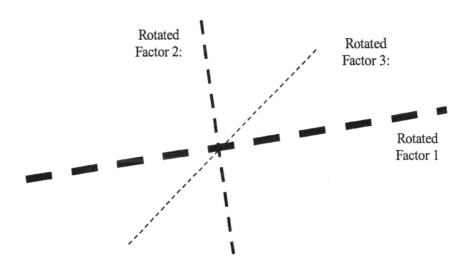

Figure 6.3 Orthogonal rotation of three initial factors

A factor loading is the extent to which each measure is related to a dimension or factor. Rotation of the axes shifts these loadings, but should make them more easily interpreted. The most common orthogonal (rigid) rotation is Varimax, and the criterion for rotation is that the variance of the loadings for each factor is maximized (hence the name "Varimax"). Since the variance of the loadings is maximized when there are more extreme positive and negative loadings, in practice this criterion causes a rotation that emphasizes a few large positive and large negative loadings, while the rest of the loadings are near zero.

In addition to Varimax and other methods for orthogonally rotating the dimensions representing factors, there is a class of methods which obliquely rotate the dimensions to find simple structure. The oblique methods result in factors that are related to each other, or correlated to some extent, rather than independent, uncorrelated factors. The critical decision that must be made in choosing between orthogonal and oblique rotations is whether the constructs underlying the covariation among the measures could actually be interdependent or correlated constructs. If the answer is yes, then oblique rotations should at least be attempted. If the answer is no, then orthogonal rotations are sufficient. Since this critical decision is difficult to make when initially exploring a set of measures, a good practice is to explore the use of oblique as well as orthogonal rotations and see which depiction of the structure is more interpretable.

Factor interpretation Having a few large positive and large negative loadings makes the interpretation of the underlying dimensions easier. The positive loadings can be interpreted as measures that are positively correlated with the underlying construct. Conversely, the negative loadings can be interpreted as measures that are negatively correlated with the construct. The construct is the continuum described by the rotated dimension.

Each factor or dimension is usually considered to be a continuum with a high end, a middle or neutral point, and a low end. A position high on this factor would be characterized by a high score on all the positively loaded measures and a low score on the negatively loaded measures. Conversely, a position at the opposite end (low end) of the factor is characterized by a high score on all the negatively loaded measures and a low score on the positively loaded measures. Middle positions represent moderate scores on all items or an even mix of high scores on positive and negative items.

To extend the aviation example presented earlier, suppose that technical and CRM teamwork performance of a crew were evaluated at each of four basic stages of flight by two types of items, one indexing positive performance and the other indexing mistakes or errors. For technical performance, one type of item would index positive technical performance and the other type of item would index technical mistakes. Similarly for CRM one type of item would index positive CRM performance while the other type of item would index CRM mistakes or errors.

If there were two separate, orthogonal or independent factors for technical and CRM performance, the resulting factor structure might look like Table 6.6 and the associated Figure 6.4 below. Table 6.6 depicts the positive, neutral, or negative loadings for each of the 16 items. The loadings are represented by the +, 0, and - entries under each factor. For Factor 1, all the loadings of the CRM items are zero; therefore, CRM assessment is not related to Factor 1. Instead, the technical performance items are systematically related to factor one. Since positive performance items have a positive relationship to this factor and errors have a negative relationship to this factor, Factor 1 represents good or positive technical performance.

Table 6.6 Factor loadings indicating technical and CRM performance factors

Phase of Flight	Item content	Item type	Factor 1 (Technical)	Factor 2 (CRM)
Takeoff	Technical	+ Performance	+	0
		Errors	-	0
	CRM	+ Performance	0	-
		Errors	0	+
Climb	Technical	+ Performance	+	0
		Errors	-	0
	CRM	+ Performance	0	-
		Errors	0	+
Descent	Technical	+ Performance	+	0
		Errors	-	0
	CRM	+ Performance	0	-
		Errors	0	+
Landing	Technical	+ Performance	+	0
		Errors	-	0
	CRM	+ Performance	0	-
		Errors	0	+

Conversely, for Factor 2 all the loadings for the technical items are zero; therefore technical performance is not related to factor two. In this case, the CRM items have strong and systematic relationships to the factor. The high end of the factor concerns CRM error items, while the low end of the factor concerns positive performance. Since high scores on the factor are errors while low scores are good performance, Factor 2 represents *poor* or *negative* CRM performance.

In the corresponding Figure 6.4, the first factor (graphically represented by the x axis) is clearly identifiable as a technical performance factor because all the items that have extremely high or extremely low loadings on the factor are technical in nature. Since the positive end of the factor has all the items related to positive technical performance and the negative end of the factor has all the items related to technical errors, the factor represents positive technical performance. In a similar fashion, the high-end of the second factor has all the CRM error items while the low end has the CRM positive performance items. Therefore, the second factor represents *poor* CRM performance. Either the tabular or graphical method can be used to isolate the content of the items most related to each factor.

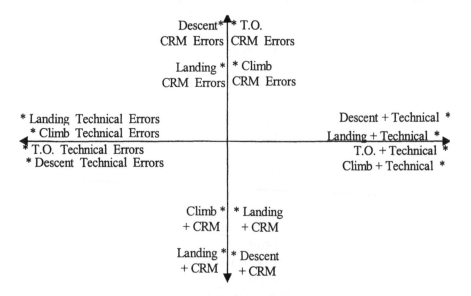

Figure 6.4 Possible item loadings (represented by *) for orthogonal technical and CRM factors of crew performance in an aviation domain

In general, to interpret a factor, the exact content of the items with high positive loadings (high end of the factor) and the items with high negative loadings (low end of the factor) must be contrasted. From this contrast, the meaning of the factor is determined. In this contrived example, the contrast was easy because the clear-cut pattern of loadings. Each end of each factor was clearly marked by all positive or all negative items. Determining precisely what is indicated by high and low positions on the factor is important since the next steps in the process will involve estimating factor scores and ultimately validating those scores.

Interpreting the meaning of factors when they are obliquely rotated is slightly more complicated. First the interpretation must take into account the degree of correlation or relationship among the factors. Factors that have very low correlations with all other factors may be considered essentially independent. Factors that are moderately related might imply common processes in the target system, second-order super-factors, or other interesting possibilities. Factors that are highly correlated might be essentially indexing the same construct and should be further evaluated in the validation step below. If the correlated factors have the *same* set of relationships with other variables they should be combined as one basic measure. If the correlated factors have distinct relationships with other variables, they should be kept as functionally distinct measures.

Second, oblique rotations create two distinct types of loadings of the original measures with the underlying factors. One is the standard loadings that represent the simple correlations of the measures on the underlying factors. This is sometimes called the structure matrix. The second type of loading has the common variance among the factors partialled out of the loadings. This is sometimes called the factor pattern matrix. Particularly when the factors are moderately to highly correlated the regular loading matrix and the factor pattern matrix may give a slightly different picture of the relevant items or measures for each factor. In such a case, the pattern matrix should be a primary source of interpretation as it removes the overlapping variance among the factors and gives a cleaner picture.

The process of interpreting the meaning of a factor should also use the information available in the SIS such as the current set of supported theories for that domain. Wherever possible, the factors should be interpreted as theoretically meaningful constructs and tied back into the theories underlying the SIS. The interpretation of a factor as representing a meaningful construct can be tested and refined by estimating factor scores and validating those scores.

Step 5: estimating factor scores   Just as in factor extraction and rotation, there are different methods for estimating factor scores which have different assumptions. Any method of estimating factor scores should produce a best estimate of the value of the underlying construct for each case. One of the most basic methods is to identify the high and low-loading measures on each factor and directly form a factor composite from these measures. This creates a unit-weighted factor score.

In the example above, the unit-weighted factor score for the technical factor would be the sum of the score on positive technical performance items minus the sum of the scores on the technical errors/mistakes items. Similarly, the unit-weighted scores for the CRM factor would be the scores for CRM mistakes/errors items minus the scores for the positive CRM performance items. Notice that the scoring for the CRM factor could also be reversed at this point so that positive scores would represent positive CRM performance. If the scoring is reversed, that must be carefully tracked in the validation step below so that the construct's meaning is correctly evaluated.

The equal-weighting approach is simple and robust to small fluctuations in factor loadings due to fluctuations in sample characteristics and so forth. However, it only uses a subset of the information available to index each factor score. It also has problems when items are highly loaded on more than one factor. In such cases, counting the item for both factors will result in some automatic degree of correlation between the two factor scores, which may not be desirable. The use of unit weighted factor scores may be reasonable when a clear factor structure with distinct, non-overlapping subsets of items for each factor is found.

Slightly more complex is the regression method for estimating factor scores. The regression method uses multiple regression to find a linear combination of the measures that will best estimate the appropriate score for each case on the underlying factor. This method is applied to each factor and results in a predicted score for each case on each factor. Generally a factor analysis program in a statistical package can automatically estimate the factor scores using this regression approach, which may be the default method.

No matter what method is used, the estimated factor score can be added to the SIS and treated the same as any other measure of a construct. Factor scores can be used to evaluate training, solve problems, or extend the theoretical base of the SIS. Since it is important to be certain of the meaning of these scores before using them for other purposes, the validation of the meaning of the factor scores should be considered an essential final step of exploratory factor analysis.

Step 6: validating the factor scores The validation of the factor scores is essentially the same as validation of any theoretical construct. Therefore, all the material on validating theories discussed in Chapter 4 is relevant and can be used in the validation process. This process can and should use the entire range of associated information in the SIS to determine if a meaningful construct is being validly measured. At the very least, the construct should be evaluated for construct or network validity and the SIS cases should be examined for natural experiments or subsets of cases that should have different values of the construct.

However, if the exploratory factor analysis finds a construct that was not expected and defined by the current SIS theories, theoretical development and integration must take place before construct validity can be empirically evaluated. The theoretical development should have the goal of elaborating the antecedents, concomitants, consequences, and relevant parameters and processes for the construct so that a distinct pattern of relationships with other SIS measures can be specified and evaluated. It is particularly important to develop and specify how the new construct will be distinguished from any other closely related constructs in the SIS. To test the unique features of the new construct, the empirical evaluation may include methods that partial out the influence of related constructs.

The principle of parsimony should be considered during this theoretical development process. If it is found that the new construct cannot be distinguished in any important way from an old construct, the new construct should either be discarded in favor of the old construct or synthesized with it. The synthesis might, for example, elaborate the basic definition of the old construct just enough to account for the residual unique features of the new construct. This conceptual pruning and synthesis is critical to avoiding an uncontrolled proliferation of theoretical constructs in the SIS.

Part of the conceptual validation should be a search for cases that should theoretically or logically have a distinct pattern of values for the construct. If natural groups of cases that should have different patterns of factor scores can be isolated in the SIS, these groups can be empirically checked for the expected differences. The expected differences might be higher or lower average values on the factor scores, or differences in the expected variation or distribution shape for the factor scores.

Following the aviation example, suppose the exploratory factor analysis showed technical and CRM performance factors. The SIS could be searched for groups with more technical training and experience that would be compared to groups with less training or experience on the technical performance factor. The more trained and experienced groups should have correspondingly higher average technical performance scores. Further, the variability in technical performance scores might be expected to be lower in the trained group but higher in the relatively untrained or inexperienced groups.

Having more than two groups for these comparisons is important as long as the expected pattern can be clearly specified. As the number of comparison groups increases, the strength of confirmation by finding the expected pattern of differences also increases. For example, if the expected mean differences in factor scores for five groups are that group A > group B > group C > group D > group E, then finding this precise pattern in the empirical results is very unlikely by chance. Finding all four differences in the expected direction gives strong support to the validity of the construct.

If the exploratory factor analysis found two or more novel constructs, the profile or pattern of scores across these constructs can also be a basis for empirical evaluation. In the aviation example, the profile of technical and CRM performance could be the basis for predicting differences among naturally occurring groups of pilots. Pilots with different kinds of military vs. civilian backgrounds might be expected to systematically vary in the degree to which they are strong in technical versus CRM performance, and these patterns of expected differences should be evident in the patterns of factor scores for these groups.

When the pattern of relationships specified for construct validity or pattern of expected differences among groups is found and cannot be accounted for by the current constructs in the SIS, adding a new construct to the theoretical base is justifiable. The new construct should help explain, predict, or control the processes and states of the target system. Since any particular sample may not represent the entire population accurately, checking on the generality of the factor analysis by cross validation is desirable before integrating the construct as a confirmed part of the SIS theory.

Cross-validation   In general, cross-validating a factor analysis involves using the same solution procedure on a separate sample and comparing the results with the results from the original sample. Cross-validation can have the goal of exactly replicating the initial exploratory factor analysis or of generalizing the results of the original exploratory factor analysis to different samples, measures, or contexts (Figure 6.5).

In a cross-validation designed solely for the purpose of replication, the same set of measures is applied to as similar a sample of the population as possible in exactly the same measurement context. The goal is to demonstrate that exactly the same results will be obtained with the cross-validation sample. This process is described in Table 6.7 for each of the major steps in a factor analysis. To ensure a fair comparison of the initial and cross-validation sample results, each step of the process must be kept as similar as possible. The initial and cross-validation results can be compared at several points as summarized in the third column of Table 6.7.

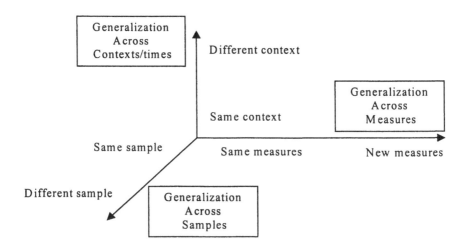

Figure 6.5   Assessing generalization of a factor structure across different modalities

The first similarity check is to compare the means and variances of the initial sample with the cross-validation sample; there should be no significant differences. The second similarity check is to compare the correlation matrices for the original sample and the cross-validation sample, which should not be significantly different. The third similarity check is to compare the number of factors and the amount of variance that is explained by these factors in the original and the cross-validation sample. Each sample should have the same number of factors explaining about the same amount of variance. The fourth similarity check is to establish that the original and cross-validation sample factors have the same fundamental meaning by checking the patterns of item loadings. The fifth similarity check is to compare the distribution of factor scores in each sample; the distributions should be shaped the same. The final similarity check is to compare the pattern of construct validity evidence for the factor scores in the original and cross-validation samples. The pattern should be similar and support construct validity in both samples.

**Table 6.7 Comparison of exploratory factor analysis and factor
analysis cross-validation**

Step	Exploratory FA	Cross-Validation of FA	Comparison
1. Selecting the measures and cases for the analysis	Which measures? Level of measurement. Which cases? Level of aggregation.	Same measures and level of measurement. Same selection heuristics and level of aggregation.	Similar sample? Same means and variances?
2. Preparing the correlation matrix	Delete Univariate or Multivariate Outliers. Assess linear relationships.	Same statistical criterion for deleting outliers.	Same correlation matrix?
3. Extracting an initial set of factors	Which extraction method? How many factors?	Same extraction method. Same number of factors?	Same amount of variance explained by the factors?
4. Rotating the factors for a more meaningful solution	Orthogonal vs. oblique rotation. What do factors mean?	Same method of rotation. Possible Procrustean rotation.	Same meaning of factors?
5. Estimating factor scores	Which method to estimate factor scores?	Same method for estimating factor scores. Cross-check coefficients.	Same distribution of factor scores?
6. Validating the factor scores	Construct validity Natural experiments.	Correlate with same external measures. Check for differences among same groups.	Same correlations and differences?

When using cross-validation to check on the generalization of the original factor structure, it is important to keep as many aspects of the original process as possible identical in the cross-validation solution. To the extent the process is identical except for the facet altered, finding the same structure demonstrates the desired stability in the factor structure. Finding differences in the factors contradicts generalization. If more than one facet has been altered, however, the interpretation of any differences may be inherently ambiguous because the observed differences could be due to any of the changes. Therefore, keeping the process as similar as possible is important.

If the goal of the cross-validation is to establish generalization across different contexts or measurement times, then the same set of measures is used on a very similar sample but in a different context. The results from the different samples can be compared in steps 1 through 6 as outlined in the table.

If the goal of the cross-validation is to establish generalization of the factor structure across different populations, then the same set of measures is used in the same context for a different population. In this case, the samples will be quite different rather than similar. Therefore, there may well be noticeable differences in the means and variances of the measures when they are compared in step 1. Further, since changes in the variances may shift the value of the correlations, the correlation matrices may be different in step 2. Similarly, the total amount of variance of the measures in the original and cross-validation samples may be quite different, and the amount of that variance explained by the factors may also be different. Nevertheless, if the factors generalize they should have the same inherent meaning and have similar external evidence for validity in both samples.

If the goal of the cross validation is to establish generality across different sets of measures then the new set of measures is applied to the same (or a similar) sample in the same context. Since the measures are different, the means and variances in the different samples cannot reasonably be compared. Similarly, it would not be reasonable to compare the correlation matrix or the amount of variance explained by the factors. If the factors generalize, however, the item loadings should imply the same basic meanings for the factors, and the factor scores should have a similar pattern of external validity evidence with the other information in the SIS.

Once confidence is gained in the generality of a factor structure by the process of cross-validation, the constructs represented by those factors can be added to the theoretical base of the SIS. These constructs now constitute an expected set of relevant information about the target system. Since these constructs are now part of the established theory, the existence and relevance of these constructs for the functioning of the target system can be precisely theoretically predicted, and should be empirically confirmed wherever possible.

Solution confirmation Testing the generality of a theoretically specified factor structure to a new sample of cases can be done with confirmatory factor analysis. The advantages of using confirmatory factor analysis for this purpose are that a precise factor structure can be tested on the new sample and that a significance test is available to evaluate whether the factor structure is significantly "wrong" for the new sample. This significance test is essentially confirming or disconfirming generality.

The essential requirement for confirmatory factor analysis is a careful and precise theoretical specification of the pattern of loadings of each of the measures on the factors. This requires, of course, the prior specification for the number and degree of relationship of the factors themselves. Essentially, the confirmatory factor analysis attempts to fit the correlation matrix in the new sample with a correlation matrix that would occur if the specified factor structure were true. The fit of the factor structure to the new sample can be tested or estimated in different ways to give an overall view of the degree of generalization.

When successful, a factor analysis gives a succinct summary of the covariation in a set of variables. Typically, however, these factors are considered as distinct entities rather than as parts of a network. For some target systems, the underlying constructs may function as an integrated whole rather than as separate entities. Structural equation modeling extends factor analysis by allowing for a network of linear relationships among the latent variables described by the factors.

Structural Equation Modeling

The basic idea of underlying factors and a structural representation of the information in a set of covariances among measures is extended in structural equation modeling (SEM). SEM includes not only the underlying factors, called latent variables, but also the idea of a structure of linear relationships among these variables. Variables are also distinguished as being either endogenous variables that interact to form the core of the modeled system, or exogenous variables that act as inputs to the modeled system (Figure 6.6).

As in factor analysis, the constructs or latent variables are assumed to underlie or cause the manifest variables or measures. The measures, however, are used to estimate what the scores on the latent variables should be in the same manner as factor scores are estimated from the measures in factor analysis. In this example, CRM skill, Technical skill, Knowledge, and Teamwork are estimated by three measures each, while Situation Assessment, SOP performance, and Safety are estimated by two measures each. In SEM, the latent variables can be linearly related to form a complex structure or model of the system.

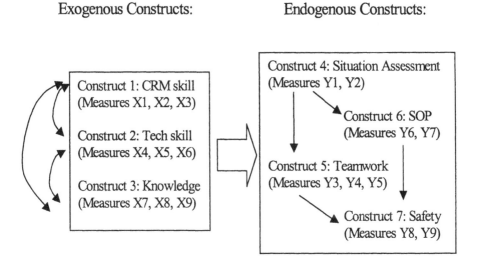

Exogenous Constructs: Endogenous Constructs:

Figure 6.6 Structural Equation Modeling (SEM)

In this figure, CRM skill, Technical skill, and aviation Knowledge of the pilots influence all four aspects of crew performance, which is indicated by the large block arrow. The double-headed arrows connecting CRM skill, Technical skill, and aviation knowledge indicate that they are correlated to some extent. These exogenous constructs represent the basic causes of the performance constructs that are the core part of the performance system.

The directional arrows in the box with four endogenous constructs indicate the predicted flow of influence or causation from one construct to another. In this example, situation assessment influences SOP performance and Teamwork, both of which in turn influence the Safety of the flight performance. Structural equation modeling tests this entire pattern of hypothesized measures and constructs.

The steps in structural equation modeling are essentially the same as the steps for a factor analysis. The level of measurement must be set, measures carefully chosen to represent the constructs, and an appropriate sample selected. If criteria like objective flight performance data are included in this analysis (e.g. path or airspeed deviations), finding a strong connection to the criteria can establish criterion validity. If many theoretically related measures are included in the SEM, finding the expected pattern of relationships can establish construct validity. This would require including the validation measures as part of the input correlation matrix and evaluating the relationships of the focal set of constructs to the validation variables in the analysis.

The capability to model a network of linear relationships among underlying constructs is the big advantage of structural equation modeling. Of course, this capability comes at the cost of increasing complexity of the analysis process and results. Where the underlying constructs of the target system are known to be or suspected to be linearly related, the technique of structural equation modeling should be tried.

STABLE DISTANCES AMONG ELEMENTS

Factor analysis and structural equation modeling emphasize covariation to determine structure. A second approach to finding structure is to examine the distances and dimensions that underlie the key elements or entities of the target system. To the extent that these entities are quite similar, they should be near each other in the structure. Conversely, to the extent that these entities are quite dissimilar, they should be far apart in the structure. If the structure can be expressed spatially, the dimensions of that space may be meaningful in the same manner as factors are meaningful in a factor analysis. Further, these dimensions may offer a concise method for summarizing the observed similarities and dissimilarities of the elements or entities.

As in factor analysis, these dimensions may ultimately represent a construct on which the elements or entities can be compared or contrasted. In turn, this construct may represent key aspects of the target system such as similar or different underlying processes or outcomes. Therefore, finding the constructs underlying the structure of elements will augment the theory of the target system with consequent better explanation, prediction, and control.

One method for examining the structure of entities arranged in a space is multidimensional scaling (MDS). This method uses similarity or difference judgments by a group of respondents as input. Using this information, MDS attempts to establish a best-fitting spatial structure for the set of entities. MDS expresses the structure in a continuous (usually Euclidean) space, typically of one to four dimensions. This method can be appropriately used for data exploration, cross-validation, and solution confirmation for structures such as sociometric structures for a group of individuals, individual cognitive structures for a certain domain of knowledge, and so forth.

Multidimensional scaling

Multidimensional scaling (MDS) establishes a set of meaningful underlying dimensions for relating a set of elements and depicts the exact configuration or position of the elements on these dimensions. Typically, pairs of elements or entities serve as stimuli for judgments about their similarity or dissimilarity. The set of similarity or dissimilarity judgments for the set of stimuli is used to establish a set of proximities that represent how close each stimulus is to every other stimulus in the minds of the respondents. The MDS analysis uses these proximities to construct a space of minimum dimensions that can capture all relevant aspects of the distances among the stimuli. Essentially, this minimum-dimension space is a parsimonious spatial model for describing the structure of the similarities and differences among the original set of entities or elements.

The steps in the process of conducting an MDS are analogous to the steps in the process of conducting a factor analysis. In both cases, care must be taken to establish appropriate data and choices must be made about the exact method for analyzing this data and obtaining the desired result. The steps in the process are listed in Table 6.8.

Selecting entities or elements The ultimate structure found by the MDS method can be influenced by the set of stimuli that are judged, the response scale, and the set of people making the ratings. The composition of the set of people making the judgments is important in exactly the same way as the cases selected for a factor analysis. These people may represent a random sample from a single population, or samples from distinct populations of interest. Similar to factor analysis, the judgments of these people may be analyzed at the individual level, a crew, team, or subgroup level, or averaged over the entire sample and analyzed at that level.

Table 6.8 Steps for multidimensional scaling

Step	MDS
1. Selecting the stimuli and measures for the analysis	Selection of representative elements or entities Select measures that represent psychological distance Respondents judge stimuli
2. Preparing the proximity matrix	Evaluate and resolve asymmetric proximities
3. Extracting an initial set of dimensions	Choose Metric vs. Non-metric MDS solution Euclidean vs. non-Euclidean space Number of dimensions
4. Rotating the dimensions for meaningful solution	Choose Orthogonal vs. oblique rotation Interpret basic meaning of dimensions
5. Validating the dimensions	Establish construct validity of stimulus structure
6. Examining individual differences	Evaluate use of INDSCAL or separate MDS for each respondent to index individual differences Establish construct validity of individual differences

The selection of the stimuli to be judged (key elements or entities) raises issues analogous to the selection of items in factor analysis. If all the key elements or entities can be included in the judgment process, stimulus selection is not an issue. However, the number of elements or entities that can be included in the judgment process will depend on the judgment method, as depicted in Table 6.9.

Free grouping or categorization methods only require that the respondent group or cluster similar stimuli. The presumption is that dissimilar stimuli are put in different categories or groupings. Since any number of stimuli can be put in the same category, this method can be used for very large sets of stimuli.

Similarity calculated from a profile of measures requires that each stimulus be rated on each measure. The number of ratings required is directly proportional to the number of stimuli and number of measures. Particularly if the profile requires many ratings, the number of stimuli may have to be limited so that the task is feasible for a human respondent. Limiting the number of stimuli require selecting appropriate sub sample.

Table 6.9 Required judgments for different MDS methods

Judgment method	free grouping or categorization of stimuli	similarity calculated from profiles over M measures	paired comparisons (similar-dissimilar)
Required judgments for N stimuli	repeated groupings of all stimuli	N * M ratings or responses	N*(N-1)/2 comparisons

Methods such as paired comparisons require many more judgments as the number of stimuli increase. Therefore the paired comparisons method is essentially limited to small sets of stimuli (30 or fewer). If the original set of key elements or entities is large and a detailed method such as paired comparisons is used for judgment, then an appropriate sub sample of relevant stimuli must be selected.

In such cases, the selection of an appropriate set of stimuli is very important. The stimuli to be judged must vary with respect to an underlying dimension in order for it to have a chance of influencing the final MDS solution. If all the key elements or entities can be enumerated, then taking a random sub sample is one way to ensure a representative sample. If they cannot be enumerated, then the sample of stimuli should be evaluated by domain experts to assess the representativeness of the stimuli for the domain.

If there is some expectation of a certain number of dimensions, the number of stimuli must also be sufficient to clearly establish all of the expected dimensions. One simple guideline is that there should be at least four to six stimuli per expected dimension. The exact number of required stimuli will, however, depend on the accuracy and reliability with which each judgment can be made and the degree of variation of the set of stimuli on the underlying judgment dimension. That is, stimuli must have values that range across the expected dimensions; they should not all cluster in the middle. MDS most commonly is used in situations where the number of expected dimensions is from 2 to 4 and the number of stimuli is correspondingly small. MDS, like factor analysis, is most useful when a few major dimensions can describe the structure of the set of elements.

Measures represent psychological distances   The paired comparisons judgment method in MDS typically uses a bipolar scale anchored by "very similar" on one end and "very dissimilar" on the other end. Each pair of elements or stimuli is judged by the respondent for the overall degree of similarity or dissimilarity. Dissimilarities represent greater psychological distances and should be coded with larger numbers.

This typical approach does not specify the exact basis upon which the similarity or dissimilarity should be evaluated, which can be an advantage or a disadvantage. The advantage is that the respondents are not being biased or primed by giving them the basis for making the judgments. Therefore, each respondent is free to choose the relevant aspects of the stimuli which he or she feels are important in making similarity and dissimilarity judgments.

This freedom to choose the basis of judgment can also be a disadvantage. Different respondents may use quite different aspects of the stimuli for grounding their similarity and dissimilarity judgments. Unless these individual differences are appropriately captured by the MDS method, the solution for the entire group of respondents may be very difficult or unclear. In such cases, a version of MDS that accommodates individual differences such as INDSCAL (individual differences multidimensional scaling) should be considered.

Fewer judgments are required when similarity is based on profiles of ratings. The respondents judge each stimulus on several rating scales or measures. The combination of these ratings form a profile for each stimulus The stimuli that have similar profiles are considered more similar while stimuli that have very different profiles are considered dissimilar. Using rating scales may influence or limit the person's evaluation of the set of stimuli, but it also helps ensure that all people in the sample are judging the stimuli on the same basis.

One common method of estimating profile dissimilarity between two stimuli is to use the sum of the squared differences for each scale in the profile. This derived measure has the property that larger scores reflect profiles for stimuli that are more different or distant, and would therefore be an acceptable choice for proximity data.

The fewest judgments are required when the respondent sorts the stimuli into categories or groups. Pairs of stimuli that are placed in the same categories are considered more similar, while stimuli placed in different categories are considered dissimilar. The nature of the judgment task is focused on groups of stimuli rather than pairs or isolated stimuli as in the other methods. The coding scheme for recording the data must convert the extent to which pairs are placed in different categories into larger numbers to reflect the psychological distances.

Evaluating asymmetric proximities Generally, MDS makes the assumption that the psychological distance from stimulus A to B is the same as the distance from B to A. Consequently, most MDS programs expect that the proximity estimates are symmetric for the reciprocal A -- > B and B -- > A estimates. Where the reciprocal estimates are available, this issue should be systematically investigated for each data set. If there are large differences in these reciprocal estimates the MDS method may not be really appropriate or may give distorted results. If the differences are small, MDS may still be appropriate and give useful information.

Some MDS analysis programs are set up to handle asymmetric proximity matrices whereas others are not. If the program does not allow asymmetry or only allows input of the upper diagonal half of the proximity matrix, the asymmetries must be eliminated prior to data input. The most common way to do this is to average the A -- > B and B -- > A estimates to obtain the best overall estimate of the dissimilarity of stimulus A and stimulus B. When the data in the proximity matrix are complete and symmetric, an MDS analysis program can be used to extract the dimensions that can describe the differences among the stimuli.

Metric and non-metric MDS MDS converts the dissimilarity estimates in the proximity matrix to a set of distances among the stimuli in a space of 1, 2, 3, or more dimensions. The dimensions and the exact position of each stimulus are calculated iteratively during the solution process. The iterative solution process adjusts the positions of all stimuli so that the distances among them are a best fit with the original proximities.

Metric MDS programs constrain the estimated inter-stimulus distances to be either proportional to the original proximities or a linear function of them. This constraint reduces the flexibility of metric MDS, and the solution for a metric MDS will typically not fit as well as the alternative non-metric MDS. However, this constraint also makes the metric MDS method less susceptible to certain types of estimation problems such as the influence of a local minimum.

In contrast, non-metric MDS programs allow the inter-stimulus distances to be any monotonic function of the original proximities. The ability to estimate different functions gives this approach more flexibility. Further, the estimated function will typically give a closer fit to the original set of proximities. One disadvantage of the non-metric MDS approach is that it may be susceptible to certain estimation problems such as falling into a non-optimal solution.

Therefore, one procedure for getting the best possible solution is to combine these methods in sequence. First, the proximity matrix is analyzed with metric MDS and then subsequently it is re-analyzed with non-metric MDS. Although the level of fit will typically be better for the non-metric MDS, the qualitative nature of the dimensions found in both solutions should be quite similar. If the dimensions found are quite similar, the non-metric MDS solution with the appropriate number of dimensions should be considered the final solution since it will have a better fit. This better fit will usually offer a better description of the psychological distances among the stimuli. If the dimensions found are quite different, the non-metric MDS solution can be checked for problems of degeneracy or local minimum and steps taken to eliminate the problems.

Euclidean space For both metric and non-metric MDS methods, the psychological space that is constructed to account for the differences among the stimuli is a Euclidean space. That is, the distance between any two stimuli is the square root of the sum of the squared distances between the two stimuli across all dimensions. Since this type of space is familiar, the resulting stimulus space is easy to understand. Generally, therefore, this is a reasonable default.

However, in certain cases the possibility of using other non-Euclidean distance metrics such as a "city-block" metric should be considered. A city block distance metric states that the distance between two stimuli would be the sum of the absolute value of the differences in their positions across all dimensions. This different definition of "distance" would potentially influence the MDS solution. Once an initial solution is obtained, the effects of different definitions of distance can be cautiously explored to see if the fit increases. The degree of fit will also, of course, typically depend on the maximum number of dimensions allowed when analyzing the proximity data and constructing the space.

Number of dimensions Each MDS analysis must specify the number of dimensions that the program should work with to establish the stimulus space. Typically, the number of dimensions is systematically increased from 1, to 2, to 3, and so forth, and the quality of the solution is evaluated at each level. The dimensionality that offers the best quality solution is chosen for the final interpretation of the stimulus space. This is analogous in factor analysis to determining the appropriate number of factors.

The best quality solution is primarily determined by the degree to which the distances among the stimuli in the space can accurately describe the original proximities. That is, how well can the original judgments be reconstructed on the basis of the dimensions and the calculated positions of the stimuli on those dimensions? In MDS, the index for this degree of fit is called the stress index (in INDSCAL, the index is the percent of Variance Accounted For or VAF index).

The stress index summarizes the squared differences between the distances in the space and the corresponding values in the proximity matrix for each pair of stimuli. Therefore, stress is high when the inter-stimulus distances do not match the original proximities. Conversely, stress is low to the extent that the distances in the space accurately match the original proximities. When the distances match the original proximities, the estimated space is a good description and summary of the psychological judgments.

To choose the appropriate dimensionality, the stress values can be plotted for the 1-dimension, 2-dimension, 3-dimension, and so forth solutions. Stress will always decrease as the number of dimensions increases, but the exact manner of this decrease will often point to the correct number of dimensions. In particular, the stress will decrease substantially as dimensions are added up to a point, and then further decreases will be minimal. This generates an "elbow" or bend in the graph, and the number of dimensions at this elbow is typically chosen for the final solution. This process is similar to choosing an appropriate number of factors in factor analysis. In MDS, however, secondary considerations such as meaningfulness of the dimensions or the stability of the solution may dictate that a solution one or two dimensions higher is chosen for the final solution.

Orthogonal vs. oblique rotation Just as in factor analysis, the underlying dimensions of the stimulus space in MDS can be rotated for better interpretability. Similar to factor analysis, the initial dimensions for the stimulus space are orthogonal. In MDS, these dimensions can be rotated while still serving as a complete and accurate basis for the positions of each stimulus. Rotating these dimensions does not change the relative distances among the pairs of stimuli and therefore does not change the stress.

The ultimate criterion for doing a rotation is the theoretical meaningfulness or practical use of the dimension. The interpretation of each dimension is based on contrasting the stimuli that have positions very high on the dimension vs. stimuli that have positions very low on the dimension. These interpretations will be checked in the validation step discussed below.

The same pros and cons of orthogonal and oblique dimensions are true in MDS as in factor analysis. A set of orthogonal dimensions makes the description of the distances between each pair of stimuli much cleaner and easier. However, oblique dimensions may be a better description of the mental model that underlies the respondents' judgments about the stimuli.

Validating the dimensions To confirm and extend the meaning of each dimension found in the multidimensional scaling, the positions of the stimuli on each dimension should be compared to external measurements. Examining the relationship of each dimension to external criteria should help confirm or disconfirm the construct validity of the interpretation of the dimension's meaning.

If the external measurement or criterion is measured on an interval scale, the relationship of the criterion to the MDS dimensions can be examined with simple correlational methods. The most basic approach is to correlate the external criterion with the positions of the stimuli on each MDS dimension. Since this correlation is across stimuli rather than respondents, the sample size for evaluating the significance of the correlation is the number of stimuli.

More typically, the scores for the stimuli on the external criterion are used as a dependent variable in multiple regression. The independent variables are the set of scores of the stimuli on the MDS dimensions. The multiple R from this analysis is an overall indicator of the extent to which the criterion is related to the set of dimension scores. If non-significant, the criterion is unrelated to the MDS dimensions. If significant, the strength of the relationship can be estimated by the square of the multiple R. This index represents the percent of the variance in the criterion that is predictable from the MDS dimension scores. If this index is high, the criterion is strongly related to one or more of the MDS dimensions. The exact configuration of the MDS dimensions that are related to the criterion is reflected in the regression weights. This detailed information helps confirm exactly how the dimensions found by the MDS are related to other aspects of the stimuli, which helps establish construct validity.

Examining individual differences Whenever respondents are used to estimate the structure of a set of stimuli, the possibility of differences among either individuals or homogeneous groups of individuals ought to be considered. This is particularly important if previous research or theory in the SIS has indicated strong individual differences for that domain. The three levels of treating individual differences are described in Table 6.10.

The assumption that individual differences do not occur leads to a simple MDS analysis based on an average or aggregate proximity matrix for the stimuli, as discussed above. In this case the tacit assumption is that all individuals in the sample use the same underlying dimensions the same way when judging the stimuli. If strong individual differences are expected, then either INDSCAL or separate MDS analyses for different individuals should be attempted.

Table 6.10 Assumptions about individual differences and appropriate MDS methods

Assumptions about individual differences	Appropriate method for MDS analysis
All individuals use the same dimensions in the same way for judging the set of stimuli	MDS based on the proximity matrix averaged across respondents
All individuals use the same dimensions, but may emphasize the dimensions differently when judging the stimuli	A common MDS analysis using a separate proximity matrix for each respondent. Individual differences scaling: INDSCAL
Individuals may use qualitatively distinct sets of dimensions when judging the stimuli	A separate MDS analysis for each individual based on his or her proximity matrix for the stimuli

INDSCAL assumes that all respondents are using the same basic set of dimensions; however, each respondent can differently emphasize these dimensions when arriving at a dissimilarity judgment. That is, some subjects may emphasize dimension 1 more than dimension 2, while other subjects emphasize dimension 2 more than dimension 1 when making judgments. Say, for example, that two dimensions for the perception of new cars were cost and size. Some consumers may emphasize cost more than size when judging the similarity and dissimilarity of new automobiles, while other consumers emphasize size differences more than cost in making these judgments.

INDSCAL uses the set of proximity matrices to calculate a set of dimensions which best describe the distance estimates among the stimuli for all respondents. During the solution process, the relative weights that each respondent uses for each dimension are also calculated. In essence, the individuals are placed on the same dimensions as the stimuli. In this case, the position of the individual in the space reflects his or her relative emphasis on each dimension when making dissimilarity judgments. The individual differences on the weighting of each dimension then become an additional focal point for the validation of the meaning of the dimensions. The weights essentially form a scale of the relative emphasis of that dimension which can be correlated with other individual difference variables.

This type of validation analysis for individual differences becomes difficult or impossible if the third approach of separate MDS analyses for each individual is chosen. Separate MDS analyses allow each respondent to have a different function that relates the proximities to the underlying distances, a different set of dimensions, and a different arrangement of the stimuli on those dimensions. In the extreme, separate MDS analyses for each respondent may result in completely different sets of judgment dimensions across individuals. This approach represents an idiographic approach to understanding structure and can very well result in incomparable structures across respondents. Nevertheless, if strong individual differences are expected, then this approach would offer the maximum amount of flexibility to estimate these differences. The associated information in the SIS offers a multiplicity of ways for validating the individual differences found by separate MDS analyses or the INDSCAL approach.

Validating individual differences If INDSCAL is used, the relative weight each person puts on the underlying dimensions can be compared or correlated with other information about the individual in the SIS. That is, the other information in the SIS concerning the respondents can be used as criteria for evaluating the construct validity of the dimensions. This validation of the individual differences may be done in several different ways.

If the associated information consists of continuous measures for these individuals, each measure can be correlated with the individual's weights on the MDS dimensions found by INDSCAL. The measure can also be used as a dependent variable in a multiple regression analysis similar to the approach discussed above for stimuli. However, in this use of multiple regression for validation, the unit of analysis is the respondent and therefore the sample size is the number of respondents rather than the number of stimuli.

Sometimes the associated information in the SIS will be a nominal scale or categorical rather than a continuous interval measure. For example, an MDS for perceiving relevant facets of aircraft operations might find basic safety and expense dimensions for making judgments. The emphasis on these dimensions may vary among pilots, managers, flight attendants, and so forth. These different groups of people are not on a linear scale, so multiple regression is not appropriate. However, the expected differences in viewpoints among these groups can still be used to validate the underlying MDS dimensions.

In this case the appropriate analysis would be a discriminant function analysis where the groups of respondents are the criterion grouping variable. The relative weights of using each of the MDS dimensions to discriminate among the stimuli (from INDSCAL) would again act as predictor variables. The results of this analysis would show two things. First, to what extent are the differences of using the MDS dimensions related to membership in the different groups? That is, pilots, managers, and flight attendants should be significantly connected to a different pattern of weights emphasizing safety and expense of operations.

Secondly, at a more detailed level, the discriminant function weights would show the exact contribution that the differences in the use of each MDS dimension makes in predicting the differences among the groups. For example, we might expect that pilots would put a particularly high weight on safety when judging different facets of aircraft operations. Conversely, managers may tend to weight cost and efficiency more when judging aspects of flight operations. If these expected group differences are reflected in the discriminant function weights, they offer strong evidence of the construct validity of the underlying MDS dimensions.

Besides multiple regression and discriminant function analysis, other statistical techniques may be appropriate for examining the construct validity of the MDS dimensions, depending on the circumstances. For example, if the scores of the stimuli on the MDS dimensions must be related to an entire set of criteria rather than an individual criterion, canonical correlations may be appropriate. The canonical correlation would relate the entire set of MDS dimension scores to the entire set of criteria scores. By using all appropriate techniques and all the information available in a SIS, the validity of the dimensions underlying the structure of stimuli can be thoroughly evaluated.

Chapter summary

This chapter examined different methods for modeling stable structures. These methods are particularly important where the stable structures concern critical aspects of the target system. However, these methods do not address the dynamic, changing aspects of the target system. These changes may be as simple as a gradual shift or drift in a structure over time. In such a case, the changes may be tracked by measuring the structure at appropriate time periods using the methods of this chapter.

In other cases, however, the dynamic changes in a target system may be far more intricate. These changes may involve shifts in processes as well as rapidly occurring shifts in structure over short time intervals. For these dynamic systems more complex methods must be used to accurately capture the changes in the target system over time. These methods are discussed in the next chapters.

7 Dynamic Modeling based on States and Events

Overview

Dynamic modeling is inherently different from the static or structural modeling discussed in the previous chapter. In all cases of dynamic modeling something is changing over time in the target system. The aim of the modeling is to explicitly capture these changes and explore the implications of the dynamics. Casti (1997), Bennett (1995), Gordon (1969), and Gilbert & Troitzsch (1999) offer readable introductions to this approach, which is sometimes called "system simulation", while Banks (1998) gives broad coverage of recent advances in his handbook.

The conceptualization of what is changing will differ depending on the approach to dynamic modeling. Dynamic modeling can be based on discrete states or events, which will be covered in this chapter, or on functions which will be covered in the next chapter. Some texts emphasizing state-event modeling in general (e.g. Law & Kelton, 1991), while other texts such as Schriber (1974) and Russell (1983) cover specific modeling languages for this approach.

A state is any discrete definable condition of the target system. Generally a state will consist of a unique set of values of relevant parameters for the target system. In dynamic modeling there will often be a starting state, intermediate states, and a target or goal state. State based modeling describes and analyzes the transitions among the states. For this approach to be feasible at all, the number of possible states in the target system must be finite or countable. For this approach to be really practical, however, the number of possible states should be relatively small.

Alternatively, the target system may be viewed as a set of events that happen over time. In this view, the timing of the events and the effects of the events on the target system are critically important. In one sense, event based modeling is the opposite side of the coin to state based modeling. That is, a state may be defined as the stable set of conditions that happen between significant events, and an event may be defined as that which makes a target system change its state.

Whether a dynamic system is modeled as a set of states or a set of events depends on the nature of the target system. This chapter will cover state based modeling and event based modeling in separate sections. Each section will provide examples of models as well as some ideas on confirming or disconfirming the accuracy of these models using the information in the SIS.

STATE-BASED MODELING

State based modeling is particularly useful when there are a small number of clearly definable states in a target system. The states should be mutually exclusive and, if possible, an exhaustive listing of possible conditions of the target system. In the initial construction of a state-based model, it may be difficult to distinctly define each of the states. In this case including a catchall state called, for example, "other" can create an exhaustive list of states. As the model is developed, however, the catchall state should be further identified and reformulated as one or more defined system states.

A state generally should represent a qualitatively distinct condition of the target system. In systems where the key variables are qualitative, the state specification should clearly list the relevant values of each defining variable. For systems where some key parameters are quantitative, however, each state should be more precisely defined by listing all critical parameters of the target system and the exact values of those parameters that characterize each state.

A state transition is anything that moves the target system from one state to another. State transitions may occur spontaneously due to processes or events inside the system, or be triggered by inputs external to the system. If transitions occur spontaneously, the timing and pattern of these transitions is the focus for understanding the system. If external inputs trigger transitions, a careful analysis of which inputs lead to which state transitions is a key to constructing an accurate state based model of the target system. State models may be classified into deterministic models for which the transitions occur in an absolutely predictable fashion, and probabilistic models for which the transitions occur with a degree of likelihood ranging from 0% to 100%.

Deterministic finite state models

If a certain pattern of variables always causes a state transition and the absence of this pattern completely prevents the transition, the process is deterministic. For deteministic models, the transition will occur all of the time when a critical set of variables has required values (100%), or none of the time if the variables are absent or have the wrong values (0%). In using a word processing program to open a file, for example, state transition variables could include the user clicking on the "open" command and typing a filename. Several other variables are necessary for this transition to occur successfully. First, the named file must be present and be an appropriate type. Second, there must be sufficient space available in the computer memory. The file will always be opened if all these variables have the appropriate values. If any of these variables does not have appropriate value, the transition will not occur (i.e. the file will not be opened).

If the target system operates in this simple all-or-nothing manner, a deterministic model is appropriate. The transition variables may depend on things inside the system (internal) or things outside the system (external). In the word processing example, the presence of an appropriate computer file and sufficient computer memory were internal system requirements, while the user clicking on the appropriate command and typing a correct filename were external requirements. Some models focus solely or mainly on internal state transitions or external state transitions. Examples of models with each type of focus will be covered in the following sections.

Internal state transitions

A simple example of state transitions is given in Table 7.1 for a system consisting of an automobile plus driver. Typically, a model of a target system will simplify the real system to capture critical states and processes while omitting unimportant ones. For this example, the target system of normal automobile driving has been simplified to essentially four states: stopped, accelerating, cruising, and decelerating. The issues of turns, hills, and traffic signals are all ignored in this simple model of driving. Each of the four states can be defined by the car's speed and the application of the accelerator or the brake. Transitions between the states are caused by pushing or releasing the accelerator or the brake.

Notice that from a given state, it is possible to transition only to a subset of possible states. When the automobile is stopped, for example, it is only possible to push on the accelerator to begin accelerating. Conversely, when the automobile is at cruising speed, it is only possible to accelerate or decelerate. This limited set of transitions from each state is a common characteristic of state based models.

This deterministic finite state model gives a certain point of view on the functioning of the target system. First, the target system may be characterized by the profile of the amount of time spent in each state. For example, some drivers might spend much of their time accelerating and decelerating while other drivers spend most of their time in cruising mode. Other aspects of system performance, such as fuel economy, might be expected to change according to this profile of states.

Table 7.1 Finite state model of an automobile

State	Critical Parameters	Transition caused by	Transitions to
Stopped	Speed = 0 Accelerator off	Pushing accelerator	Accelerating
Accelerating	Speed increasing Accelerator on	Releasing accelerator Pushing brake	Cruising Decelerating
Cruising	Speed = constant > 0 Holding accelerator	Pushing accelerator Pushing brake	Accelerating Decelerating
Decelerating	Speed decreasing Accelerator off or brake on	Pushing accelerator Releasing brake Speed = 0	Accelerating Cruising Stopped

Second, the target system may be characterized by the pattern of transitions from state to state. A tailgater, for example, might be characterized by a transition pattern of accelerating up to the rear bumper of the car ahead and decelerating to cruise at exactly the same speed as that car. This pattern of frequent accelerate-decelerate-cruise transitions might be associated with poor fuel economy and other aspects of poor system performance such as increased rates of accidents, especially rear end collisions!

Since the brakes and accelerator are key controls of the automobile system, the transitions for this model are caused by things occurring inside the human-automobile system. The cause of these transitions can be alternatively viewed as either an internal change of a system parameter value or an internal event. These state transitions are diagrammed in Figure 7.1. "Release accelerator" is defined as easing off on the accelerator just enough to reach a steady cruise speed. Similarly, "release brake" is easing off the brake just enough to reach a steady cruise speed.

The issue of causality may become more complex as the model is extended to more completely portray the target system. If this model were expanded to include external inputs such as traffic signals, for example, the ultimate cause of an internal transition such as braking could be the occurrence of a red light or stop sign in the driver's perception. The internal, proximal cause of the change in state to deceleration could be tied to the external, distal cause of the traffic signal or sign. For such an extended model, causality would have at least two levels.

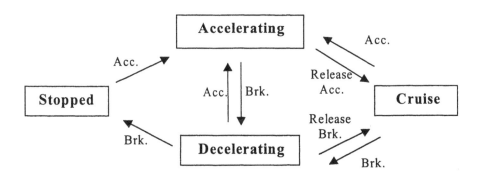

Figure 7.1 State transitions for finite states of an automobile
("Acc" is pushing the accelerator; "Brk" is pushing the brake)

Using finite state models in an SIS This type of finite state model can be used either descriptively or analytically with the information in the SIS. The descriptive use of this model would involve measurement of the time spent in each of the states and the frequency of state transitions for real drivers in some driving situation. The profile of time spent in each of the states, for example, could be illustrated by a pie chart with the proportion of time in each state being one segment.

Alternatively, the profile of time spent in each state could be listed on a figure by simply putting a percentage in the box for each state as illustrated in Figure 7.1. Since the states should be mutually exclusive and exhaustive, the percentages should sum to equal 100%. For example, a driver may spend the majority of driving time accelerating (40%) and decelerating (40%), and relatively less time cruising (10%) or stopped (10%).

This type of summary information can be used analytically by incorporating it in the SIS. Specifically, the amount of time spent in each state and the relative frequency of state transitions for each driver can become data in the SIS. To be comparable across persons, for example, this data should be based on a standard set of driving tasks or a representative sample of driving tasks in the natural domain. This data would, of course, be subject to the same scientific scrutiny as all other data in the SIS. That is, this data would be examined for sensitivity, reliability, and validity as discussed in Chapter 3.

After establishing the basic scientific quality of this data, the state and transition information can be analyzed either separately or together with the other data in the SIS. One of the first analyses may look for types or groups of drivers that are characterized by different data patterns. For example, the profile of time spent in each state could be cluster analyzed to find distinct types of drivers. The cluster analysis might identify some drivers who frequently shift between accelerating and decelerating ("Speed Oscillators") and other drivers who spend the majority of their driving time cruising at a steady speed ("Steady Drivers"). Other types may also be identified based on the relative amount of time spent accelerating, decelerating, and cruising. For example, some drivers may avoid stopping by accelerating to run yellow lights and decelerating while approaching a red light in order to give it more time to turn green. These drivers would have a pattern of less time stopped and relatively more time accelerating and decelerating.

Similarly, the pattern of transition frequencies among states could be analyzed for distinct clusters or types of drivers. Sequences or chains of the states could also be coded for analysis. A cluster analysis of the sequences, for example, might find a cluster of drivers with a high frequency for the transition sequence "Accelerate-->decelerate-->cruise" which would typify tailgaters.

The connected data in the SIS offers a wide variety of possible avenues for additional analyses. These analyses would connect the state or transition profiles to other data in the SIS. For a SIS in the driving domain, external information could include other information about the driver, the vehicle, and typical driving conditions. Connecting this state and transition data from the model analysis to the other SIS information can extend the knowledge available from the SIS.

The group of "Speed Oscillators", for example, might be found to be mostly teenage drivers who have not fully developed a steady control of the vehicle, or who enjoy rapid acceleration and deceleration. Similarly, the group of "Steady Drivers" might be found to be older drivers who have vehicles with cruise control. Finally, the group of tailgaters might be found to be drivers who drive typically in rush hour traffic and have personalities characterized by impatience and aggression.

The theories in the SIS should be extended to accommodate this new form of data. How well, for example, do the theories of driver behavior in the SIS account for either the profile of time spent in the driving states or transitions among states? Do drivers with more tickets and accidents on their driving record actually drive in a different manner that can be objectively identified by their profile of states and likelihoods of state transitions? Can driver training really change the pattern of driving behavior? If so, exactly how does it change driving behavior? Would training in defensive driving, for example, lead to more emphasis on deceleration to avoid accidents?

Asking and empirically analyzing these questions forces the theories in the SIS to address the *processes* in the target system as described by the state model. Broadening the scope of the SIS theories to include processes in a target system in addition to outcomes is an important extension because it broadens the set of relevant data for making predictions. A broader set of predictions can be pragmatically useful in solving problems (see Chapter 5) as well as allowing stronger confirmation or disconfirmation of SIS theories. Since the processes may be quite distinct from results or outcomes for the target system, this extension of the domain of relevant data allows a much stronger confirmation or disconfirmation of relevant theories in the SIS.

Further, if the SIS theories can be successfully tied to system processes, the control utility of the theories is greatly enhanced. Ties to system processes allow many more systematic interventions to alter or control system outcomes by altering system processes. In particular, the processes impinging on the profile of states or the relative frequency of transitions among the states can be targeted for change. Using the state model and ties to SIS theories, the exact expectations of the changes in either the profile or transition frequencies can be calculated. These expected outcomes from process changes can be predicted and compared to the real outcomes after an intervention such as training.

The automotive domain is, of course, a relatively simple domain with one operator in the system. More complex domains with multiple operators or more complex tasks would require correspondingly more complex models of states. Further, the transitions among states may depend on external triggers or stimuli in the environment as well as the internal events and processes in the target system.

External state transitions

The transitions among states of a finite-state model will often depend on events or processes happening outside the target system. Many dynamic systems are strongly influenced by input from the system context. The target system may change its fundamental state as a result of these inputs.

An example of a model with external state transitions from the aviation domain is given in Table 7.2. This set of states is a complete and exhaustive description of the states of the aircraft during a routine leg of a commercial flight. As in the automobile example, each state is characterized by a set of critical parameters. The set of allowable transitions is less than for the automobile example due to the tight scripting of commercial flights by air traffic control. This tightly scripted control leads to a more constrained transition among states than for the relatively unconstrained domains such as driving.

As in the automobile example, the transitions between the states are caused by particular, definable events. Unlike the automobile example, however, these transitions are caused by external as well as internal events. For this example, internal events are events having to do with the aircraft and crew only. External events would be events due to air-traffic control, company dispatch, the weather, and other factors. For example, all the transitions caused by clearances from air traffic control are external events. Examples of state transitions caused by internal events are the transitions due to conditions of the aircraft such as "Positive rate, gear up" and "Leveling out at a specified altitude".

The basic states for modeling the flight of a typical commercial aircraft are shown in Figure 7.2. For clarity, the triggering events that lead to the transitions among the states have been omitted from the figure. At this high level of analysis, the states and transitions are determined largely by the execution of the flight plan by commercial aviation pilots in accordance with air traffic control directives and company standard operating procedures. Therefore, much less variability in the state profiles and the transition probabilities would be expected in this model than in the automobile model. However, even at this high level of analysis there is the possibility that differences among crews could yield informative data for the SIS.

Although the time required for most states is tightly controlled by external factors, the time spent in the pre-departure and post-flight states is more flexible. Pre-departure or post-flight time could, therefore, be influenced by differences among crews or pilots. For example, some crews might spend more time at the gate or prior to the flight briefing the flight plan and expected cockpit interaction during the flight while other crews would not. Similarly, some crews may systematically debrief the good and bad aspects of the flight at the end of the flight while other crews would omit this process.

Table 7.2   Finite state model of an airplane flight

State	Critical Parameters	Transition caused by	Transitions to
Predeparture (At the gate)	Speed = zero	Departure clearance and pushback	Taxi
Taxi for take off	Speed = ground speed, path to a runway threshold	Takeoff clearance	Take off
Take off	Throttle TOGA, Accelerate to Vr	Positive climb rate, Gear retracted	Climb
Climb	throttle climb, gear up, flaps up, altitude increasing	Level out at specified altitude	Cruise
Cruise	Throttle cruise, altitude constant	Clearance to higher or lower altitude	Climb or Descent
Descent	Throttle reduced, altitude decreasing	Clearance to land on runway	Cruise or Landing
Landing	Throttle at idle, speed and altitude decreasing	Clearance for taxi after touchdown and rollout	Post-flight taxi
Post-flight taxi	Speed = ground speed, path to gate	Clearance for gate	Post-flight briefing
Post-flight briefing (At the gate)	Speed = zero	(Departure clearance for new leg)	(taxi for takeoff on new leg)

Time spent in the pre-departure and post-flight briefing states may help produce more effectively coordinated crews. Therefore, the profile of time spent in these two states could be analyzed with other information in an aviation SIS. If the time spent in these activities actually helps crew performance, pilots and crews that have increased time in the pre-departure or post-flight briefing segments should also have better performance indicators in the SIS. The SIS theories should be able to help predict the expected relationships between better briefings and other aspects of crew performance.

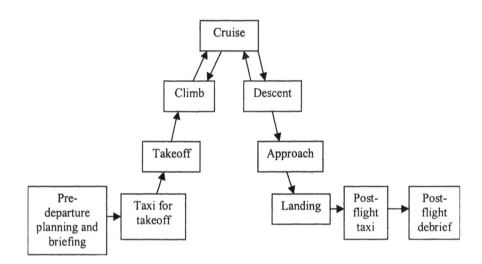

Figure 7.2 Finite state graph for an airplane flight

Conversely, good SIS theories should also help explain the antecedents of these briefing patterns. That is, why do some crews put in this extra time while others do not? Appropriate SIS measures should predict differences in the time spent in the pre-departure or post-flight briefing states. For example, Captains who emphasize a smooth, efficient working relationship with their First Officers may spend more time in the pre-flight stage on the first flight of the day with the new First Officer. Similarly, Captains who place a high-value on mentoring and training First Officers might be found to take more time for post-flight debriefs.

Since the transitions among flight states are also tightly controlled by air-traffic control, the only differences that might indicate crew qualities are those transitions over which the crew has some control. In the cruise phase of flight, for example, the crew may elect to request higher or lower altitudes to avoid turbulence, to increase fuel efficiency, or to obtain more favorable wind conditions. Crews with Captains putting a higher value on passenger comfort might have a higher number of these transitions as they seek flight levels with a smoother ride for the passengers.

A more fine-grained models may be more informative than this upper level model. For example, the sequential execution of the steps of a checklist is largely under the control of the crew. Therefore, the transitions among the steps of a checklist might be more diagnostic about the nature of the crew and their interaction. A crew that puts a high priority on standard operating procedures, for example, might have a very consistent pattern of transitions across checklist items that is precisely according to the official flight operations manual for that aircraft. Such a crew may also resist any deviations from, or interruptions in, their checklist procedures. The information in the SIS could be used to understand the training antecedents of such a crew and the long-term performance consequences of this pattern.

Probabilistic finite state models

In contrast to deterministic models, probabilistic models assume more gradations in the likelihood of state transitions in the target system. That is, the relative likelihood of a transition depends to some degree on the occurrence of a pattern of inputs, but may occur anywhere in the interval between 0% and 100%. That is, a certain pattern of inputs may make a state transition more likely (e.g. 90% likely), but the transition may occasionally occur even without the inputs (e.g. 10% likely). This pattern of conditional likelihoods of transitions may be more appropriate for target systems that do not operate in a simple all-or-nothing manner. As for deterministic models, a critical distinction for probabilistic models is whether the likelihood of the transitions is dependent only on internal system factors or is dependent on external triggers or variables.

Internal state transitions

Some systems spontaneously or probabilistically change states without any external input. An example is traffic lights changing color. For traffic lights that are not controlled by sensors, the state of the traffic light does not depend on the presence of traffic or on any other external factor. Instead, the light changes color based entirely on internal events or processes. The changing of most traffic lights is typically determined by a timer, but these timers are set differently for each light. From the point of view of a driver approaching the light, the sequence of changes is somewhat unpredictable or probabilistic. This type of situation where all changes occur probabilistically over time is often described by a Markov model. A Markov model assumes that the shifts among the states of a system are due to a purely probabilistic process which is not affected by internal or external inputs.

Despite the timing differences among traffic lights, they all change in a known and predictable pattern. Also, it is typically true that the yellow light is shorter than either the green or the red. A Markov model of the probabilistic change in states of a traffic light is given in Table 7.3.

Table 7.3   Probabilistic transition model for a traffic light

		State of Traffic Light in 5 seconds		
		Green	Yellow	Red
Current State of Traffic Light	**Green**	80%	20%	--
	Yellow	--	--	100%
	Red	20%	--	80%

This type of model is simple because there is no need to list the internal or external causes of the transitions. Rather, the critical thing to list is the set of probabilities of transitions in states from one time to the next. This is sometimes called a transition matrix and is illustrated by Table 7.3. In this table, if the traffic light is currently green the likelihood of it staying green in the next five seconds is 80%, while the likelihood of it turning yellow is 20%. If the traffic light is currently yellow, the likelihood that the state will still be yellow in 5 seconds time is zero because yellow lights are almost always shorter than 5 seconds; instead, the yellow light will certainly change to red (100%). If the traffic light is currently red, there is an 80% chance of it staying red and a 20% chance of it turning green in the next five seconds. In the U.S. at least, the red light will transition directly to green rather than to yellow and then to green.

Although in the aviation domain examples of systems that change in a purely probabilistic manner may be rare, one example might be the weather. At a very gross level, a destination airport may have three qualitatively different conditions of weather: visual meteorological conditions (VMC), instrument meteorological conditions (IMC), and airport closed due to weather. Since weather changes spontaneously, the transitions among these three conditions may be considered a Markov model. Although there are three states as in our traffic light example, the relative likelihood and transitions among the states are different. For example, although weather conditions are very unlikely to change from VMC to airport-closed conditions in one-hour, it is possible. For example, a severe and quick moving front with severe weather might cause an airport to shift from VMC to closed in one hour. A possible set of transition probabilities is given in Table 7.4 for the weather example.

Table 7.4 Probabilistic transition model for airport weather

		State of weather in one hour		
		VMC	**IMC**	**Airport closed**
Current State of the weather	**VMC**	90%	9%	1%
	IMC	20%	77%	3%
	Airport closed	5%	45%	50%

In this example, the relatively high probabilities of staying in a state are due to the normal slowness of change in the weather. The typical transitions are from VMC conditions to IMC conditions and back. Transitions into closing an airport due to the weather are rare and typically occur after the weather has already deteriorated to IMC. Once an airport is closed, it may remain closed for a few hours but the likelihood of an airport being reopened quickly is quite high due to the fact that every effort is made to facilitate airline traffic by reopening closed airports.

External state transitions

Since systems with these types of pure, internal probabilistic transitions are rare, most state-based models will have some external determinants of how the system changes state. Returning to the finite state model of driving an automobile, the probability of state transitions for a given driver could be illustrated as in Figure 7.3 by an appropriate proportion or percentage attached to each of the arrow links.

As in the stop light model discussed above, the total percentage or probability of transitioning out of each state is 100%. For example, from Decelerating the transitions are 10% to Stopped, 10% to Accelerating, and 80% to Cruise, for a total of 100%. The hypothetical pattern of transitions and time profiles presented in Figure 7.3 might be descriptive of a tailgating driver.

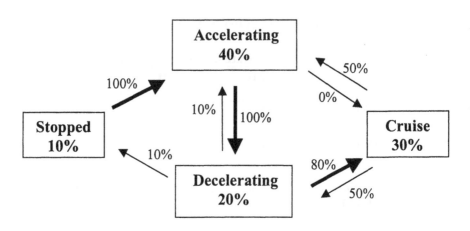

Figure 7.3 State transition likelihoods for a typical tailgater

In this example, the tailgater spends most of his time accelerating up behind the car in front (40%), and then decelerates in order to match speeds. Another 30% of the time is spent cruising directly behind the car in front. This tailgater shows the characteristic accelerating-->decelerating-->cruise pattern that was described earlier. The transition probabilities along the arrows on that path are particularly high (thicker arrows). The noticeable lack in this pattern is a lack of transitions from accelerating to cruising by releasing the accelerator. This type of transition from acceleration to cruise might be more frequent for normal drivers.

The external inputs that can change the transition probabilities of this system are illustrated in Figure 7.4. In this example, the presence of a red light changes the likelihood of decelerating to a full stop to be 100% instead of the normal 10%. Similarly, the presence of an open space in the lane ahead changes the likelihood of switching from decelerating to accelerating to be 100%. Even the transition using partial braking can be facilitated if the external input is the presence of a police car! In this fashion, the effects of external inputs on the transition probabilities of the finite state model can be included.

The stage profile and transition probabilities of a probabilistic finite state model are a concise way to descriptively summarize model information. This summary information can be used to evaluate the model as discussed below. If, however, there are very large number of states, this approach becomes unwieldy due both to the large number of states and the even larger number of potential transition probabilities. A more feasible approach in that case is to use event based modeling that is covered in the next section.

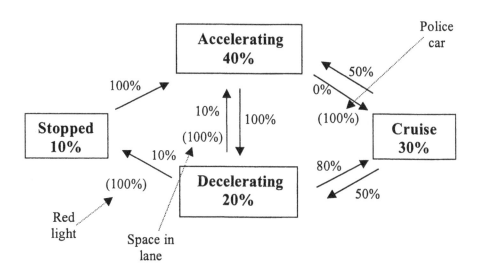

Figure 7.4   Effects of external inputs on transition likelihoods

Evaluating State Based Models

Although a state based model represents a simplified version of reality, it is at the same time a form of theory of the target system. Therefore, it can and should be evaluated on the same basis as other theories in the SIS. That is, a model should enhance the explanation, prediction, and potential control of what is happening in a target system. Both the qualitative and quantitative data in SIS can be used to assess these functions of the model.

Explanation The plausibility of the model's account of the target system should be evaluated. However, to the extent that the model is a simplified version of reality, it cannot be expected to be a completely plausible account. Nevertheless, the key aspects of the states and transitions that are postulated by the model should be assessed for plausibility.

The states should be evaluated to see if they are the best possible mutually exclusive and exhaustive set of states for the target system. This may involve the qualitative evaluation of whether the states occurred in a target system in the manner postulated. For example, subject matter experts could evaluate whether all of the postulated states would occur in a target system, and whether other qualitative states that are not postulated should be added to the model.

Quantitative analyses should also check on the plausibility of the postulated states. For example, a cluster analysis using the set of key or critical state variables should find clusters corresponding to the states. Similarly, domain experts should be able to categorize examples of real system states quickly, easily, and reliably into the set of states postulated by the model.

The plausibility of the state transitions postulated by the model should also be examined by qualitative and quantitative methods. Subject matter experts should evaluate which transitions are possible and which not, the relative likelihood of the transitions, and the causal factors involved in the transitions. Case studies of typical and critical incidents in the SIS should help clarify the important transitions of the model.

Quantitatively, if longitudinal records for some target systems are available, the relative frequency of state transitions observed in the SIS can be compared to the model's hypothesized transition probabilities. If discrepancies are noted, the reasons for the model having too high or too low a probability of state transitions should be explored. Based on these analyses, the model should be adjusted as necessary before proceeding to the evaluation of prediction in the next step.

Prediction To evaluate predictions, the sequence of states predicted by the model should be compared to the sequence of states observed on a sample of cases in the SIS. It is equally important to evaluate transitions that are expected to happen and do not, as to evaluate transitions that happened but were not expected. Deriving the set of expected states and transitions, however, will depend on the complexities of the model.

The derivation of expected states and transitions is simplest for Markov models. This is because the transitions between states in a Markov model are dependent only on probability and time, but not on other external or internal events. Therefore, predicting transitions among states for a Markov model can be done with two essential elements: a vector of the distribution of starting states and the transition matrix discussed earlier. This is illustrated in Table 7.5 for the traffic light example.

Table 7.5 Predicting distributions of a set of traffic lights over time

Vector of likelihood for initial states of a set of Traffic Lights			Traffic Light transition matrix (based on 5 second intervals)			
				Green	Yellow	Red
Green	Yellow	Red	Green	80%	20%	--
45% or .45	5% or .05	50% or .50	Yellow	--	--	100%
			Red	20%	--	80%

Suppose the sample consisted of 100 traffic lights, and at the beginning 45 of them were green, 5 of them were yellow, and 50 were red. To obtain a predicted distribution of traffic light states in the next five seconds, the vector of initial likelihoods is multiplied by the transition matrix. This results in a predicted distribution of lights five seconds from now. That distribution, in turn, can be multiplied by the transition matrix to predict the distribution of lights 5, 10, 15, and 20 seconds from now. Table 7.6 gives the results of using this procedure to make predictions from the model for 5 to 30 seconds in the future. In practice, this matrix multiplication process can be continued indefinitely to make predictions for an arbitrarily long time from the current time. The matrix multiplication process is simple enough to be included in popular spreadsheets such as Microsoft Excel.

Even though this is a very simple Markov model, the predictions it makes could be tested in several different ways. First, the distribution of a random sample of traffic lights over time could be predicted and compared to a real sample. The predictions in Table 7.6 are that the sample should stabilize at about 45% green lights, 9% yellow lights, and 45% red lights. These precise predicted percentages can be checked against the results from a real sample of traffic lights.

Second, more specific subsamples could be selected and examined over time. For example, a selection of all green traffic lights could be checked over time to see how the distribution changed. Making predictions for a sample of all green traffic lights would just involve changing the initial likelihood vector to be 100% (green), 0% (yellow), and 0% (red) and doing the matrix multiplication. Similarly, samples of all yellow and all red traffic lights could be checked to if they change in the predicted manner over time.

Table 7.6 Distribution of traffic lights over time

	Distribution of Lights		
	Green	Yellow	Red
Start:	45.00%	5.00%	50.00%
5 seconds	46.00%	9.00%	45.00%
10 seconds	45.80%	9.20%	45.00%
15 seconds	45.64%	9.16%	45.20%
20 seconds	45.55%	9.13%	45.32%
25 seconds	45.51%	9.11%	45.38%
30 seconds	45.48%	9.10%	45.42%

Finally, the expected transition likelihoods should be compared to the relative frequencies of transitions in the sample. Particular attention should be given to transitions that should never occur, such as transitions from a green light directly to a red light. If any of these transitions occur and cannot be otherwise explained (by, for example, a burned-out yellow light), the model is disconfirmed.

Establishing predictions for more complex state based models may require the running of a computer simulation. Any internal or external events that affect the transition from one state to another must be correctly input to the simulation at the appropriate time. In the simple driving simulation, for example, the internal events of pushing or releasing the accelerator and brake would have to be accurately input to the simulation. Since the behavior of the target system over time is being modeled, the correct timing of each input is critical.

As in the traffic light example, the predictions of the simulation will generally be compared to the results for a sample of cases in the SIS. Typically, the simulation is run repeatedly to generate a complete set of expected results. The timing of each input, such as the pushing of the brake pedal, can be estimated from the sample of cases and this distribution of times used as input for the simulation. The results of the simulation are evaluated both qualitatively and quantitatively.

Qualitatively, behavior of the simulated system should be compared to the sample of real cases. One way of doing this is to have a sample of experts evaluate a mixed set of simulated and real cases. If the simulated results resemble the real cases, the experts should be unable to distinguish the real from simulated cases at beyond a chance level. If the experts can reliably distinguish simulated from real cases, however, this simulation must be qualitatively distinct in some manner. The manner in which the simulated cases are systematically distinct from real cases must be examined further to see if this indicates a serious flaw in the simulation or is an unimportant distinction.

Quantitatively, the simulation will give an expected distribution of values for each key parameter. The results predicted by the simulation should match the distribution of results for the sample of real cases as precisely as possible. Particular attention should be paid to the aspects of the sample of cases that were not used to estimate simulation parameters. If the brake presses from the sample were used as input for the simulation, for example, the average frequency and timing of brake presses in the simulation will of course match the sample. An aspect of the sample cases that was not used for estimation, such as the average cruising velocity of the automobiles, would be a good basis for comparing the simulation results to the sample.

The change in system states over time should be one focal point of the comparison of the simulation results to the sample. For any particular time or time interval, the distribution of states found in a simulation should match the distribution of states found at the corresponding time in the sample of real cases. If the states are qualitatively distinct, the profile of the expected probabilities of each state should match the relative frequencies of the states in the sample. If the states are defined by values on continuous variables, the comparison of the expected and obtained distributions should be a thorough quantitative comparison. That is, the mean, variance, and overall shape of the distribution predicted by the simulation should be compared with the corresponding distribution of real cases. Any statistically significant difference between the distributions may indicate some degree of inaccuracy in the model.

Control If the state-based model is plausible and predicts outcomes of the real target system reasonably well, it may be usable for altering or controlling the outcomes of the real system. Control options are less likely for simple Markov models since the state transitions do not depend on any internal or external events that could be changed or manipulated. For Markov models, the only control options are to alter the balance of starting states or to try to find system parameters that will change the likelihood of transitions in the transition matrix. If such parameters can be found, they can be tested by finding a real sample of cases with the same system parameters. The results from the sample should be compared to the results of the simulation with the modified start vector or transition matrix.

State based models with transitions triggered by internal or external events lend themselves more to control of the target system. If the model is accurate, controlling the sequence of states in a target system should be possible by controlling the events that lead to state transitions. A real world example of this is anti-lock brake systems in the driving domain. Such systems pulse the brakes and prevent locking the wheels during heavy braking. Essentially such systems change braking inputs and prevent an undesirable state of loss of directional control.

Similarly, in the aviation domain stall proof airplanes have been developed. Again, a technological fix has been applied to the system in order to prevent the system from entering a certain undesirable state. Preventing these states, of course, may have undesirable side effects. For example, to make a full stall landing a pilot may wish to have sufficient control to stall an airplane.

Alternatively, a control effort may be focused on adding possible new states for the target system. Suppose, for example, that aviation crews are having trouble completing checklists when interruptions occur. A new state of "Checklist pending" could be added to the standard operating procedure for the aircraft. This new state could be associated with electronic or physical means of reminding the crew to complete the checklist.

No matter what the nature of the control interventions, their effect should be modeled using the simulation procedure and the results of this simulation compared to appropriate real data in the SIS. This may, of course, require making an intervention for a sample of cases in the SIS. If the expected results are obtained from the intervention, the validity of the model is strongly supported.

If the expected results are not found, the possible faults of the model should be investigated in the same way as possible weaknesses in a theory. In some situations, the model will not work correctly because it is too simple. In such a case, a more complex model may have to be developed. At some point, the number of possible states in state based modeling becomes unwieldy. At that point, further development may require a shift in focus to event-based modeling to more easily accommodate the increased complexity.

EVENT-BASED MODELING

Event based modeling can be as simple as a list of events and their scheduled occurrence in the target system. For example, an almanac lists a set of events such as tides or phases of the moon and the expected time for each event. The expected times for tides are the hour and minute of flood tide and ebb tide during each day. Similarly, the phases of the moon are specified by days of a monthly cycle. However, event based modeling can also be considerably more complex.

Since the occurrence of any event technically can be defined as something that changes the state of the system, the event-based models are formally equivalent to the state based models covered previously. If, however, the focus is on a target system and the events that impact the target system, then the event based modeling approach may be more natural. Certainly when the number of states is very large, the event-focused approach may be more feasible than a state based approach.

Finite event models

For a simple system with few possible inputs, a graph can represent finite events and the states which occur between these events. To illustrate this, the event-based model for the simple automobile system is illustrated in Figure 7.5. The focal events for this system are either pushing or releasing the accelerator or the brake. The states discussed previously in the finite state model for the system occur between the events and depend on the value of critical parameters.

Although this finite event model is essentially equivalent to the finite state model for the automobile presented earlier, it tends to emphasize the events rather than the states of the system. The events of pushing or releasing the accelerator or the brake have now assumed the central role in this model. The states defined in the finite state model for the automobile have essentially become transition periods between events. Further, this presentation may clarify that certain events, such as simultaneously pushing the brake and accelerator, should not occur in the system.

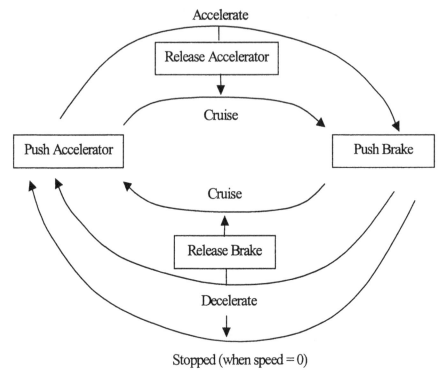

Figure 7.5 Finite Event model for driving an automobile

Finite event simulation can be made more powerful by differentiating events that have qualitatively or quantitatively different effects on the simulated system. In the driving example, stamping on the brake or applying the brakes gently has the same net effect of decelerating the car, but are qualitatively distinct. The finite event model could be elaborated with separate events for "normal braking" and "maximum braking". For simulation focused on the quality of the driving behavior, therefore, preserving different types of throttle or brake application as distinct events might be necessary.

Script based models One type of event based modeling uses scripts. Scripts are systems in which particular events are defined to have certain effects. If, for example, the script involved eating at a fast food restaurant, the sequence of events given in Table 7.7 might be expected. This script specifies the actors, actions, and events of each scene in the script. This type of script together with the timing information can be used to construct a dynamic model of the target system.

Table 7.7 Script for eating at a fast-food restaurant

Time	Person 1	Person 2	Server	Event
T-1	Pair enters restaurant			Enter System
T-2	Pair wait in line			Queuing
T-3			Asks for order	Start Service
T-4	Orders		Takes order	
T-5		Orders	Takes order	
T-6			Obtains food	Hamburgers, French fries
T-7	Pays			
T-8			Makes change	
T-9			Gets food	End Service
T-10	Selects table		Serves next customer	
T-11	Eat	Eat		
T-12	Pair exits restaurant			Exit System

Focus on events vs. focus on processes For a simple event based model, the focus on the information derived from the simulation can either be information about the objects in the simulation or the processes acting on those objects. In the above example, the objects or entities are the couple visiting the restaurant. If the simulation is focused on information about these objects or people, then the relevant information from the simulation is the average time required to serve the people at each stage of the process. This average time required for service might be correlated with other relevant reactions of the persons recorded in the SIS, such as the satisfaction with the dining experience.

Alternatively, the simulation focus might be on some aspect of the process itself. In this case, the major aspect of the process is the server behind the counter. Simulation focused on the process, therefore, would emphasize information about the relative amount of time that the server spends on each task and the relative amount of time that the server is not busy doing any task. This information on server utilization could be very important for fine-tuning the staffing of the restaurant and other managerial concerns. If several servers have periods of no utilization, for example, the staffing of the restaurant might be reduced without any negative effects on average service times. Alternatively, if there is a bottleneck of servers all trying to access the French fries at once, for example, optimizing the system may require providing another French fry cooker. One way to focus on process is to use a form of event based modeling that describes each step in a process as an event. An example of this type of approach is using production systems to model each step in a process.

Production systems One way of modeling each step in certain types of processes is a production system. This approach uses a set of simple rules to model human experts or mental processes such as human cognition and action. A production rule is a simple atomic "if-then" statement. When applied to humans, the if-then statements concern elemental cognitive operations such as a recall from memory, a perception of an external element, or a simple step in the cognitive process. The execution of each of these productions can be considered a cognitive event, and thus this type of modeling is event based modeling of cognition.

Since this level of detail is quite fine, modeling of even moderately large cognitive tasks requires a large number of productions. Therefore, this level of modeling is appropriate when it is necessary to gain a more complete analysis of a specific, important task. An example of these types of tasks in the aviation domain is using the radios to communicate to air traffic control (ATC) and using the automation to plot and fly the course of the aircraft. A simple example of using production rules to model a change in radio frequency is given in Table 7.8.

In this example, the productions are numbered to help track the execution of the model. The firing of each production is a discrete event in the model. Productions do not, however, fire in any particular preset order. Rather, a production fires when the conditions on the "if" side are satisfied. When a production fires, the products on the "then" side are added to the system. These additional products may help satisfy the "if" side of other productions in the system and allow them to subsequently fire.

In this simple example, production 1 is triggered by approach control giving the pilot the frequency for the tower. After that, each production is triggered by an internal set events that include the results of the firing of previous productions. That is, production 2 is fired as the pilot recalls the first digit of the frequency and enters that digit on the keyboard. Entering that digit removes it from memory. If there are still digits in memory that must be entered, the pilot recalls those digits one at a time and enters these numbers using productions 3 and 4. Once all digits in the assigned frequency have been entered, production 5 fires which closes the goal of entering the frequency and sets the goal of switching the frequency to be the active frequency. Finally, production 6 is fired and the pilot presses a key that activates the new frequency in the communication radio.

Table 7.8 Example of a production system for changing radio frequencies

Relevant fields for productions that the production system			
If (conditions that must be satisfied for the production to fire)	**then** (all cognitive and behavioral results of the production firing)	**time** (time required to execute production)	**other effects** (any other effects caused by the execution of the production)
1. Pilot hears ATC give tower frequency	pilot sets goals of entering numbers and switching to active frequency		Interrupts other tasks of pilot not flying until task completion
2. Pilot recalls first digit of frequency	pilot presses digit on keyboard	500 milliseconds	possible key entry errors
3. More digits are in the assigned frequency	pilot recalls next digit in the frequency	200 milliseconds	possible recall errors
4. Pilot recalls next digit in the frequency	pilot presses digit on keyboard	500 milliseconds	possible key entry errors
5. All digits in the assigned frequency have been entered	Satisfy goal of entering frequency. Set goal of switching to active frequency	200 milliseconds	possible premature closure errors
6. Goal is to switch active frequency	pilot presses key to activate new frequency	500 milliseconds	possible termination error

Since the firing of each production is an event, other important aspects of the event, such as the time required for the cognitive operation, can be modeled along with the production. These associated effects of a production firing should include the change in any system variables that are relevant to the overall model. In the example above, the other effects that are listed for the execution of each production focus on possible errors. In other cases, however, the possible effects of a production firing could include many different effects such as changes in the cognitive workload or cognitive capacity of the pilot, changes in motivation, mood or emotions, etc. If the production system is embedded in a higher level model, the relevant direct and indirect effects of the production system should be passed back up to the higher levels of modeling. The relevant direct effects would include the successful or unsuccessful execution of the frequency switch. The indirect effects of success or failure would be changes in other relevant variables such as ATC responses and pilot motivation or workload.

Evaluating event-based models

The evaluation of event-based models involves the same basic criteria of explanation, prediction, and control as the evaluation of state-based models. The evaluation process is essentially the same for each facet of evaluation, but the focus is shifted to be nature, timing, and sequencing of system events rather than system states. For example, a Markov model of changes in a traffic light should accurately predict the distribution of times for changes in the traffic light over time. The plausibility of the explanation and control potential of the model are also evaluated in the same basic manner as for state-based models.

Advanced state/event dynamic modeling

Complex, multi-faceted systems may require more than one type of approach to adequately capture the richness of the events and processes in the target system. The basic approaches of state-based and event-based modeling can be extended in at least two distinct ways to create models with more complete coverage of the target system. First, state-based models may be combined with event-based models to create hybrid models that cover more aspects of the system. Combining approaches in this manner may increase the breadth of covering different aspects of the target system. Second, models may be developed at different levels of generality or grain-size and combined into a multi-level model. Combining general and detail-level approaches in this manner increases the depth of coverage of selected aspects of the target system.

Hybrid models

Hybrid models combine state based models with event based models. Different states of the system may imply that a qualitatively distinct set of events and processes should occur. Within each general state of the system, the modeling proceeds due to certain events (event based modeling).

An example of hybrid modeling would be a model required for a complex script. Complex scripts can be modeled as scenes in which actors play a role and accomplish certain tasks between their entrances and exits. Each scene would be a different general state of the system. A "first date" script, for example, may consist of a scene for the initial greeting, a scene for driving to a restaurant, a scene for eating at the restaurant, a scene for driving to a movie, and a scene for seeing the movie.

The events and processes in these scenes may be similar or quite different. For example, the "driving to a restaurant" scene and the "driving to a movie" scene might have quite similar events and processes. On the other hand, the events and processes of the driving scenes might be quite different from the "eating at a restaurant" or the "seeing a movie" scenes. Similarly, the restaurant and movie scenes might have in common some eating processes but be quite different in the expected conversation processes.

In the aviation domain, the finite state model of an airplane flight presented earlier could be augmented by event based modeling of each phase or state of the flight. The usefulness of modeling a specific state or phase might depend on the degree to which a detailed model was necessary. If more problems are being encountered in the takeoff and landing phases, for example, then these phases may have more detailed event models developed. An example of a more detailed event model for a phase of flight is given in Table 7.9 for the landing phase.

In this example, notice that there are several critical external events consisting of the clearances, but there are also other internal events occurring as the crew carries out necessary tasks. To flesh out this model, the average time required for each of these tasks or segments would be specified in the computer simulation. The distribution of times for each task might come from SIS information from simulations of landings or from data recorded during observations of actual commercial flights.

Table 7.9 Script type event model for the landing phase of flight

Time	Captain	First Officer	ATC	Event
T-1			Contact by Approach Control	Cleared for approach
T-2	Crew performs landing checklist and configures aircraft appropriately			
T-3			Specifies shift to Tower on frequency X	
T-4	Pilot not flying enters the tower frequency on radio # 1			
T-5	Pilot not flying contacts tower and asks for permission to land			
T-6			Contact by Control tower	Cleared for landing
T-7	Pilot flying establishes correct air speed and rate of descent	Pilot not flying calls out appropriate altitudes and airspeeds		
T-8	Pilot flying initiates reverse thrust and braking as appropriate			
T-9			Contact by Control tower	Cleared for taxi to gate
T-10	Controls aircraft	Monitors traffic		
T-11			Specifies shift to Ground Control on frequency Y	

As it currently stands, this model depicts normal performance. The times or events of the model can be altered to depict performance and more unusual or abnormal situations. For example, the presence of unusual wind conditions could require maintaining a higher airspeed than normal during landing. In such conditions, the distribution of average times for T-7 would be shorter than normal.

Similarly, the set of events depicted in this normal landing model would be different if a go-around was required. The cause of such a missed approach could be either internal or external to the system. Internal causes could include failure to see the runway at the decision altitude or an approach that was too high and fast. External causes could include a failure of ATC to provide a landing clearance or the perception of an obstruction on the runway. These causal events would have to be explicitly modeled in this simulation. The go-around procedure itself could be modeled as a variation on the normal approach and landing, or as a separate scene in the script for the entire flight.

Multi-level models

Modeling may take into account an analysis of the target system at different levels of detail. Naturally, as the grain of the model becomes finer and finer, the events and processes that constitute the model become correspondingly smaller. The real power of modeling of complex systems in this manner lies in the ability to integrate models at any level of detail. These models may range from very course grained accounts of large-scale tasks or even entire jobs to the very fine-grained models of cognition or neuronal activity in individuals.

These models are integrated during the computer simulation of the target system by appropriately linking the inputs and outputs from each distinct level of modeling. Typically, the higher level models will set the general context, goals, and upper-level parameters for the execution of the more detailed models. The more detailed models, in turn, will execute when triggered by the upper-level models and will feed the relevant results of their execution back up the chain. The models at different levels do not, however, have to be of the same type. In fact, it will often be the case that these models will be distinct types of models since they have to model different levels of the target system, and different levels may be naturally suited to qualitatively different forms of models.

An example of multilevel modeling in the aviation domain is given in Figure 7.7. The upper-level task is flying an aircraft from one city to another. The upper-level model is the state-based model of the entire flight discussed earlier in this chapter. In multilevel modeling, this upper-level task and corresponding model will activate and set the relevant contextual variables for the middle and lower-level models. In particular, the triggering event of the ATC approach clearance would activate the more detailed middle level model that focuses solely on the approach process.

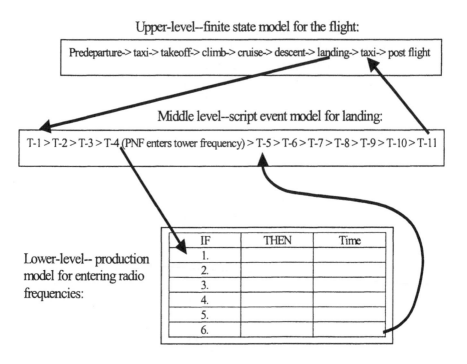

Figure 7.6 An example of multilevel modeling in the aviation domain

The middle level task concerns the final approach to the airport at the end of the flight, which is only one segment of the upper-level model. The corresponding model at this level is the script model for the approach tasks that was presented earlier. Among the tasks included in the script is the task for the pilot not flying to shift over from the approach control frequency to the tower frequency for that airport. The execution of this particular task would activate the more detailed lower-level model of pilot cognitions and behaviors.

The lowest level task in this multilevel model is the detailed cognitive and behavioral events that the pilot not flying must do in order to appropriately change radio frequencies. The corresponding model at this level is the production rule model discussed above for the cognitive and behavioral activities for this task. The execution of the set of productions will take some predictable model time and have an expected result. In most cases, this expected result would be a successful entering and activation of the new radio frequency. In some cases, however, the expected result would be an error somewhere in the process such that the correct frequency is either not entered or not activated.

Whether the process concludes successfully or unsuccessfully, the conclusion of the execution of the lower-level model will re-activate the middle level model and provide it with the set of relevant outcomes from the lowest level. In this case, the relevant outcomes would certainly include the fact that either the correct radio frequency has been entered and activated or possible errors of incorrect entry or non-activation. The relevant outcomes may also include workload and other side effects of the completed process.

The middle level script model would then continue executing from that point. It should be clear, however, that the script would take a different branch in the case with a radio frequency that has been correctly entered from the case where it is incorrect. The script branch for an incorrect frequency might include several unsuccessful calls to the tower followed by a switch back to the approach control frequency to confirm and re-enter the radio frequency for the tower. This branch would, of course, take extra time and result in extra workload and confusion for the pilots during the critical approach phase of flight. The net time remaining before the runway threshold and the total workload on the pilots might be an outcome of the middle level script model that would be input to the upper-level state model of the flight. In any case, when the middle level script model has completed its execution, the control of the simulation is transferred back to the upper-level state model along with relevant outcomes from the script execution.

Finally, the upper-level state model of the flight proceeds to the next state. In some cases, the transition to the next state will depend on the results of the middle level model. For example, a missed radio frequency on approach may have required so much extra time and generated enough extra workload and confusion that the pilots decide to do a go around instead of a normal landing. In this case, the transition of the upper-level state model would be from the approach to a go around or missed approach rather than to a normal landing.

The advantage of multilevel modeling is exactly this interconnection between the models of the target system at different levels of detail. It is critical for multilevel modeling that the essential information for the lower-level models is appropriately communicated downward from the execution of the upper-level models. It is equally important after the lower-level models finished execution, they communicate all relevant results upward and that the upper-level models have paths for all results including error conditions. By combining models at different levels it is possible to go into much more detail for exactly those parts of the upper-level models that are most important and critical to modeling the operation of the target system.

Evaluating advanced models

In order to evaluate a model fully, the simulated results should be analyzed both qualitatively and quantitatively. Initially, these analyses should be focused on establishing whether or not the model is an accurate description of the target system and its processes at some level of granularity. Subsequently, the focus of the evaluation can shift to a more general assessment of the model's adequacy for explanation, prediction, and control. That is, after initial validation, the model should be evaluated in the same way as any theory. The initial evaluation should include examination of both the qualitative and quantitative modeling results.

Qualitative analysis of modeling results The qualitative analysis of modeling results should compare critical qualitative aspects of the simulated results with what is known about the target system. The critical qualitative aspects of the simulation include such features as the normal operations of the target system and the qualitatively distinct ways in which the system performs better than normal, worse than normal, or noticeably different from normal in some important respect. For performance that is better than normal, the specific conditions under which this performance is known to happen should be reflected in the simulated results. For worse than normal performance, the specific timing and modes of failure when the system does not perform normally, should be reflected in the simulated results. Early in the development of a SIS, this qualitative comparison of the "known facts" about the target system with the simulation results must rely largely on the opinion of subject matter experts as relevant comparison cases may not be available.

Later, when the SIS is more fully developed, the database information in the SIS should also be used as a source of qualitative information about the target system. Critical incidents that illustrate the good, bad, or merely ugly functioning of the target system should be qualitatively compared to the simulated model results. Qualitatively unique patterns of measured data about real cases of target system performance should be compared to simulated data to see if the same pattern will occur under the same or similar conditions.

Using real cases in this matter augments the basis of comparison beyond the memory and expertise of a single subject matter expert. The set of cases in the SIS may be broader than any single expert would experience and therefore give a greater potential scope for the qualitative comparison. The qualitative comparison of the model to the information in the SIS should be acceptable before the more detailed quantitative analysis is performed. That is, qualitative faults in the model should be identified and corrected before investing the time in conducting precise quantitative evaluation.

Quantitative analysis of modeling results   To analyze the model results quantitatively, the key or most important model parameters must first be selected.   The key parameters may include the timing, sequencing, antecedents, or consequences of the events or states in the model.   For quantitative analyses, these key parameters must be explicitly defined in terms of one or more measured variables.   Quantitatively analyzing the model may involve analyzing model results for the key variable means, standard deviations or distribution shapes, the covariation of two key variables, or the interrelationship of sets of key variables. In all cases, the results from an appropriate sample of data from the SIS should be compared to the simulation results for the same or similar system conditions. Executing the simulation multiple times to obtain a distribution of results for the variables in the model is called the Monte Carlo method.

Monte Carlo analyses of model   The focus of a Monte Carlo analysis is to obtain the distribution of key model results under particular conditions (e.g. Rubinstein, 1981).   These overall contextual conditions should match the contextual conditions from the comparison sample in the SIS.   In particular, the starting conditions should be simulated to have the same amount of variability as the starting conditions in the real systems in the SIS. Further, the nature, timing, and quality of inputs during the simulation, such as state transitions or events, should match the real sample.   Only when the contextual conditions, starting conditions, and inputs during the simulation are essentially the same as the real systems, can the simulation results be sensibly compared to real data in the SIS.

The computer simulation of the model should be executed repeatedly to obtain a distribution of results for each key variable. These results would include the means, variances, covariances, and so forth for the set of key variables. These quantitative results should be analyzed both internally as well as externally with the real SIS data to evaluate the model's accuracy. One form of internal quantitative evaluation of Monte Carlo results is the assessment of model sensitivity.

The core idea of model sensitivity is how much the simulated set of model outcomes will change for a specified change in model input.   The sensitivity of model outcomes to changing parameters is assessed by calculating the extent to which model outcomes change with small changes in key parameters. This concept is similar to the concept of measurement sensitivity covered in Chapter 3. If sensitivity is high, the model outcomes will change a great deal with small changes in the parameter. If sensitivity is low, the model will not change, or change only in a very small way, with even moderately large changes of the parameter.

In general, the simulated results should be sensitive to changes in parameters that are already known by previous research or theory to be strong, influential variables in the target system. Conversely, the simulated results should not be sensitive to parameters that are already known to be weak influences or minor players in the target system. Occasionally, the sensitivity analysis will show a surprisingly high or low sensitivity for particular parameter. The origins of unexpectedly high or low sensitivity in the simulated results should be explored and resolved in order to bring the model into better compliance with known facts about the target system.

For external evaluation, the major focus of the Monte Carlo analysis is the comparison of the simulated results to the real results in the SIS. The values of key parameters can be extracted from the simulated results and compared in the appropriate manner to the real data. This may involve the comparison of means, variability, distribution shapes, or relationships among sets of variables. For all comparisons, the degree of match or mismatch must be appropriately and carefully assessed.

One aspect of mismatch between simulated vs. real results is whether or not they are statistically significantly different. The statistical power of these tests will depend on the number of real and simulated cases. The simulated cases can be increased at low cost by re-executing the computer simulations. The comparison cases from the SIS, however, will be fixed unless resources are expended to add data to the SIS. Since that may incur high cost, and would certainly take time, the power of the statistical comparison of the model to the real data will generally be limited by the availability of appropriate SIS data.

Given a sufficiently large sample of real and simulated data, the significance tests will have adequate power. A "not significant" difference between simulated and real data gives at least some confidence that there are no large, systematic differences between the simulation and the real data. That is, non-significant differences imply that the simulation is reasonably accurate. Conversely, statistically significant differences between simulated and real results imply that there is some non-chance discrepancy or difference between the model and the real system.

However, the statistically significant difference may be a small, unimportant difference, particularly if the sample sizes are very large. Therefore, a necessary addition to the statistical evaluation of the differences is some indication of the size or strength of the differences between the simulated and real data. If the differences between the simulated and real data are both statistically significant and also quite large, then the model is inaccurate either in whole or in part. If the differences are statistically significant but quite small, the model may be slightly inaccurate but still be essentially correct. Even if the model is not perfect, using an essentially correct model to solve problems is more likely to be effective than not using any model or relying solely on expert opinion.

Cross validation of models For many state or event based simulation models, some of the information on real cases in the SIS will be used in constructing and fine tuning the model. For example, the average time required to execute certain tasks may be estimated by averaging the times for appropriate sets of real cases in the SIS. Whenever this is the case, the validation of the model should be made as independently as possible from the SIS source data. If only one sample is available, this principle implies that the validation should focus on aspects of the SIS source data that are as unrelated as possible to the parameters that have been estimated from this data. If a sufficient number or cases are available, however, then techniques such as cross validation should be used in model construction and validation.

The basic technique of cross validation can be adapted to modeling and Monte Carlo simulation. In broad outline, a large sample of appropriate SIS cases would be randomly divided into at least two samples (Figure 7.8). The entire modeling procedure would be carried out separately for each sample. That is, the distributions of key parameters as well as qualitative information on the conditions and sequencing of key events, would be used to independently construct a model for sample 1 (model 1) and a model for sample 2 (model 2). Each model would be initially validated on the qualitative and quantitative information in the sample upon which it was constructed.

The subsequent cross validation step would involve the execution of each model with the appropriate set of inputs for the other sample, followed by a comparison of the model results to the results of real systems in that sample. For example, the model for sample 1 would be executed with the inputs that would characterize the real systems in sample 2. The results of these inputs would, however, be simulated without using any further information from sample 2. The simulated results for sample 2 (using model 1) would be qualitatively and quantitatively compared to the real results for sample 2 cases in the SIS. Similarly, model 2 would be executed with the inputs that characterize the sample 1 cases, and results of this simulation compared to the real results for sample 1. The combined information from both of these comparisons would constitute the double cross validation of the models.

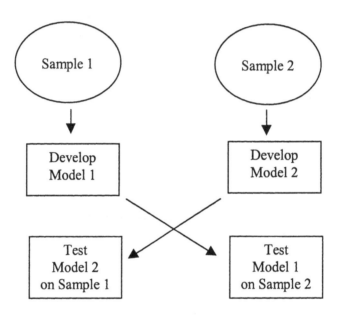

Figure 7.7 Cross validation of system models

There is, of course, nothing magical about the use of exactly two samples. Conceptually, this approach can be generalized to any number of initial samples that would be randomly derived from the set of cases in the SIS. The time required, however, for the construction of multiple models and for the cross validation of each model onto all the other subsamples would greatly increase as the number of subsamples increases. Therefore, most often the cross validation process will be carried out with a small number of sub samples such as two or three.

Modeling systematically different subsamples In cross validation, the goal and intent of the process is to find no differences; that is, to find very similar models that will give very similar results when applied to the different sub samples. When systematically different sub samples are suspected, however, the goal and intent of the validation process is to explore and confirm these differences. One example of different sub samples would be models describing the decision processes of novices vs. experts. In this case, the expectation is that the model based on one sub sample will *not* accurately characterize the other sample. That is, a model built to describe novice decision making would not describe expert decisions very well, and a model built to describe expert decision making would not describe novice decisions very well.

In fact, if appropriate theory exists on the essential differences in the two samples, a set of predictions or expectations can be specified for how each model's results should differ from the real results for each sub sample upon cross validation. That is, the expected results should clarify how a model built to fit sample 1 will be systematically incorrect when cross validated on to sample 2. Similarly, the expected mismatch in results when the model built on sample 2 is applied to sample 1 should also be specified. In this example, novice decisions might be based on superficial understanding of the problem and the use of weak decision heuristics while expert decisions might be based on extensive, organized knowledge about the problem and strong decision rules tied to domain content.

However, the process of developing a model in each sub sample and applying that model to a different sub sample, is the same as for cross validation. Since these two sub samples may be very different, the models constructed for each sub sample may be both qualitatively and quantitatively different. They may, for example, use different sets of inputs, different key parameters, and different processes for state transitions or events. To the extent that the models are very different, the basis of the evaluation may be quite different. For example, each model may specify different subsets of key outcome variables. One important principle for evaluating each model is that the model should be initially evaluated on its own terms, that is, using its own set of key outcome variables.

For example, suppose the model developed on novices uses set A of variables for input and has the key output variables of set B, while the model developed for experts uses set C for input and set D for output. To the extent that set B differs from set D, the evaluation of each model can be quite different. That is, if the decision processes of novices seem to emphasize a different set of outcomes than the decision processes of experts, the evaluation of each model can be based on either the novices' output variables or the experts' output variables. The most typical outcomes of the evaluation process in this case are shown in Table 7.10.

Table 7.10 Typical results expected for models developed on very different subsamples

	Model developed on novices (A -> B)	Model developed on experts (C -> D)
Accounts for output variables considered important by novices (set B)	Very well	Poorly
Accounts for output variables considered important by experts (set D)	Poorly	Very well

The most typical result when models are developed on very different subsamples is that the model will account very well for the results of the sub sample with which it was developed. Conversely, when this model is applied to a very distinct subsample, it will only account poorly for the results in that subsample, particularly if the subsamples are really best described by different variables and processes. In this case, the model developed for novices accounts for novice outcomes very well but does not account very well for the outcomes emphasized by experts. Conversely, the model developed for experts accounts for expert outcomes very well but does not account for outcomes emphasized by novices. If these results are found, the SIS theories must preserve the distinctions underlying the different models.

Chapter summary

This chapter covered state based and event based models. These modeling approaches have a common assumption that the essential dynamic aspects of the target system can be captured in a finite number of states or events. Usually the finite state models assume a relatively small number of possible states, whereas finite event models may include quite a large number of possible events. In either case, the dynamic change in the target system may depend on variables internal to the target system or external to it. Different levels of the target system can be modeled with different approaches and tied together in a hybrid, multi-level model. Such models can be quite powerful computational simulations of the target system and very useful for explanation, prediction, or control of the events and outcomes of the systems. The approaches do not, however, focus on system processes. The approaches covered in the next chapter emphasize the simulation of system processes and should be considered whenever the target system is known or suspected to have very important, complex processes that influence system states, events, and outcomes.

8 Dynamic Modeling based on Functions

Overview

Dynamic modeling based on functions has a different emphasis than the dynamic modeling based on states or events discussed in Chapter 7. In general, functions describe a mapping or conversion from one set of variables to another set of variables in the target system. Typically, the functions form an interconnected network that describes the dynamic processes of the target system. For each function, the mapping is from input variables to output variables. However, in the network of functions the outputs from some functions may serve as the inputs for subsequent functions. Therefore, the network as a whole may have functions representing early, middle, or late sets of events and processes in the target system. Some modeling texts explicitly cover this form of modeling (e.g. Morrison, 1991, and Fishwick, 1995), and some computer modeling languages are explicitly designed to facilitate functional modeling (e.g. Franta, 1977).

Target systems in different domains require very different types of variables. Some domains will have relatively concrete, physical variables. One example would be modeling fluid flow in the pipes of an oil refinery. The quantification in this case will concern the rate of flow of the fluids in the respective pipes in the network. Other domains will concern target systems that have more abstract or symbolic variables. The domain of economics, for example, quantifies variables such as monetary units, work hours, and so forth.

Different domains may also require different types of functions. Domains that are strongly quantifiable, for example, may make use of differential equations. Qualitative domains that follow logical rules may require logical functions. Other domains, such as human cognition, may make use of quantitative functions to describe variables such as reaction time, and qualitative functions to describe variables such as symbolic cognition.

For a domain to be appropriate for this type of modeling there should be a definable flow in the target system that helps identify the input and output variables for each function. In modeling the target system, focal emphasis can be given either to the functions themselves, to the variables that are affected by these functions, or to constraints that limit the system. The relative emphasis on functions, variables, and constraints should be determined by a careful examination of which facet is most important in describing and modeling the target system. If the functions are complex and the set of variables rather simple, the natural focal point for the modeling is the functions. Conversely, if the functions are simple linear functions but the set of variables is complex, the natural focal point for the modeling is the set of variables. If constraints are a dominant force in the system dynamics, they must be a focal point of the modeling.

Function-focused models

The functions should be the focal point of the modeling whenever they are the natural point of reference for summarizing the processes in the target system. A typical first step in modeling is to draw a graph that emphasizes the inputs and outputs of the interconnected set of functions. The elements of this graph can then be transformed into the set of equations or computer executable functions that perform the simulation. These modeling steps will be illustrated with a simple model of household finances.

Household finance simulation

An example of a functional system involving the flow of material is the flow of money in a household. For most households, money enters the system at a relatively constant rate due to salaried employment. This money tends to fill a "storage tank" called the checking account. From this storage tank, the money tends to flow into savings or family expenses. On a monthly basis, some of the money outflow is relatively constant such as mortgage payments. Other forms of money outflow, such as the monthly food bill, will vary somewhat but tend to have the same average value over time. Still other forms of money outflow, such as the monthly utility bill, will have quite different values depending on the time of the year.

The savings account may act as a buffer or resource for months with unexpectedly large expenses. A second such buffer would be a credit card which adds a resource for discretionary spending besides the salary. Thus, in months where the net outflow is greater than the net inflow from salaries, savings or the credit card can be used to make up the difference. However, there is a credit limit to the reservoir of money represented by the credit card, and a limit on the savings pool defined by its current balance.

The interconnected functions of this model should encapsulate the spending strategy of the family. First, the relative amount of money available and the amount of money needed for monthly bills may affect the use of resource pools such as savings or the credit card. Furthermore, the spending patterns of a family may alter as the supply of money changes. For example, if there is a net surplus on a given month, the family may increase spending on discretionary items such as entertainment. Conversely, if there is a net deficit on a given month, the family may decrease discretionary spending or economize in other ways.

Model graph A good model of family finances should accurately describe the flow of money through this system. This description should include the source and final destination of the money as well as the interdependent linkages of the inputs and outputs of these money transfer functions. A graphical form of this family money model is presented in Figure 8.1.

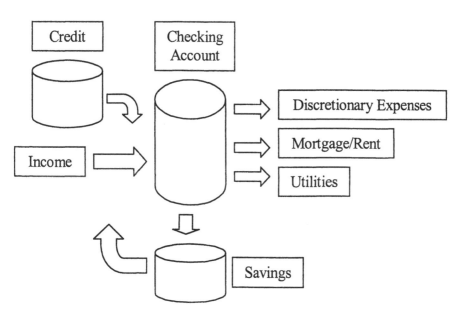

Figure 8.1 Graphical depiction of money flow during a typical month

As indicated in Figure 8.1, there are three basic "storage tanks" or pools for money in this simplified household financial model. The amount of money in the checking account pool is highly dynamic and is one major focus of the model. The amount of money in the savings pool is less dynamic, but is an equally important focal point. The amount of money available in the credit pool may be static or dynamic, depending on the family's use of credit. Families that do not systematically use credit or pay off the bill each month would have a relatively static pool of available credit. Conversely, families that accumulate credit charges would have a dynamically changing credit pool.

To some extent, this model can be seen as an extension of the finite state or event models covered earlier. Monthly or biweekly paychecks, for example, could be equally well thought of as events that constitute money inputs. The continuous flow of money through this system, however, makes a functional analysis more appropriate. The checking account, for example, could change its level on a daily basis across the month. Further, defining all possible states of this system would involve defining all possible monetary values for each of the subsystems, and clearly that is impractical. To actually simulate this system, the graphical representation above must be converted to a precise set of equations or to a set of computer functions that can be executed.

Model equations Converting this graph to a set of equations requires a more detailed specification of exactly how the inputs are transformed into outputs for each function. Setting up the set of equations for even such a simple system as this one requires careful thought of how the target system operates. Suppose, for example, that there is money left over in the checking pool after the payment of mortgage/rent and utilities for the month. In some families the surplus might be simply carried over to the next month, in other families this surplus might be saved, while in still other families the surplus might be spent on discretionary expenses. To model this, "spend it" percentages and "save it" percentages can be added to the equations of this model as in the example below. These percentages fix the percent of a surplus that will be saved and the percent of a surplus that will be spent. Clearly, the sum of these two percentages for a particular family cannot exceed 100 percent, and would be less than 100 percent to the degree that money is simply left in the checking account at the end of the month.

In this example, income per month and mortgage/rent are constants. Utility expenses are treated as a random variable; that is, an expense that will vary in unpredictable ways across the months. During the simulation, the values for utility costs for each month would be specified by randomly selecting appropriate values from a defined distribution of reasonable values. The average value and overall shape of the distribution for utility expenses in a given month could, for example, be drawn from the distribution of utility bills for this family in the given month over previous years. Alternatively this distribution could be estimated from typical utility expenses contained in a SIS focused on family finances for similar families and living conditions. One set of equations that would represent the functions of this model would be the following:

1. Income (I) = constant

2. Mortgage/Rent (MR) = constant

3. Credit Limit (CL) = constant

4. Credit available = CL - credit advanced + credit repaid

5. Utilities (U) = random variable (average value, distribution)

6. Discretionary Spending (DS) = spend % * (Checking + I – MR –U)
    + credit advanced - credit repaid

7. New Saving (NS) = save % *(Checking +I – MR –U)

8. Savings (S) = old savings +interest + new saving - savings to checking
    transfer

9. Checking (C) = old checking balance + (I + savings transfer)
    - (MR + U + DS + NS)

To this set of nine basic equations must be added a final equation that describes to what extent a family will augment the monthly income when necessary by using either a withdrawal from savings or credit. This equation can effectively summarize a family's fiscal strategy or customary pattern for this issue. One strategy might be to use savings until the savings are exhausted, and only then to use credit. To encode this strategy, a formula with logic that acts like a switch in withdrawing from savings vs. using credit should be added to the original equations. Using the abbreviations listed above, one possible formula representing this strategy is given below.

10. If (MR + U + DS) > I, then transfer necessary $ from savings to
        checking
    If (Savings < necessary $ and credit is available) use credit to pay
        residual difference,
      Else if (savings + credit < necessary $) do Bankruptcy

Different families may, of course, have different strategies for handling
their finances. These different strategies can be represented by different
equations for the family financial model. In particular, some families may
have the strategy of using credit first before savings when income does not
meet the required expenses for given month. Modeling families of this type
would require a different formula:

10. If (MR + U + DS) > I, then transfer necessary $ from credit to checking
    If (available credit < necessary $ and savings are available) use savings
        to pay residual differences
      Else if (credit + savings < necessary $) do Bankruptcy

If a population of families were being simulated, appropriate variations
in these strategies as well as in the appropriate values for the input
parameters for the model would have to be included. For example, the
simulation would have to include the correct proportions of families with
different amounts of credit available and some families with no credit cards
at all. Similarly, the starting distribution for savings for the simulated
families would have different amounts of savings available and include
some families with no savings at all. Distributions for different amounts of
family monthly income and necessary expenditures for mortgage/rent and
utilities could be obtained from a government statistical almanac or from a
SIS focused on family finances.

Depending on the goals for the simulation, the simulation program
could be executed on a monthly, weekly, or daily basis. If the goal of the
simulation were to understand the long-term accumulation of either savings
or debt, the natural interval for the simulation would be months and years. In
this case, all the relevant money variables would be summarized on a net
monthly basis. For validation, the results for this relatively coarse time grain
simulation should be appropriately compared to relevant financial data in
the SIS summarized on a monthly and yearly basis.

Other modeling goals might require a finer grain for the simulation. For
example, if the goal of the simulation were to model day-to-day cash flow
and the occurrence of things like bounced checks or savings withdrawals
during the month, the natural interval for the simulation would be days and

weeks. This fine grain for time may require other modifications to the simulation to preserve accuracy. For example, a very detailed day-to-day modeling of things like bounced checks might require building into the model the appropriate time lags for deposits being credited to the checking account and time for checks to clear.

Model evaluation Just like the state and event based simulation discussed in Chapter 7, both the qualitative and quantitative outcomes of the simulation should be compared to data in the SIS. Qualitatively, this simulation should describe families that get into trouble by over extension of credit, families that ultimately declare bankruptcy, families that successfully amass savings over time, and families that simply live from month-to-month without either accumulating savings or over extending their credit. Quantitatively, this simulation should describe the time course of the flow of money for each type of family accurately. Further, the distribution of the key parameters of the model such as the amount of credit, the amount of savings, and the amount in a checking account, should match the distribution of the same parameters in the SIS data.

Aircraft control simulation

An aviation example of functional modeling might be the set of functions that describes how a pilot physically controls an aircraft during an approach and landing. Since an aircraft is controlled in a three-dimensional space, the nature of this control process is much more complex than decelerating and stopping an automobile. Further, there is a clear and inherent trade-off between speed and altitude that does not occur for automobiles. For example, the pilot dives the aircraft to lose altitude, the air speed will inherently increase and may become unacceptably high. To keep this description manageable, we will focus solely on the aspect of aircraft control that involves air speed and rate of descent while ignoring the other facets of aircraft control such as course and distance which further complicate this task in reality.

The time scale for this simulation would be quite distinct from the time scale for the family finance simulation discussed previously. In the case of aircraft control during descent, the processes are occurring across the span from milliseconds to seconds and minutes. This changing time span does not change the fundamental nature of the modeling, but it does change the required time span of the data that is collected in the SIS which would be used to validate the model.

Model graph A simple model for a pilot controlling an aircraft on descent is presented in Figure 8.2. Once again, the flow of inputs and outputs through the functions of the system is clear. The process begins with a function in which the pilot compares the desired air speed and rate of descent with the indicated air speed and rate of descent. Based on this comparison, the pilot adjusts throttle and elevator settings. These actions are input to the hydraulic system for the elevator and the engine fuel control system, which actually change the elevator setting or engine power setting, respectively.

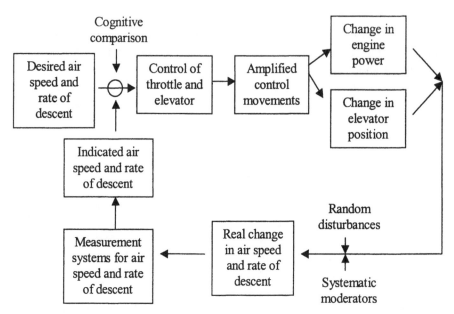

Figure 8.2 Functional model for pilot controlling an aircraft during descent

The effect of these changed settings on the real air speed and rate of descent of aircraft is moderated by both systematic factors and random disturbances. The systematic moderators consist of such factors as the aircraft's gross weight, the altitude, and the air density. The systematic moderators will have large, stable, and predictable effects on how much the changed settings actually change the status of the aircraft.

In contrast, random disturbances have more minor, transient, and unpredictable effects on how much the changed settings change the aircraft status. Examples of these random disturbances would be the effects of turbulence or small updrafts and downdrafts on the pitch and air speed of aircraft at a particular moment. Although these momentary disturbances may be small and not strongly affect the actual flight instrument readings, in extreme cases such as wind shear the momentary disturbances will be severe enough to strongly affect the status of aircraft.

Finally, the real changes in the air speed and rate of descent of the aircraft are measured by sensors, and this information is communicated to the pilot. Different physical and electronic systems may measure this information. These measurements vary in the sensitivity, accuracy and time lag of displaying the information to the pilot. Part of the pilot's mental comparison process at the beginning of the cycle may include adjusting for the known qualities of each measurement system.

To fully specify this model, the exact way that the inputs to each of these functions are transformed into the outputs used by the next functions in the process would have to be specified. In particular, the way that the moderated inputs affect critical functions would have to be clarified. For example, exactly how the systematic moderators would affect the translation of control inputs to real changes in aircraft status would have to be specified. This part of the modeling is largely in the physical domain.

More complex and varied relationships might be found in the part of the model that involves human cognition and crew teamwork. For example, the function of the pilot flying comparing the desired air speed and rate of descent to the indicated air speed and rate of descent also has to be accurately specified. Similarly, in a typical approach the pilot monitoring calls out critical airspeeds and altitude. These calls would be additional sources of relevant input to the cognition of the controlling pilot. Modeling the cognitive function accurately could be critically important for the entire model to work well. One aspect of modeling this component is the fact that different pilots may attempt to control the aircraft in systematically different ways.

For example, if given a choice some pilots may prefer to bring aircraft in rather low and slow while making glide path adjustments primarily with the throttle. In contrast, other pilots may prefer to bring the aircraft in higher and gradually trade the altitude for air speed while using low thrust settings. These different mental prototypes for optimal approach could have a strong effect on the desired air speed and rate of descent. It is easier to focus on the pilot aspects of the control to develop appropriate equations if we simplify the model.

A simplified model that focuses only on the pilot control aspects of this system is illustrated in Figure 8.3. This model tacitly assumes that many of the functions in the preceding model are direct 1:1 translations. Also, the random disturbances and systematic moderators of the previous model are assumed not to make a critical difference in this model. These simplifications allow us to focus on the direct effect of pilot inputs in the control process.

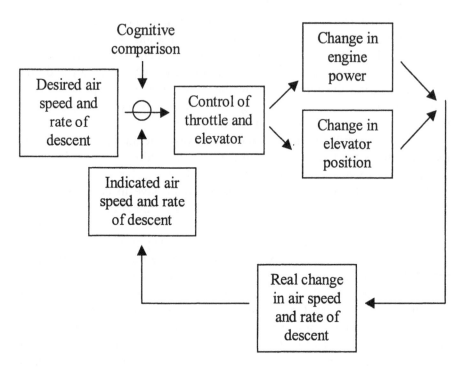

Figure 8.3 Simplified model for pilot controlling an aircraft during descent

Model equations This graphical representation must be converted to equations for the simulation. The equations should correctly represent the basic dynamic nature of the system. For this system, one important feature of the dynamic nature of the system is the interdependence of rate of descent and air speed. That is, increasing pitch to decrease the rate of descent tends to reduce air speed, and decreasing pitch to increase the rate of descent tends to increase air speed. Some basic equations that would illustrate this interdependence are the following:

1. Indicated air speed = real air speed + measurement error

2. Indicated rate of descent = real rate of descent + measurement error

3. Real air speed = old air speed + change in air speed

4. Real rate of descent = old rate of descent + change in rate of descent

5. Change in air speed = (delay X) (change due to thrust changes + change due to elevator changes)

6. Change in rate of descent = (delay Y) (change due to elevator changes + change due to thrust changes)

In converting the graphical form of the model into a set of equations, one critical issue will be the relative emphasis that the pilot places on maintaining a desired air speed or a desired rate of descent. Simultaneous control of two distinct aspects of a dynamic system can be quite difficult, and early in the training, pilots may tend to concentrate on one aspect at the expense of the other. Some pilots during training, for example, may target the desired air speed and control that quite well using the thrust while letting the rate of descent vary. Conversely, other pilot trainees may target the rate of descent and control that using the elevators quite well while letting the air speed vary. These differences would be illustrated by the following two strategies:

7. If (desired air speed not = indicated air speed) then adjust thrust,
   Else if (desired rate of descent not = indicated rate of descent) then adjust elevator.

OR

7. If (desired rate of descent not = indicated rate of descent) then adjust elevator,
   Else if (desired air speed not = indicated air speed) then adjust thrust.

At the most detailed level, the precise control movements made by the pilot in response to a discrepancy between a desired and indicated air speed or rate of descent must also be modeled by a function. That is, exactly how the pilot makes the control movements is also important to the dynamic behavior to of the system. Suppose, for example, a pilot in training tries to separately control the air speed using the thrust and the rate of descent using the elevators. The simplest method of thrust control is the decision of "too slow-increase thrust" or "too fast-decrease thrust". Similarly, the simplest method of controlling the rate of descent is to increase the elevator setting (pitch) if the descent is too fast and to decrease the elevator setting if the descent is too slow. This is illustrated in equations 8 and 9 below.

8. If (desired air speed < indicated air speed) then increase thrust setting,
   Else if (desired air speed > indicated air speed) then decrease thrust
      setting.

9. If (desired rate of descent < indicated rate of descent) then increase pitch,
   Else if (desired rate of descent > indicated rate of descent) then
      decrease pitch.

Given the interrelationship of changes in air speed and rate of descent discussed above as well as the inherent time lags in the response to the control movements, this method of control will almost always lead to over control. That is, the aircraft's air speed will oscillate from too slow to too fast, and the aircraft's rate of descent will oscillate from too high to too low. Many beginning pilots suffer from this form of "pilot induced oscillations".

In a physical system such as a house heating and cooling system, this type of control would produce continuous temperature oscillations. A simple thermostat with this type of control would cause sudden and repeated shifts from heating to cooling and vice versa. Clearly, these control strategies are rough and inefficient and should be changed.

The key to avoiding oscillations in either system is to more gradually adjust the controls. To avoid pilot induced oscillations, the pilot should take into account the size of the discrepancy between the desired air speed and descent values, the current rate of change in both these parameters, and the cross-link between thrust and pitch changes. That is, decisions to change a control input should first take into account the current rate of change in the parameter and then use adjustments that are proportional to the required change. As the difference between the desired state and the current state of the aircraft decreases, the control input should become correspondingly smaller. In essence, to avoid oscillations pilots must change what they are focusing on and how they are transforming the information inputs into control outputs.

This change.in the pilots' decision-making can be illustrated by revised equations 8 and 9. Let DAS designate the difference between the desired air speed and indicated air speed, and DVS designate the difference between the desired rate of descent and indicated rate of descent. Examples of the revised formulas 8 and 9 are:

8. If (DAS > [rate of positive air speed change * time]) then *increase thrust* proportional to the difference
    Else if (DAS < [rate of negative air speed change * time]) then *decrease thrust* proportional to DAS.

9. If (DVS > [rate of positive altitude change * time]) then *increase pitch* proportional to the difference
    Else If (DRC < [rate of negative altitude change * time]) then *decrease pitch* proportional to the difference.

These revised formulas would simulate a smoother set of adjustments to both thrust and elevator settings during descent. This set of adjustments should better converge to the desired values and produce a more stabilized descent. To be fully accurate, however, these simulation equations should also take into account the pilot's intuitive judgments of how much a change in the pitch angle will affect the air speed and vice versa. That is, the cross linkages between pitch and thrust adjustments would also have to be included in the equations.

Model evaluation As with the previous simulations, the results of this simulation of pilot control of an aircraft would be compared to appropriate data in the SIS. Qualitatively, the simulation should be able to reproduce the known important problems that pilots have in this regime of flight such as under-shooting or over-shooting control goals, or pilot induced oscillations. Quantitatively, the evaluation of this simulation will be stronger the more completely it can be compared with measured data. In this case, more complete data could be obtained by a more detailed and dense time record for the precise status of the aircraft or simulator.

Detailed and dense time records for aircraft status can, for example, be obtained from the simulator during flight training. For critical aircraft parameters and control inputs, the data can be measured at a frequency of once per second or more. That type of extensive, dense database gives a sufficiently rich set of information to validate or invalidate the results of this type of functional simulation.

The actual validation should at least compare the distributions of values for critical variables such as air speed and descent rate at all critical points in the descent profile. This would involve comparing mean values and variances of variables like air speed and rate of descent for critical points such as 1000 feet above ground level, 500 feet above ground level, flare, and touchdown. To be completely validated, the distribution of simulation results should not significantly differ from the distribution of observed data from simulator flights recorded in the SIS for any critical variable at any point in the descent.

Further validation analyses could compare the sequential dependencies of variables in the simulation with the corresponding measured variables in the SIS. That is, would the simulation results for being too high and too fast at 1000 feet on the remainder of the descent be comparable to the database results for training flights in which the crews were also too high and too fast at 1000 feet? Finding congruent sets of consequences from similar antecedent conditions supports the validity of the model.

Once validated, the results of this type of simulation should guide the selection and analysis of other relevant variables in the SIS databases. Using this example, the analysis of the approach profiles of crews in training should be guided by the simulation. For example, an appropriate cutting point for determining when a measured control input oscillations exceeds the normal range and becomes pilot induced oscillations could be determined from the simulation. This cutting point could then be applied to the actual training data to give better feedback to the pilots concerning the occurrence of pilot induced oscillations.

A second example would be determining objective criteria for defining a destabilized approach. The simulation results could be examined to see how much variability in air speed, rate of descent and deviation from the descent path are "normal". Again, cutting points set around the normal limits could be used to classify the results of training flights and give more effective feedback about destabilized approaches to pilots. This simulation-guided analysis of approach data could also be correlated with other information in the SIS to extend knowledge about the causes and consequences of good vs. poor approaches.

This extension of the SIS data can utilize both qualitative and quantitative analysis. For example, the occurrence of qualitative problems such as pilot induced oscillations should be correlated with relatively inexperienced pilots or with pilots transitioning to a new aircraft with more sensitive controls. Similarly, such quantitative results of the simulation as the total number and size of control inputs required during the descent or variation around the desired flight path, should also be related to aspects of pilot experience, training, descent control strategies, or other relevant variables in the SIS. Thus, models focused on dynamic functions can significantly add to the theoretical base of the SIS. In some situations, however, the focus of the modeling should be on the variables rather than the functions.

VARIABLE-FOCUSED MODELS

Variables should be the basis of functional modeling whenever they are the natural focal point for summarizing the processes of the target system. This is a matter of relative emphasis--the same simulation model can potentially be done either by emphasizing functions or by emphasizing variables. Nevertheless, some systems will be more gracefully modeled by systems of equations that emphasize the variables rather than input - output functions.

Linear systems

One way that this can occur is if the system has a set of functions that are simple linear functions. In such a case, the emphasis may be on the variables with only a basic statement of the linear dependencies and parameters in the system. Many statistical methods such as factor analysis and structural equation modeling assume linear relationships and focus on the variables. Although these methods are typically used to capture system structure (Chapter 6), they can also be used as dynamic models of simple linear processes.

In this type of use, the linear relationships are assumed to be an adequate approximation of system processes and the focus is on the variables. One example of a domain in which true linear relationships may occur is chemistry. The conversion of one chemical to another by a catalyst or reaction is often proportional to the amount of each reagent present. If all steps in the reaction occur in this manner, the entire system may be modeled as a set of simple linear equations. Alternatively, the linear system may be used as a first approximation to processes that are actually non-linear. to the extent this is a good approximation, the linear system may suffice. To the extent that the target system has strong non-linear dynamics, a shift to a non-linear model may be required.

Non-linear systems

Complex real-world systems will often require complex sets of non-linear equations. In many natural systems, positive and negative feedback loops are a critical part of the system and cause strong non-linear effects. A positive feedback loop consists of a downstream variable having a positive or augmenting affect on variables earlier in the causal process. A negative feedback loop, conversely, consists of the downstream variable having a dampening or negative effect on variables earlier in the causal process. These feedback loops should be explicitly represented in the initial graph of the system in order to construct an accurate model.

Certain qualitative aspects of the behavior of a dynamic system can be inferred already with this initial graph. If, for example, a system has all or a preponderance of positive feedback loops, it may tend to "runaway" into extreme positive conditions over time. Conversely, if the system has a preponderance of negative feedback loops, the system's behavior may tend to dampen over time and converge on a stable state.

Therefore, it is very important that the graphical representation of this type of dynamic system include the depiction of all positive and negative feedback loops in the system. This causal graph is then converted into a set of equations for the simulation. This set of equations should faithfully capture the interdependencies among variables, any positive or negative feedback loops, and any rates of growth or decay in key system parameters due to the passage of time.

The set of equations may either be differential equations or difference equations depending on how time is treated. If time is assumed to occur in a specific and indivisible unit, such as years, then time is incremented as an integer during the simulation and the equations are called difference equations. Alternatively, if time is assumed to be a continuum that is infinitely divisible, then the simulation may occur with very fine time slices and the equations are differential equations.

College simulation

An example where the natural time increment is years is the college system in which students are recruited and graduated on a yearly basis. All administration and student records are based on the academic year as a unit. Therefore, the appropriate focus for a dynamic model is the number of students enrolled in the college each academic year and changes in this number. Since time is incremented as a whole number, the equations for this model will be difference equations.

Model graph The graphical model for this system is presented in Figure 8.4. In this model, a pool of available students is given offers for entrance into a particular university. A subset of these students enrolls in the university and occupies seats until they graduate or drop out.

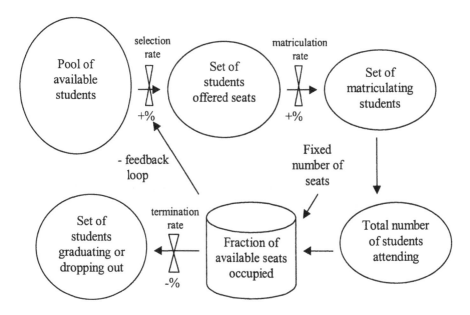

Figure 8.4 Dynamic model for number of students at a college or university

Model equations The equations that correspond to this figure concern either the rates of student selection, matriculation, and termination and the levels of students and the fraction of seats occupied. In converting this graph to equations, the critical point is that there is a feedback loop between the fraction of seats that are occupied and the selection criteria used by a college. College administrators attempt to fill all the available seats, but also attempt to do so with the best qualified students. If few seats are available, the selection criteria are raised in order to fill the seats with the best possible students. If many seats are available, the selection criteria are lowered in order to be sure of having enough students to fill the seats. A set of equations that might represent this dynamic system is the following, where the letter "f" represents an appropriate function that combines the information on the right side of the equal sign.

1. Net change of students each year = f (matriculation rate, graduation rate)

2. Matriculation number = f(population size, selection rate, acceptance rate)

3. Termination number = f (number of students, average length of degree)

4. Selection rate = f (population size, acceptance rate, seats available)

5. Number of students = f (old number of students, net change of students)

6. Fraction of seats occupied = f (number of students, seats available)

For each of these equations, the generic function indicated by the letter f could be any function that satisfactorily combines the information on the right to yield the number for the variable on the left of the equal sign. In order to simulate this system, specific functions have to be specified for each generic function. Initially, these functions should be kept as simple as possible while still describing the processes in the target system. For example, simple linear functions can be tried as a basic starting point and later modified to non-linear functions where necessary. An example of equations 1-6 with simple linear functions inserted is given below:

1. Net change of students
    each year   = matriculation number - termination number

2. Matriculation number = population size * selection rate * acceptance rate

3. Termination number = $\dfrac{\text{number of students}}{\text{average length of degree}}$

4. Selection rate = $\dfrac{(\text{number of seats} - \text{number of students})}{\text{population size} * \text{acceptance rate}}$

5. Students = old number of students + net change of students each year

6. Fraction of seats occupied = $\dfrac{\text{number of students}}{\text{number of seats}}$

With these functions specified, the simulation of this system would proceed on a yearly basis. For each simulated year, a set of students from the simulated pool of available students would be given offers depending on the selection rate. A subset of the students given offers would accept these offers and matriculate as students depending on the acceptance rate. These students would add to the student body at the college and take up available seats. During the year, some fraction of the students attending the college would either graduate or dropout, making more seats available for the next incoming class.

Model evaluation Qualitatively, this simulation should account for the gross behavior of the real system. The pool of available students, for example, will change each year depending on the demographics in the population. A "baby boom" would have the predictable effect of tightening the admission standards for the college. Conversely, a "baby bust" should reduce the available pool of students and ultimately lead to a lowering of admission standards.

Quantitatively, the accuracy of the numerical results of this simulation should be compared to real data from an appropriate set of colleges. The accuracy of this simulation would be enhanced by including changes in the year-to-year values of relevant rates in the model such as the acceptance rate and graduation rate. The acceptance rate, for example, may go down in the years of a baby bust due to the fact that students can choose among more universities and colleges. The average length of time required for the degree may change due to changes in economic conditions or changes in the mix of college majors selected by the students. Similarly, other critical parameters in the flow of students through the college will be changing over the years, and capturing the changes in these parameters will increase the accuracy of the results.

To increase the quantitative accuracy of the simulated results, changes in the equations must also be considered. The simple linear versions given above, for example, may have to be modified to have non-linear components. One form of nonlinearity would be potential interactions among the coefficients. In this example, loosening of the admission standards in order to fill seats may result in selecting a less qualified cohort of students who will take longer to graduate and potentially have a higher dropout rate. That is, a higher selection rate may be tied to a lower graduation rate and higher dropout rate 4-5 years later.

A second modification might be to define different pools of students that might have distinct values for the model parameters. If, for example, a college is attempting to recruit a certain number of women or particular minorities, these populations should be separately modeled. Modeling separate pools of students allows for different selection, matriculation, and graduation rates to be applied to each population. If these rates are noticeably different, capturing the differences by using multiple pools of students and different parameter values for each type of student will increase the overall accuracy of the simulation.

Obviously, making these changes requires careful consideration of both the students and how the target system operates. Further, there is also a point of diminishing returns beyond which adding further complexity to the model yields only minor increases in the accuracy of the simulated results. Nevertheless, when costs of inaccuracies are high an extensive effort to increase the accuracy of the model may be justified. An example is the potential legal costs of a situation that has adverse impact of the selection rules on minority students. In such a case, alternative selection rules should be one focus of the modeling.

Passenger service simulation

An example from the aviation domain of a functional model focused on variables is serving passengers during a flight with beverage and meal service. For this model, the processes are simple serving processes and the focus is on the number and rate of passengers served. The flight attendants must make at least one pass through the passenger cabin to serve all the passengers beverages, a second pass to serve meals, and a third pass to clean up. Essentially, the passengers are a queue that must be served by the flight attendants at a certain rate. Further, a certain percentage of the passengers will need to use the restrooms and may have to queue for that facility. Therefore, this model will be a good example of functional models that include queues.

In constructing a functional model, it is important to remember that the model is a simplification of reality that still preserves the most important aspects of the target system. Unimportant aspects of the target system, however, may be ignored with little impact on the simulated results. Therefore, the model will often diverge from the real system in certain respects.

For this model, the real physical system is the cabin of an aircraft. Passengers are sitting in their seats waiting to be served food and beverages, and flight attendants move up and down the aisles performing these services. For this simulation, however, the passengers will be conceptualized as a queue of entities waiting to be serviced, and flight attendants will be conceptualized as fixed servers. The passengers will "flow" from one server to the next to obtain food and beverages. Although the conceptual model presented in Figure 8.5 does not directly match the physical system, it makes the queues for the required services more explicit.

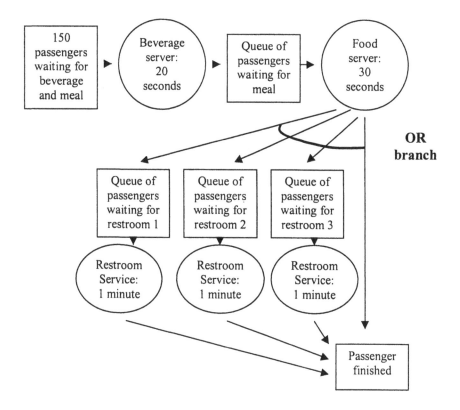

Figure 8.5 Passengers queuing for services during a flight

Model graph As Figure 8.5 depicts, the passengers on the flight essentially form a queue for beverage service and for food service. Typically, the beverages are served first and the food or meals are served second. For this simulation, the average time for serving a beverage is set at 20 seconds while the average time for serving a meal is set at 30 seconds. Since the beverages are being served more quickly, a queue of passengers who have been served beverages but are waiting for their meal will accumulate in the cabin.

After being served food, there will also be a delay for eating the meal that is not represented in this graph. After finishing the meal, however, there is one final choice point for the passengers. That choice is whether or not to use a restroom. In this model, an aircraft with three restrooms is depicted. For the OR branch representing the passenger decisions at this point, the relative probability of a passenger trying to use restroom 1, restroom 2, restroom 3, or no restroom at all would have to be specified. The simplest way of modeling this decision is to specify that the passenger is equally likely to randomly pick among restrooms 1, 2, or 3 if the passenger decides to use a restroom at all.

Model equations The equations for this model are simple but must include the concept of a queue for each required service. Different types of queues act in a different manner. A first-in-first-out or FIFO queue operates such that the first person into the queue is the first person served. Examples of FIFO queues are the waiting lines in banks or fast food restaurants. A last-in-first-out or LIFO queue operates such that the last person to enter a queue would be the first person to be serviced. An example of this is the way luggage is handled on a commercial flight. Passengers arriving first and checking their luggage early have it stored deeper inside the airplane with the result that their luggage is the last retrieved at the end of the flight. Conversely, passengers embarking at the last minute have their luggage stored on top where it is retrieved first. Sample equations for cabin service are below.

1. If (beverage queue not empty and beverage server available)
      then serve beverage with time = 20 seconds
      remove passenger from beverage queue

2. If (meal queue not empty and meal server available)
      then serve meal with time = 30 seconds
      remove passenger from meal queue

3. If (restroom queue not empty and restroom available)
      then passenger uses restroom with time = 1 minute
      remove passenger from restroom queue

   Although very simple, this simulation model' could give several very important results. The "throughput rate" is the number of passengers served per unit time. The "mean service time for passenger" is the average amount of time the passenger spends being served as opposed to waiting in a queue or in a delay status such as eating. This service time can, of course, be subdivided into service time for beverage, service time for meal, and service time for restrooms. Finally, "server utilization" is the percentage of time the server is busy serving passengers as opposed to idle. Server utilization is particularly important for the flight attendants who are serving the beverages and food. In a smoothly functioning crew, the tasks of the flight attendants will be arranged in such manner that their time is efficiently utilized.

Model evaluation This model would, of course, be evaluated by comparing the simulated results to real flight crews in both a qualitative and quantitative matter. A critical qualitative result, for example, would be whether the food and beverage service is completed by the scheduled end of the flight, or at least in time for the passengers to complete eating their meals. Similarly, having long queues of passengers at the restrooms when the seat belt light for the final descent is turned on would also be a qualitative result that is undesirable in terms of passenger comfort.

Quantitatively, the average values for the throughput rate, mean service time for passenger, and server utilization over a set of simulations would be critical values. The validation should match the mean and overall distribution of these critical values for the set of simulations to the observed distributions of real values recorded for actual cabin crews in the SIS. This simple model could be adjusted in many different ways for a better qualitative and quantitative match with real data.

First, the relevant parameters and input distributions for the model can be adjusted to better reflect reality. For example, it is very unlikely that the service time for beverages takes precisely 20 seconds each time, or that the service time for meals takes precisely 30 seconds each time. The simulation can be improved by substituting the distribution of times that typifies real flight attendants on actual flights. When running the simulation, these distributions of service times are randomly sampled for each simulated passenger. Doing this will add the real variation of the time required for the flight attendant's work into the simulation, and should make the simulation results more accurate.

Second, the basic structure of the model can be altered to see if it better conforms to the observed results. For example, some attendant crews may attempt to serve the meals first and then the beverages. Modeling these attendant crews would require shifting the food service process to occur before the beverage service process. Simply by looking at Figure 8.5, we can see that if this change were made in the model it could have important qualitative and quantitative effects. By putting the food service ahead of the beverage service, the size of the queue for beverage service will decrease to zero because foodservice takes longer than beverage service. In this case, the flight attendants serving beverages can always keep up with the flight attendants serving the food.

While this might be thought of as a good effect of such a change in the serving process, other side effects of the change are not good. Because the foodservice is slower than beverage service, the flight attendants serving beverages will necessarily spend some of the time idle. Further, the throughput of how many passengers can be served in a given amount of time is lower. Therefore, the quantitative results of reversing the order of food and beverage service could be 1) the average size of the queue of passengers who have neither drink nor food increases, 2) the idle time for the flight attendants serving beverages increases, and 3) the number of passengers served per unit time decreases.

Certain aspects of the qualitative results may also change with this reversal of the serving process. For example, the decreased throughput increases the total time required to serve the entire set of passengers. For shorter flights, this may result in some passengers not being served or not being served early enough to finish their meals by the end of the flight. Similarly, the longer time to service the entire set of passengers would push the set of queues for the restrooms to occur later in the flight, and this may result in some passengers still being in restroom queues when the seat belt light is turned on.

Finally, some aspects of the model could be elaborated for better simulation results. This may involve distinguishing among the passenger entities and the services. For example, suppose the airline has a first-class and a coach class, each with a distinct set of restrooms. The first-class passengers can probably use either the first-class restroom or the coach restrooms, but the coach passengers cannot use the first-class restroom. Distinguishing the two types of passenger in the simulation could lead to a more accurate portrayal of how the queues build up for the respective types of restrooms.

Similarly, the passenger's decision-making about whether or not to queue for restroom could be included in an elaborated model. For example, the model could specify that a passenger would typically choose to queue for the restroom with the shortest line. In this way the model can be made a more accurate depiction of events occurring during a real flight.

Even simple models may, however, have basic contextual constraints. The passenger service model is limited by the number of beverages and meals aboard the aircraft as well as by the total time of the flight. For the above simulation, these constraints are boundary conditions that limit the simulation but are not the focal point. For some functional models, these constraints become the focal point of the modeling.

Constraint-focused models

Some forms of functional modeling place an emphasis on the constraints on the dynamic behavior of the target system. If the dynamic behavior of the target system is strongly constrained, it is critically important for the equations to accurately express these constraints. If the changes in the target system occur in discrete shifts or jumps, the constraints may be expressed as a set of difference equations using the appropriate time interval (milliseconds, seconds, minutes, hours, days, months, years, etc.). If, on the other hand, the target system changes in smooth or gradual ways across time intervals, the constraints may be expressed as a set of differential equations across time.

Any domain that has a tightly interlocked set of variables may be appropriate for constraint- based modeling. Mechanical or physical systems emphasizing force, velocity, momentum, distance, power, and energy might be suitable for constraint modeling because all of the relevant parameters are tightly linked by appropriate equations. Similarly, electrical systems emphasizing voltage, current, electrical flux, charge, and the magnetic or electrical effects of a circuit would also be suitable for constraint modeling. Finally, biological systems with tightly interlinked entities and overall limits on the entire ecosystem may also be suitable for constraint modeling.

In modeling the processes of complex human-physical systems, the applicability of constraint based modeling may not be immediately apparent. However, the above domains should be kept in mind as the underlying dynamics of the target system are explored. If aspects of the target system begin to resemble components of a mechanical/physical, electrical, or biological system, then constructing and modeling dynamic equations based on those analogous systems should be explored.

Predator-prey model

One type of biological system with strongly linked variables and overall constraints is the interaction of predators and prey within a closed ecological system. Consider, for example, the interaction of wolves and moose on Isle Royale in Lake Superior. Although large, the island is limited in size and the populations of wolves and moose are isolated from mainland populations unless Lake Superior freezes over during a particularly harsh winter.

Model graph The populations of moose and wolves wax and wane in an interlinked fashion over the years (Figure 8.6). When wolves are very prevalent, the population of moose declines. When the population of moose declines, the population of wolves that depend on the moose must sooner or later decline.   Subsequently, a decrease in the predation caused by the wolves allows the population of moose to recover. The purpose of the dynamic modeling is to describe and gain some understanding of how the population shifts occur.

A natural time unit for measuring the changes in sizes of populations in a temperate climate is the year since that time interval includes one breeding cycle for both species.   Therefore, this model may be expressed in difference equations that summarize the changes in moose and wolf populations over the span of several years. The equations should summarize the core factors influencing the births and deaths of each species, and the interaction between wolves and moose.

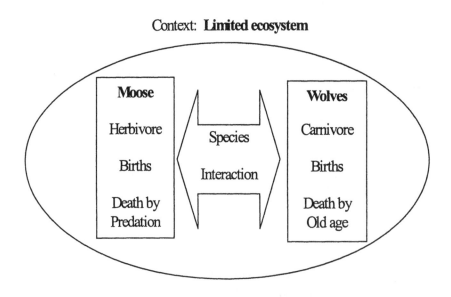

Context: **Limited ecosystem**

Figure 8.6 Conceptual graph for a dynamic predator-prey model

Model equations   In particular, the model equations should reflect the constraining effect that the level of the moose population has on wolves, and that the level of the wolf population has on moose. The interaction between the two species depends on the relative number of each species that are present, but interaction has a distinct effect on each species. The more interactions between the two species, the more moose will be killed. Moose being killed, however, allow the wolves to have larger litters each year and thereby increase their numbers. For simplicity, we will assume that old, feeble moose are the first victims of the wolves. Moose, therefore, do not typically die of old age, although wolves usually die of old age. A verbal form of the equations that might model the dynamic changes in wolf and moose populations over years is the following:

1. Change in number of moose = moose births – Interaction (moose killed)

2. Change in number of wolves = Interaction (wolf births) - wolves dying

3. Moose births = Moose fertility (births per female) * number of female moose

4. Wolf deaths = old age constant * number of wolves

5. Interaction = constant * number of moose-wolf contacts

These verbal equations must be mathematically specified to perform the simulation. The first two equations above emphasize the interdependent nature of the moose and wolf populations. The potential number of interactions between the moose and wolf populations depends on the numbers of each species. The more interactions there are between the moose and wolves species, the more moose will be killed, and this food supply helps wolves reproduce. Assuming female moose are one-half the moose population, the difference equations for the simulation become:

1. Change number of moose = moose born – Interaction

2. Change number of wolves = Interaction – wolves dying

3. Moose born = Fertility constant * (1/2 * number of moose)

4. Wolves dying = Old age constant * number of wolves

5. Interaction = predation constant * number of moose * number of wolves

Condensing all five equations into the first two equations summarizes this simple predator prey system in a very concise manner. Using M for the number of moose and W for the number of wolves, the simplified and condensed formulas look like:

1. Change M = Fertility constant * (1/2 M) – (predation constant * M * W)

2. Change W = (predation constant * M * W) – (Old age constant * W)

One important result of a constraint system such as this one is that the relative levels of animals in each population can oscillate over the years. Suppose we start with a lot of moose and only a few wolves. The wolves should kill enough moose to increase their numbers gradually over years. At some point, however, the increased predation due to the larger number of wolves will overcome the natural fertility of the moose. At that point the moose population starts to decline. Inevitably, the decline of the moose population will have a subsequent effect of decreasing predation and the associated new births in the wolf population. With the constant percent of wolves dying from old age, the decline in new wolf births will ultimately reduce the wolf population. With a reduced wolf population, the fertility of the moose population can now overcome the losses due to predation and the moose population will increase.

The two difference equations above can capture the oscillatory dynamics of this predator-prey system. In the extreme, this type of oscillatory behavior can lead to catastrophic outcomes such as population crashes. In this simple example, one extreme is that wolves die out, leaving moose to reproduce unchecked. The other extreme is that moose die out, leading to a subsequent extinction of wolves.

Notice that the role of causality becomes blurred in this type of interdependent and strongly constrained system. For example, it is clear that an increasing number of wolves is the proximal cause of a decline in the number of moose. However, the increasing number of wolves was in turn caused by a previous increase in the number of moose, which constitutes a more distal cause. Therefore, assessing causality in mutually constraint systems quickly becomes a problem. In such a case, the mutual dynamic causality is best described by the system of equations and the mutually interdependent nature of the simulation results for the key variables. Thus, although a simple, direct causality may be easy to assign with some dynamic systems such as those with clear, unidirectional flow from initial inputs to final outputs, in other dynamic systems the question of causality becomes inherently more complex. For those systems the causal picture must be conceptualized, described, and modeled in a more sophisticated manner.

Model evaluation Regardless of the change in the view of causality, the results from this type of complex dynamic model would be both qualitatively and quantitatively compared to relevant information in the SIS just as for simpler models. Given its extreme simplicity, this version of the model may not predict every nuance of the changes in the populations. Depending on the purpose of the modeling, however, even a rough prediction of population changes may be sufficient. It may be sufficient, for example, to know that rather extreme oscillations in the populations of predators and prey can naturally occur. If more accuracy is required, these equations can easily be modified or augmented by other factors to increase the model's accuracy.

Modifying the current content of the model would have to focus on either adjusting the constants or fine-tuning the births and death expressions. The old age constant, for example, may be fine-tuned by keeping track of the entire distribution of ages of the simulated wolves in the population. Only the proportion of these simulated wolves that exceeded a specified old age limit would die of old age. Similarly, predation could differentially effect male and female moose. If so, the simple fraction of 1/2 for moose females may be inaccurate and the gender composition of the moose may have to be more precisely linked to predation and tracked during simulation.

Adding other factors to the increase/decrease equations of the model is another way to increase the model's accuracy. For example, as the population of moose declines due to over predation, some subset of the wolves may die due to starvation rather than old age. The number of wolves dying due to starvation could be calculated from a relatively small size moose population compared to a large size wolf population. Adding this component to the model may enable a more accurate prediction of the decline in the wolf population after the decline in the moose population. Most real dynamic systems are at least in part constrained by some limiting aspect of the context or system. Even in aviation, some aspects of the flight task have the qualities of a constrained dynamic system.

Airplane climb model

Model graph In the aviation domain, the forces acting on aircraft during climb form an interlocked set of variables that may be modeled by a constraint system. Since the aircraft is changing its position in a smooth and gradual way during climb, the set of equations appropriate for modeling this process would be differential equations.

The main forces acting on aircraft are drag and gravity. Opposing these forces are the forces of thrust and lift, respectively. Figure 8.7 gives a graphical depiction of the balance of forces at any particular moment during climb.

This figure illustrates that the force of gravity must be counterbalanced by the lift of the aircraft, and the force of drag must be counterbalanced by the thrust acting on the aircraft. The total of the positive forces acting on the airplane to keep it steady on a flight path must equal the total resistance coming from drag and gravity. A net acceleration along the flight path will occur to the extent that the total positive force exceeds total resistance. Further, this acceleration will be inversely proportional to the mass of the aircraft; that is, given the same net positive force a light aircraft will accelerate more quickly than a heavy one.

The acceleration along the X-axis (horizontal speed) is jointly constrained with the acceleration along the Y-axis (vertical speed). The joint constraint works through the desired angle for the path of the aircraft. The steeper the angle, the more the total positive force will be acting to increase vertical speed (gain altitude) rather than horizontal speed. The shallower that angle, the more the total positive force will work to increase horizontal speed rather than vertical speed.

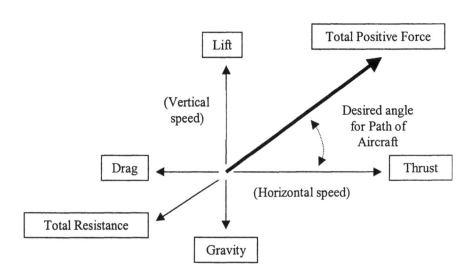

Figure 8.7 Forces acting on an aircraft during climb

This trade-off is exploited by pilots to adjust a climb path to different requirements for airport departures. Some airport departure patterns require a steep climb out to avoid noise sensitive areas. In this case, the climb angle is increased to maximize altitude gain during the initial part of the climb. In other departure conditions, the emphasis might be on quickly gaining sufficient flying speed to insure a safe margin above stall speed at the beginning of a flight. For those conditions, the departure angle might be kept shallow to rapidly increase air speed for the initial part of the climb.

Model equations Using the X and Y coordinate system, this system can be condensed into a set of differential equations. One equation represents the velocity and acceleration along the X-axis (change in horizontal speed). The other equation represents the velocity and acceleration along the Y-axis (change in vertical speed). The velocity of the aircraft along its flight path (air speed), depends on each of these two components. One set of equations which could approximate this system is:

1. Air speed = square root (vertical speed ^{2} + horizontal speed ^{2})

2. Vertical speed = air speed * sine (climb angle)

3. Horizontal speed = air speed * cosine (climb angle)

4. Change in rate of climb = (lift - gravity)/mass of aircraft

5. Change in ground speed = (thrust - drag)/mass of aircraft

6. Total applied force = square root (lift force^2 + thrust force^2)

Model evaluation This set of equations represents one possible simplified and idealized model for the climb situation. The changes pilots make to effect the behavior of a real airplanes during climb can be compared to this idealized model. One important use for comparison is to consider the idealized model as a normative model of what the pilots should be doing for a given situation. If the model can generate a mathematically optimal solution for the situation, then deviations of actual pilot behavior from this mathematical optimum signify poor performance. For example, the flight path chosen by a crew for an airport requiring a steep departure angle could be compared to see how much it deviates from a simulated optimum flight path. The extent of this deviation could be used as training feedback for the pilots.

To accurately model the complete departure sequence from an airport, however, additional components would have to be added to the model. For example, the maximal thrust setting used for the initial takeoff is often reduced during climb. This reduction lowers the total positive force acting on the aircraft, and may require a change in the trade-off set by the pilots. Further, different pilots may have different priorities for adjusting the climb trade-off, for example, prioritizing altitude gain vs. air speed gain. If so, accurately modeling real departures may require simulating those pilot priorities and the process of making decisions in a dynamic situation.

Chaotic dynamic systems

Many systems that can be represented by difference or differential equations change in a smooth fashion as depicted in the above examples. Other systems, however, may exhibit more extreme, turbulent, or catastrophic changes in system behavior. The sudden, discontinuous changes in the processes of the system are often summarized under the label of "chaotic behavior". For such systems it is critical to accurately model the nature of the system results for the areas in which the chaotic behavior is observed.

For example, when heating water the initial stage of gradual heating and convection currents that transfer the heat smoothly throughout the water is supplanted by the chaotic stage of boiling. The boiling process disrupts the smooth convection currents and qualitatively changes the nature of the heat transfer process. When boiling, much of the input heat will be used to change water from liquid to steam. Further, the exact manner and sequence in which the bubbles form while boiling is exceedingly unpredictable. Sudden, qualitative shifts in a process and a high degree of unpredictability may characterize a chaotic process.

A second example is water dripping from a tap. Smooth, steady states of the water leaving the tap would be a steady stream of water or a steady drip. If the faucet is adjusted slowly below the level of a steady drip, however, the dripping will become irregular and unpredictable. The transition from the steady drip to the irregular drip occurs quite quickly. The sudden qualitative changes in the dripping process as well as the unpredictability of the resulting state both illustrate the chaotic process.

In the aviation domain, one physical illustration of a chaotic physical process is the occurrence of flutter for wing surfaces. Certain airfoils under certain wind conditions will experience a rather sudden onset of degenerative oscillations. If not immediately damped, these oscillations will quickly become strong enough to destroy the airfoil. Chaotic behavior may also occur in the human elements of a complex system.

Human interaction with even a simple device may lead to chaotic system behavior. An example is speed control on my treadmill. My treadmill has a 3-4 second delay between changes I make on the speed lever and resulting changes in velocity of the belt. This interval is just long enough for me to notice that the speed has not changed and make further adjustments on the speed lever. When the cumulative impact of all these adjustments finally occurs on the belt, the speed is much too high and I have to quickly make adjustments downward. Adjustments downward are equally uncertain due to this 3-4 second time lag in the effect of the speed adjustment, and the net effect is extreme changes in treadmill velocity.

One important locus for modeling chaotic processes would be crew decision-making during abnormal or emergency situations on the flight deck. Under low levels of workload and low stress, decision-making might be smooth, thorough, and nearly optimal. In abnormal or emergency situations, however, the workload and stress may be very high. At some point, the smooth, reasonable decision-making flow may change to a turbulent and erratic mental process. If this is true for some crews under certain conditions, it would be very important to model this chaotic process and understand it well enough to design special training for those crews.

Chapter summary

This chapter has explored the modeling of dynamic systems using functions. The focus of the modeling may be on the functions themselves, the variables affected by those functions, or the constraints on the system that those functions represent. Developing an appropriate set of functions for modeling key aspects of dynamic systems is one important goal in a SIS. Descriptively, these functions can be a succinct summary of key aspects of the target system. Empirically, these functions can be used to precisely predict behavior were possible or show the conditions under which system behavior becomes difficult to predict or chaotic. Finally, when validated these functions can be used to specify the effects of changes in key system variables or components.

9 Personnel Selection and Evaluation with a SIS

Overview

The preceding chapters provided examples of how a SIS can be constructed for a complex system. The next chapters cover the use of the SIS for understanding and managing several critical aspects of complex human systems in an organizational context. The first focus will be personnel selection and evaluation.

Understanding and improving the performance of a complex system requires very careful and complete analysis of the system. Some systems will be complex because several persons have to coordinate in accomplishing job tasks (e.g. team sports). Other systems will be complex because of the intricate and sophisticated devices or tools required for the tasks (e.g. flying a multi-engine commercial aircraft). Still other systems may be complex by the inherent nature of the task itself (e.g. writing a book). In all complex systems, the nature of the complexity should be identified and theoretically modeled and tested as described in preceding chapters.

This theoretical framework sets the basis for any interventions designed to improve the performance of the target system. Two major approaches to improving performance in complex systems are changing the personnel or changing the device, tools, or nature of the task. The focus on changing personnel can be either directed at selecting personnel or training personnel. These approaches are illustrated in Figure 9.1.

This chapter covers personnel selection including choosing and evaluating selection instruments, choosing and evaluating performance criteria, and validating personnel selection instruments using a SIS. Subsequent chapters will address personnel training and changes in job tasks, procedures, artifacts, or devices. Personnel selection is a critical "upstream" determinant of the performance of a complex system. Where possible, selecting appropriate personnel may have advantages over alternative approaches to improving system performance such as training, job redesign, or changes in the work tools and devices. Resources required to test a job candidate may be far less than resources required to adequately train a candidate or redesign the job such that a candidate can be an effective worker. Further, selecting knowledgeable and skilled personnel can save resources by either shortening or eliminating required training. Similarly, selecting personnel that can successfully adapt to the current job context may eliminate the need for changes in tools or devices.

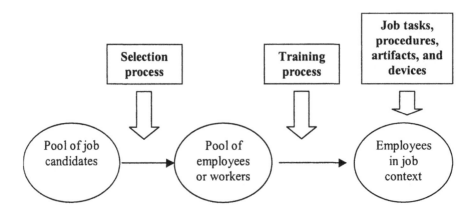

Figure 9.1 Illustration of three methods for improving system performance

This chapter will start with a brief review of job analysis. Job analysis is a critical preliminary step for selecting appropriate performance criteria and for determining appropriate measures of personal qualities that will become part of the SIS and form the basis of the selection system. Subsequent sections will cover in more detail the definition and measurement of job performance, definition and measurement of personal qualities, and the validation of personnel selection instruments. The chapter will close with a review of important practical and legal issues in personnel selection that can be delineated by the information available in the SIS.

Job analysis

A job may be defined as any coherent set of tasks or activities that are performed in an organizational setting. Job analysis is any systematic method that details the nature of these tasks and activities and the knowledge, skills, abilities, or other characteristics of the persons who perform them (KSAOs). A detailed job analysis is important because it provides the basis for designing and implementing job performance criteria and for choosing appropriate selection methods for personnel. This process is graphically depicted in Figure 9.2.

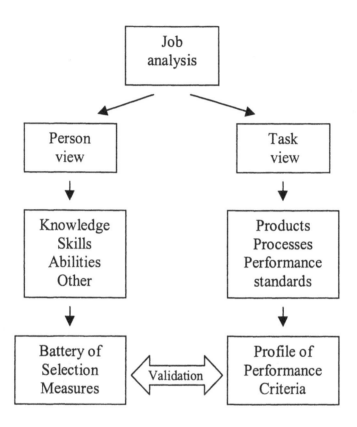

Figure 9.2 An overview of the development and validation of personnel selection measures

Several different methods can be used for job analysis such as functional job analysis, task inventories, the critical incident method, etc. Although these methods differ in their exact steps and processes, the end results of determining critical factors about the job activities are often quite similar. Therefore, this section will briefly describe the generic steps of task analysis and determining KSAOs in this process with specific reference to the role of the SIS. More detailed coverage of specific job analysis methods is available in specialized texts (e.g. Schmitt, Borman, & Associates, 1993). In the aviation domain, Seamster, Redding, & Kaempf (1997) thoroughly cover cognitive task analysis and related issues while Hunter and Burke (1995) cover a broad range of issues related to pilot selection.

Task analysis

Task analysis is typically an interactive process that successively refines descriptions of the job tasks using incumbents or subject matter experts (SMEs) and a job analyst. The initial statements should describe required job actions using explicit and concrete verbs, specify the objectives of the task, and list all equipment or other job resources that are necessary to perform the task. Where possible, task descriptions should be simple declarative sentences about the actions of the worker on the job.

Either the job analyst, a group of SMEs, or current job incumbents can be the source of the initial task descriptions. These task descriptions are iteratively refined by having the subject matter experts or incumbents judge their adequacy. During this process the judges add necessary tasks, redefine ambiguous tasks, and merge or delete tasks as appropriate. This refinement cycle stops when no significant alterations are being made to the task listing and the task list is judged to be complete and accurate by the incumbents or SMEs.

Hierarchical decomposition In addition, these tasks may be decomposed into smaller sub tasks or aggregated into larger scale tasks. Generally, there will be a "natural" level at which the tasks of a given job can be best described. However, aggregating these tasks into larger scale tasks can be valuable for offering a more condensed general picture of job performance. Conversely, the decomposition of the natural level tasks into more detailed sub tasks can give a more fine-grained view of the required activities for the job. A general picture of hierarchical decomposition is given in Figure 9.3.

For a specific example, Figure 9.4 illustrates hierarchical decomposition for the first stages of a commercial flight. This illustration is greatly over-simplified but will give an example of the job, task, and sub task decomposition process. In this decomposition, the basic tasks of pre-flight planning, aircraft inspection, and preparation for the departure of the flight are decomposed into sub tasks. In practice, the sub tasks would be further decomposed into relevant job elements where necessary.

The adequacy of this hierarchical decomposition can be cross-checked by using a new group of SMEs and a statistical method such as cluster analysis. That is, a separate sample of job incumbents or SMEs should be able to cluster or categorize the tasks in the same manner as the original sample. Any discrepancies in the hierarchical decomposition of the task between the task analysis sample and the cross-validation sample should be resolved before the task list is used further. The further use of the task list will often involve extra judgments about the tasks such as the performance standards for the task, frequency of the task on the job, criticality of the task, and so forth.

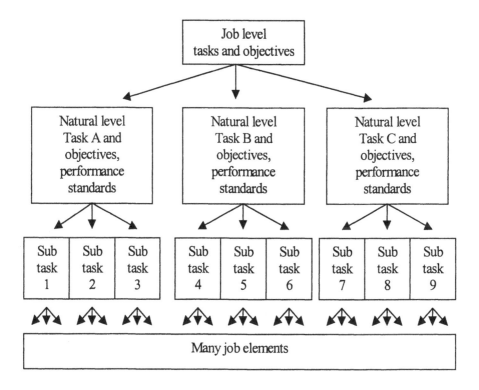

Figure 9.3 A graphical depiction of the hierarchical structure of tasks

Performance standards At some level of detail in the hierarchical decomposition of the job tasks, the SMEs should be able to determine appropriate criteria for good and bad performance. These performance criteria should be attached to the task descriptions as performance standards. In the aviation domain, these are known as "qualification standards".

Wherever possible, the performance standards should specify some potentially observable or measurable facet of job performance. For example, the performance standard could specify a minimum or maximum time interval for performing the task, a desired quantity of task outcomes, a desired quality standard for task outcomes, or some other discrete and measurable aspect of good or bad performance. These performance standards should help determine the definition, selection, and measurement of job performance in a subsequent steps.

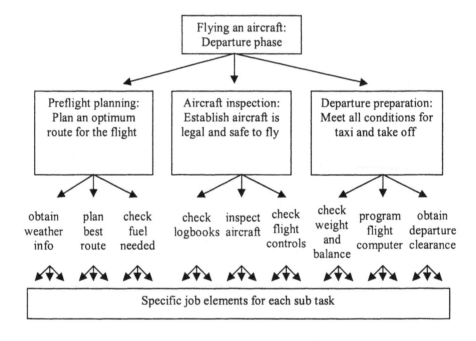

Figure 9.4 An aviation example of the hierarchical structure of tasks

Other task judgments Once the basic task list has been assembled, each task can be judged on other relevant characteristics. These other characteristics should be chosen either to directly help the personnel selection process or to more generally provide relevant information about these tasks in the SIS. One type of judgments that may help personnel selection is the judgments of the KSAOs necessary for good job performance. An example of judgments that provide other relevant SIS information are more detailed assessments of the job and job context.

Examples of task judgments that may reflect necessary KSAOs are assessments of the extent to which each task requires dealing with information, other people, or artifacts and devices. Tasks that require dealing with information may suggest certain intellective KSAOs. Tasks that require dealing with other people may suggest certain social KSAOs. Tasks that require dealing with artifacts, devices, or tools may suggest certain psychomotor KSAOs.

Task judgments that are more relevant to the description of the job include any descriptive judgments that point more to the job context than to the person. For example, judgments about how much time is required for a task or how frequently the task must be performed on the job are clearly more relevant to the job context. Similarly, how significant, important, or critical the task is to overall job performance is referencing the job context. Finally, the relevance of the task or sub task to other critical job parameters such as safety may also be judged.

Establishing job requirements

At some point, the information about the job must be used to develop a list of requirements for the persons that will be selected for the job. These job requirements are generally summarized as requirements for knowledge, skills, abilities, or other characteristics (KSAOs). The distinctions among these categories may occasionally be fuzzy at the boundaries, but the core definitions imply qualitatively different constructs:

- knowledge: the required set of information needed to do the job
- skill: the sequential, procedural performance of an integrated set of activities
- ability: a physical or mental capacity or capability (usually assumed to be stable)
- other characteristics: anything other than the above KSAs including personality, motivation, etc.

Establishing KSAOs A list of potential KSAOs must first be generated by either the job analyst, the SMEs, or incumbents. As before, the process of development of an adequate list of KSAOs is an iterative process. The initial development of the list of KSAOs may be started with a representative sub sample of the tasks from the task analysis. However, the final version of the list of KSAOs must be judged against all of the tasks for the job to ensure that the KSAOs adequately cover all tasks.

General KSAOs will be listed for a wide variety of the tasks and sub tasks of the job. These general KSAOs will be obvious by inspecting the lists at the end of the judgment task. Specific KSAOs, conversely, may be listed for only a small subset of required tasks for the job. For example, the job of salesperson may include many tasks, one of which is making change for the customer after a sale. The ability to do mental arithmetic might be required for this specific task but none of the other sales tasks.

Usually, the generality of a KSAO will reflect its relative importance for either selection or training. However, sometimes a very specific KSAO will be critical for overall job performance although only influencing a small set of the tasks. In the aviation domain, for example, a pilot does many tasks among which is the use of the flight automation computer. The specific KSAO of being able to make error-free computer inputs may be related only to the task of using the flight automation system. Nevertheless, since the correct use of the flight automation is critical to flying modern aircraft, this KSAO may be critical for overall job performance. Whenever the KSAO is critical for overall job performance, it is also a candidate for selection or training. Once the KSAO list is finalized, it can also be used as input for other important judgments.

One particularly important judgment is whether the KSAOs can or should be trained. If the KSAO cannot be trained and is critical for performance of a wide variety of tasks for the job, then it must be included as a goal for the selection process. If a KSAO can be trained, the decision still must be made on whether it should be a focus of selection or a focus of training. This decision may be based on the relative cost effectiveness of training vs. selection, legal considerations such as adverse impact, contractual or union obligations, or other considerations. In the end, however, all of the job-critical KSAOs should be represented either in the selection plan or in a training plan (discussed in Chapter 10).

Directly assessing job requirements   Using KSAOs derived from the job analysis to choose and construct selection instruments is an indirect method. Alternative methods would involve directly linking the required actions on the job to a selection instrument without deriving the KSAOs. One approach to doing this is the job sample or work sample approach.

A job sample is a representative set of the tasks required for a job. In this approach, the set of tasks is transformed into selection tasks as directly as possible. The most direct transformation is to have a candidate perform actual job tasks in a realistic setting. Putting a candidate on the job for a short probationary period could be considered one method for directly assessing potential job performance.

This very direct approach is often not feasible due to practical considerations such as availability of workspaces, contractual limitations, liability concerns, or safety issues. More typically, a subset of job tasks is transformed into versions that can be administered in an off site selection setting. When fidelity to the tasks of the original job is preserved, this method may be useful for selection.

However, there are at least two important disadvantages for a direct link method in the context of a SIS. First, because the KSAOs are not developed in a direct link method, they are not available for the development of training. Unless the selection process is comprehensive enough to provide candidates who can immediately and competently perform the job, training will be necessary. Without KSAOs, there is no precise guidance to the development and validation of this training.

Second, because the tasks for a specific job form the basis of the tasks used in the direct method, the direct link method results in a selection procedure that is quite job specific. This usually prevents the generalization of selection beyond a single job or at most a cluster of closely related jobs. If a selected employee will be confined to one job for an entire career (e.g. a sweatshop) this selection may be adequate. If, however, in the normal course of employment an employee will change jobs as his or her career develops, the selection information will become increasingly irrelevant as the performance of the employee is tracked over time in the SIS. This issue is further discussed in the performance criteria section below.

Therefore, although direct methods may form a part of the ultimate selection battery, the KSAOs developed in the job analysis process are very important for of the selection and training functions. These two approaches are not, of course, inherently incompatible. Evaluating relevant KSAOs may be more feasibly done by a direct job or work sample than separate, stand-alone measures. The important point in integrating these approaches would be to specify exactly which KSAOs are assessed in this direct work-sample manner and which KSAOs are assessed by the other measures in the selection battery.

Extended job analysis

Although the job analysis is usually tightly focused, extending this focus to include other products may be advisable. Other possible products of job analysis can contribute to designing a more effective selection system. This associated data can increase the range of relevant information in the SIS for training and performance appraisal. Although these other possible products will require time and effort on the part of the SMEs or job incumbents who are doing the job analysis, the information gained in the SIS may outweigh the incremental costs.

One additional product would be the analysis of the job in the full organizational context. That is, the organizational goals accomplished by the job can be connected to the tasks and functions of the job position. Analyzing the organizational goals can help ensure goal alignment between what is expected at the organizational level and what is expected for a job incumbent. For example, if the organization emphasizes low-cost of a product or service rather than high quality or proficiency, the definition of good or bad task performance should reflect this emphasis. Further, candidates should be selected whose performance will most naturally fit the expected profile for the organization.

An extended job analysis could also determine the organizational, divisional, or working group climate for the job. These climates may have important impact on job performance particularly as they help set the expectations for contextual job performance. Contextual job performance includes any aspects of the expected activities of a job incumbent that are not covered by a formal task analysis (Borman & Motowidlo, 1993). The climate information can also be used in selection to help assess the fit of particular candidates to the organization, division, or work group.

This organizational goal and climate information can be used in other ways that will affect the selection process. For example, the candidate can be given a realistic job preview concerning the job context and overall expectations for job performance in the organization. This more realistic information may help the candidate select an organization, which is the converse side of the selection process. Among other possible effects, having realistic expectations may help prevent later attrition of selected job candidates due to unmet expectations and disillusionment.

More generally, the organizational goal and climate information will be valuable additions to the SIS. This information greatly expands the potential analysis of moderators of job performance. Moderators such as a cooperative or competitive working group climate may substantially alter the relative impact of KSAOs on job performance. For example, social skills such as negotiation and conflict resolution may be much more important in a competitive workgroup than in a cooperative one. Therefore, wherever possible the job analysis process should be extended to include relevant organizational factors.

A second example of useful SIS information from an extended job analysis would be the analysis of job rewards and incentives.

Understanding the rewards and incentives attached to a job can help determine the match between a job and the motivational profile of a candidate. Further, understanding the job's rewards and incentives may clarify how candidates applying for the job are self-selecting. For example, candidates for a high-salary position may be self-selecting on the basis of monetary need or greed. To the extent that needs and motivations play a key role in short-term or long-term job performance, job rewards and incentives should be analyzed and motivational assessments should be one aspect of the selection battery.

Once the task analysis, analysis of personal qualities, and other job relevant information have been determined, the selection process can proceed to the next phase. It would be tempting to move directly into the phase of selecting or developing measures of personal qualities for the selection of job candidates. However, choosing selection measures without first carefully considering the performance criteria against which these measures will be validated is a mistake. Therefore, a more optimal sequence is to first carefully define the criteria for performance and only then evaluate and choose selection measures. In this way measures may be developed that accurately and validly relate to the job tasks and performance standards and correctly reflect the real influence of the KSAOs.

Performance criteria/metrics

For even a simple task, there are many different aspects of performance that could serve as criteria or metrics. A production task, for example, could have a quantity index of performance as well as several different aspects of the quality of production. For more complex tasks, the variety of possible performance criteria is even wider. The challenge is to define, select, and implement an appropriate set of performance measures. This process is essentially defining a theory of the job, including job processes and outcomes and the best indices or measures for the important facets of job performance that will become part of the SIS.

Defining performance criteria/metrics

A good definition of job performance is essential for defining performance criteria. In part, the definition of good performance is contained in the objectives and performance standards of the job analysis. This definition must be carefully and systematically extended to determine performance criteria that are potentially measurable and can serve as performance metrics. An appropriate team for defining performance metrics would include an SME or job incumbent, a person with a background in scientific measurement in the targeted domain, and a person familiar with the contents and function of the SIS.

The process of defining and measuring performance criteria should be an iterative process similar to the basic job analysis. An initial set of performance criteria and appropriate metrics should be developed by this team and then reviewed by all relevant stakeholders. After review, the proposed metrics should be revised as appropriate until they are satisfactory to both the development team and other stakeholders. This iterative process must address several aspects of the performance metrics such as the precise composition of the metric from SIS variables.

Univariate performance metrics A univariate metric is a single, unitary measure of job performance. For a sales position, for example, a single univariate metric might be the average dollar amount of sales. Even for this simple job, however, alternative metrics such as the number of repeat sales are possible. Therefore, for most jobs included in a SIS the performance metrics would have to include more than one relevant variable. However, these relevant performance variables can either be combined into one composite performance index or treated separately as a performance profile.

Composite performance metrics For some corporate purposes, information about performance has to be reduced to a single number. An example is assigning salary increases to workers based on performance. To assign salary increases in an objective manner, the performance information must be combined into a composite index that is then tied to the salary increase.

A composite index is usually formed by a weighted linear combination of different performance variables. Each performance variable should tap a separate aspect of job performance such as the quantity of objects produced, the quality of objects produced, and so forth. Forming a composite index is tacitly assuming that poor performance on one aspect of the job can be compensated for by extremely good performance on some other aspect of the job. If a composite job performance index is formed, the theoretical justification should be made clear. In practice, the weighted performance composite is calculated by multiplying the value of each separate performance variable by its corresponding weight and adding these products.

The set of performance variables that is used for this weighted composite should be a minimal spanning set. That is, care should be taken to avoid performance criteria that duplicate or overlap with each other, but care must also be taken to completely cover the domain of relevant performance variables. If they are not eliminated, duplicate or overlapping performance criteria will play a much stronger role in the weighted composite then they really should. In effect, duplicate or overlapping criteria will bias the result of the composite toward those criteria.

Conversely, not completely covering the domain of performance variables also biases the resulting composite. In effect, relevant aspects of performance are ignored or omitted from the composite. Forming a good composite, therefore, is a balancing act between including enough performance variables to have a reasonably complete description of performance but not so many variables that they unnecessarily overlap or duplicate each other.

The weights on each performance variable must also accurately reflect the contribution of that aspect of performance to overall job performance. To the extent these weights are not accurate, the composite performance index will be misleading and may create adverse effects. For example, if an incorrectly weighted index is used to assign salaries, employees may be dissatisfied with their relative salary increases. Scientifically, if an incorrectly weighted index is used to validate the personnel selection battery, the resulting selection battery may be biased and not pick the best set of candidates.

Multivariate performance metrics The problem of appropriate weights is avoided if the aspects of performance are recorded separately in the SIS. When the performance variables are measured and recorded separately, they can be used as a set of separate performance indicators or as a performance profile. This approach is essentially a multivariate approach for analyzing job performance. The advantage of this approach is that multi-faceted aspects of performance can be captured and no *a priori* decisions need to be made about the relative weight of each performance variable. The disadvantage of this approach is that the profile is a complex set of performance metrics that may be difficult to work with and may make selection and validation analyses more complex.

For some corporate purposes such as salary assignments, the performance profile may not suffice because a final unitary assessment is required. However, for selection, training, and performance appraisal, the performance profile may give a richer and more complete set of information about a person's performance. If the entire set of performance variables is preserved in the SIS, they can be used as a profile or as a composite. When used for some functions such as salary increases, these variables can be formed into the necessary weighted linear composite. When used for other functions such as selection, training, or job appraisal, these variables can be used as a performance profile.

Outcomes vs. processes   Most commonly, when people think of performance criteria they think of the outcomes of the job. For example, they may consider the "bottom line" of performance in terms of number of widgets produced, dollar amount of sales, and so forth. However, the processes by which the job tasks are performed are also critical performance indicators and may become relevant job performance criteria.

In many cases, the process criteria of performance will be superior to an outcome performance criterion. That is particularly true when the outcomes for job performance are constrained by factors not related to the person's KSAOs. For example, the person's bottom line job performance may be constrained by poor quality machinery, uncooperative coworkers, or other organizational constraints. In such cases the process of job performance may be a far better performance indicator than job outcomes.

Moreover, focusing on process opens the door for an even wider variety of potential job performance criteria. Processes can be measured at any stage in the task performance, and at each stage different measures can reflect different facets of the process. This large potential set of job performance measures must be evaluated to find a truly diagnostic subset. This evaluation can be accomplished objectively if the task processes have been validly modeled in the SIS as discussed in Chapters 7 and 8.

If the dynamic modeling methods discussed earlier have been applied to modeling performance on a particular job, that model can be used to objectively select a diagnostic subset of process indicators. If the model is in the form of an executable computer simulation, this simulation can be tested with changes of the process at different points to see which of the processes are truly critical to job performance. Once the critical processes or subprocesses are identified, appropriate measures for these processes in the real task situation can be developed. These process measures will serve as criteria for selection, training, performance evaluation, and so forth.

For example, in the aviation domain the crew interaction process may be a critical determinant of job performance, particularly in abnormal or emergency situations. The first step would be to construct and validate a dynamic model of the crew interaction process using the information in the SIS. After this validation, the simulation of this interaction process during the accomplishment of necessary flight tasks can be used as the basis for developing specific, task-relevant process measures. These measures, in turn, can be the basis for validating selection instruments or training methods focused on crew interaction processes.

Selecting performance criteria

The set of potential performance criteria should be carefully winnowed prior to implementation as performance metrics or measures. This initial evaluation will help ensure that the time spent in developing performance measures in the next step is not wasted. Further, the evaluation may raise issues that require a rethinking of the basic set of potential measures found in the previous step. This initial evaluation should include a check on the potential suitability and measurement qualities of the proposed criteria. One aspect of this evaluation is whether the proposed performance criteria are general and stable or changing dynamically due to changes in the job or job context.

Dynamic (changing) performance criteria Changes in organizational goals, working conditions or the organizational context over time can change the criteria for job performance. If, for example, an organization shifts from emphasizing high-quality, high-priced goods or services to high-volume, low-priced goods or services, the corresponding performance criteria for many jobs will change.

Similarly, organizational restructuring or changing economic conditions may change working conditions and cause a corresponding change in job performance criteria. In these cases, workers selected and trained to perform according to one set of performance criteria may not perform well when the criteria change. Therefore, if these changes can be anticipated, they should be used to select the relevant performance criteria.

An example in the aviation domain is the trend for some domestic carriers to set up low-cost subsidiaries to compete economically in the aviation marketplace. These low-cost subsidiaries may have different salary schedules, equipment, working conditions, and other factors compared to the parent organization. For example, the low-cost subsidiary may emphasize low operating costs of the aircraft more than on-time arrivals as performance criteria. If this difference can be anticipated, the different relative weights should be reflected in the job performance criteria used to select, train, and evaluate pilots for the low-cost subsidiary.

Although changing performance criteria due to organizational changes may be difficult to anticipate, changing performance criteria due to different stages in the career path of employee should be predictable. That is, in many if not most domains an employee typically progresses through a known set of stages in his or her career path. The stages would include the training stage, the early job performance stage, and subsequent stages that depend on advancement, promotion, or other forms of career progress.

Training performance is clearly relevant to the training costs of the organization, and can be included as relevant performance criteria. However, it is very important that performance criteria not be limited to the training stage because training performance criteria may not generalize to subsequent day-to-day performance on the job, much less to future job performance in the later stages of the career. In fact, the extent to which training performance generalizes is a very important empirical question that should be addressed by the information in the SIS as it becomes available.

As employees advance up the career ladder, they may shift into management positions that require more interpersonal skills and abilities, planning, and strategic decision-making skills than the initial job position. Therefore, different performance criteria that cover at least the training stage and the initial job performance stage should be included in the final set of criteria for personnel selection. If criteria can be identified that are appropriate for distal aspects of job performance such as long-term job performance or promotions, these should also be included in the final set. To be theoretically analyzable in the SIS, each criterion must be clearly related to an appropriate job performance construct.

Overlap of criterion and construct   The underlying construct of job performance should be defined or specified from existing theory for each measured criterion. The performance construct is the conceptually meaningful aspect of job performance measured or indexed by each criterion. Performance speed, for example, can be indexed by the number of widgets produced per hour. Defining the performance construct clarifies the theory underlying performance measurement and will help theoretically integrate the criterion metrics into the database of SIS information.

However, the construct validity of any particular criterion of job performance will be imperfect. For example, in the aviation domain the safety of the flight may be an important criterion for job performance. Since the safety of a typical line flight cannot be directly estimated, the criterion of job performance that indicate safety may be the number of errors or the level of performance on a test flight in a simulator. The performance on the simulated flight for that crew may be an imperfect reflection of the safety on a typical line flight.

To the extent that the measured criterion does not reflect the entire intended construct, it is deficient in construct validity. In this example, criterion deficiency means that the number of simulator errors may not completely capture the real safety of the crew's performance. This criterion deficiency might occur if the typical pressures for on-time performance, air traffic control demands, and other aspects of a normal line flight were not adequately captured in the simulator evaluation scenario. If so, the simulator performance could be somewhat deficient in reflecting the safety of the crews on line flights.

To the extent that the criterion systematically measures something other than the intended construct, it is contaminated or biased. For example, suppose a simulator assessment context with company evaluator tends to elicit "textbook" flying behavior rather than the crew's natural flying behavior. In this case, the simulator assessment could be measuring an unnatural style of performance rather than natural performance. High fidelity simulators and realistic flight scenarios administered in a natural fashion would tend to minimize this bias.

To the extent that the measured criterion does accurately and fully capture the intended performance construct, it is relevant and should have construct validity. However, the extent of either criterion deficiency or criterion bias will correspondingly decrease the usefulness of the measure for selection, training, and other uses. In this example, to the extent that the simulator evaluation is either deficient or biased, the performance measure would be less useful for pilot selection or training.

Measuring performance criteria

After the important aspects of good and bad performance have been defined and an initial set of performance constructs selected, these constructs must be tied to appropriate measures. In some cases, appropriate measures may already exist in the SIS, or measures in the SIS may be adapted to measure the desired aspect of performance. In other cases, appropriate measures must be developed and added to the SIS databases. Clearly, the latter option is more expensive and time-consuming than the former. However, the extension of the SIS databases to cover more relevant aspects of performance could be a valuable enhancement of the information available in the SIS. The implementation costs should be weighed against a broad spectrum of potential information gains in the final decision to adopt a specific set of measures.

Objective measures versus judgments   One basic issue in measuring performance is the choice between more objective measures of performance, such as task outcome measures, and judgment-based measures. Commonly, outcome measures such as the number of widgets produced or the dollar amount of sales are considered objective measures of performance, while supervisor ratings or other person-based evaluations are considered subjective measures. While simple, this division of performance measures can be very misleading.

A more solid basis for evaluating potential measures of performance is the scientific viewpoint covered in Chapter 3. That is, proposed measures may differ in the degree to which they directly or indirectly index the desired performance construct. Additionally, any proposed measure should be evaluated according to its sensitivity, reliability, and validity of measurement.

Although objective measures may be more sensitive, reliable, and valid than subjective measures, this will certainly not always be the case. Therefore, this issue must be considered for each proposed measure. Objective measures can have statistical problems such as poor distributions, bias or contamination by confounds, or deficiency in indexing precisely the desired performance construct. An example of a statistical problem for a proposed objective measure may be a ceiling effect such that the great majority of workers perform with almost maximal scores. Almost all candidates for an entry-level position could, for example, mop a floor successfully. The restricted variability of the resulting assessment scores may preclude adequate sensitivity, reliability, and validity of measurement.

An example of an objective job performance measure that could be biased and deficient is sales dollar volume for the job of sales clerk. Sales dollar volume may be biased by the salesperson's schedule of days on the sales floor or by the region or store to which the salesperson is assigned. Salespersons who are regularly scheduled to work on heavy volume days or who are employed in high traffic stores will have an artificially better performance record. Sales dollar volume may also be deficient in not reflecting the long-term sales outcomes of a salesperson's actions. Actions such as taking time to form a good relationship with the customer and thereby bringing in long-term, repeat sales may actually reduce the immediate dollar volume of sales and therefore not be accurately represented by that measure.

Once the potential defects in a proposed objective measure have been identified, it may be possible to adjust the measure to minimize or eliminate the biases and deficiencies. For example, sales dollar volume could be adjusted by region, store, and day of the week to eliminate the average effect of those confounds. Further, a secondary measure of repeat sales could be devised for situations where customers are identified in each transaction. This repeat sales measure could be combined with the adjusted dollar sales volume to form an objective sales performance index that would be less biased and deficient. For example, a composite performance measure using appropriate weights for dollar sales volume and repeat sales could be calculated for each salesperson.

If objective performance measures with sufficient scientific quality cannot be developed, alternative subjective measures of these facets of performance must be considered. Further, there may be some aspects of performance, such as contextual job performance, which are not reflected at all in the objective performance record and must be evaluated subjectively. Facilitating the social relationships in a workgroup, for example, may not be directly reflected in any objective job performance measure. If this type of facilitation were a relevant aspect of job performance, however, a subjective method of assessing this performance variable should be constructed.

Since subjective measures are typically more indirect than objective measures, each step of the assessment process should be carefully considered to ensure the scientific quality of the measurement. Typically, this will involve a careful consideration of who, what, when, where, and how the subjective evaluations are made. The scientific quality of the resulting measure may suffer if any of these basic components of the process are ignored.

Who The basic choice for who will make the ratings is very important. Supervisors, coworkers, subordinates, clients and customers may all have a different view on the performance of an employee. Co-workers and subordinates, for example, may value the technical expertise of a person while supervisors value the interpersonal skills or managerial competence of a person. Further, clients and customers are interacting with a person on a very different basis and have their own agendas for evaluating performance. The choice for who will make the ratings should be based on a careful analysis of the content of the rating and which group of potential respondents has the experience and motivation to rate the performance most accurately.

What Exactly what is being rated must be made clear to the respondent. This involves clearly defining and specifying the aspect of performance to be rated. It also, however, involves specifying the absolute vs. relative or comparative nature of the judgment. That is, is the performance of the person to be evaluated against an absolute scale of performance? If so, the anchor points for making these judgments must be clearly defined in terms of job performance and meaningful to the respondent. Alternatively, is the performance of the person to be evaluated in a relative or comparative matter? If so, the comparison group must be clearly defined to the respondent. Further, the anchor points for making these judgments must be clearly defined in terms of the relative performance for that group.

When When judgments are made is also critical for accurate evaluations. For example, the typical end of the year performance evaluation by a supervisor requires that the supervisor recalls and appropriately weighs relevant performance incidents over the last year. Since recall for past events is biased in several ways such as recency effects, making this judgment accurately may be an almost impossible task. Wherever possible, the judgment should be made as close as possible to the performance in order to avoid memory loss or bias. Immediate judgments would also be advisable if the judgment goal is absolute judgments of performance rather than relative judgments. The immediacy of the performance context and sole focus on the person's performance would be desirable in this case.

If, however, the judgment goal is a relative judgment rather than an absolute judgment, the immediate judgment may not be the best procedure. That is, for relative judgments the evaluator has to carefully consider the observed performance in the context of his or her recall about the performance of the comparison group. The recall of comparative performance could be inhibited in the immediate judgment context as the observed performance is such an overpowering focal point. Therefore, comparative judgments might be best made after observing a range of relevant performances and reflecting on the comparison group.

Where Where a judgment is made includes both the physical context or situation for making the judgment as well as the mental context for the judgment. The absolute vs. relative nature of the judgment plays a similar role in determining where as well as when a judgment should best be made. An absolute judgment, for example, might best be made in the performance situation itself. This helps focus the judges on the job and relevant tasks to be performed. Conversely, a relative judgment might best be made outside of the performance situation and focus the judges on the distribution or variability in performance typically observed for the comparison group.

How How the judgments are made concerns the exact process of evaluating the specified aspect of performance. Clearly, the specified process is different for absolute vs. relative performance evaluations. Absolute performance evaluations focus the judgement process on the exact performance observed and the performance standards for each evaluated performance component. Relative performance evaluations focus the judgements on the performance for the particular job incumbent in the context of what is typically observed for the comparison group.

However, the process of judgment is also influenced by the nature of the rating scale or rating format on which the judgments are made. Many different types of rating scales have been developed. Each type of rating scale makes a certain assumption about the best way to obtain the desired judgment. Further, each type of rating scale may require distinct types of scale preparation or training for the judges to use it well. The important point is that some systematic development process is followed which will typically result in a good quality judgment.

After the performance constructs have been defined, selected, and transformed into appropriate job performance measures, the selection process can focus on defining and measuring the personal qualities that should predict good or bad performance. The transformation of the KSAOs from the job analysis into appropriate measures of personal qualities should be done as carefully as the transformation of the task analysis into measures of job performance. As in defining appropriate measures of performance, several important issues must be considered in the transformation of KSAOs to personnel selection instruments.

Defining and measuring personal qualities

In a SIS, defining and measuring personal qualities that are related to job performance requires theory that elaborates how individual characteristics influence either job process or outcomes. This theory should, of course, integrate with the theory of the job that was elaborated while defining, selecting, and measuring aspects of job performance. The integrated theory of the job and the persons who perform it should be made explicit and included as part of the theoretical structure of the SIS. This integrated job-person theory is the essential justification of the personnel selection process and will be the focus of the validation process discussed in the next section.

The critical issues for measuring personal qualities are the basic definition of the individual differences, the level of specificity for measuring individual differences and connecting them to job performance, and the final choice of an appropriate selection battery of tests or measures. Addressing these issues requires careful elaboration of the job-person theory.

Individual differences

There are fundamentally different views on the relevant individual differences underlying job performance. These different views influence what individual differences are measured, when they are measured, and how they are used to predict job performance. One way of summarizing these viewpoints is using the dimensions of inherent vs. learned and stable vs. unstable in Figure 9.5. In this figure, each quadrant represents a distinct viewpoint.

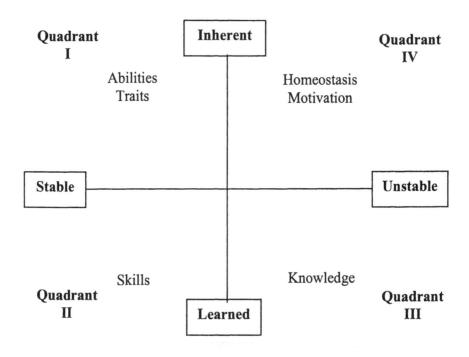

Figure 9.5 Different viewpoints about individual differences

The theoretical view of Quadrant I is that the relevant individual differences are inherent rather than learned and stable rather than unstable. In this view, the main differences in job performance are due to stable, inherent differences such as mental or physical abilities, personality traits, and so forth. In this view such abilities as intelligence or such personality traits as introversion-extraversion will influence performance over a long span of time and across a wide variety of jobs. Since abilities/aptitudes are assumed to be stable and relatively invariant, at least in adulthood, the emphasis is on selection rather than training. Under this view, the assessment of abilities, capacities, and traits is the focal point for analyzing the effects of individual differences on job performance. Since they are assumed to be inherent and stable, these differences may be measured in any context and at any time prior to or during employment.

Quadrant II emphasizes individual differences that are stable but are largely influenced by learning. A good example of this is skills. Such skills as learning to ride a bicycle are learned rather than inherent and, once learned, are relatively stable. Of course, skills may rely in part on underlying abilities or aptitudes, but they are assumed to be most strongly influenced by training and experience. From this point of view individual differences in prior training and experience may be strongly related to job skills. Job skills would predict performance over a long span of time on a particular job. A specific set of skills may not, however, predict performance across a variety of jobs. Current skill assessments would be relevant for selection, but could also be focus for training since skills are assumed to be learnable.

Quadrant III emphasizes individual differences that are learned but unstable. A good example of this is task-related knowledge. Clearly, appropriate knowledge must be learned for each specific task. However, once learned the knowledge is unstable due to the effects of interference, forgetting, and problems in selective activation of the correct knowledge for the task. Due to this instability, the best measures of task related knowledge may have to be administered immediately before or during task performance. Further, the appropriate task context may be necessary to activate and accurately assess relevant task knowledge.

Quadrant IV emphasizes individual differences that are inherent to the person but can be expected to change over time. One example of this may be motivation, which is included under the "other" category in KSAOs. Basic motivation may be due to homeostatic factors or the cumulative effect of early experiences, but it is usually assumed to be inherent in the person and depend on underlying motivational constructs that are relatively unchangeable. However, different job contexts or task situations may activate different motivation constructs and therefore elicit different motivational states over time. Although possibly unstable over time, motivation to do the job may be as important in predicting job performance as knowledge, skill, or ability. Like knowledge measures, best measures of task related motivation may have to be administered immediately before or during task performance and in an appropriate task context.

As shown by the above examples, although Quadrants I to IV have different theoretical assumptions, they each may contain relevant individual differences for predicting job performance. Therefore, relevant individual difference measures from each of the quadrants should be considered for the personnel selection battery. For most complex jobs, relevant abilities or aptitudes, skills, training or experience, knowledge and motivation should be considered. Besides this basic set, other individual differences might be relevant for specific jobs or organizations and should be included in the selection battery as appropriate.

Level of specificity

The level of specificity of a measure refers to the measurement focus. At the most specific level, a measure may be tightly focused on the performance of a specific job. At the most general level, a measure may be intended to predict the job performance of employees in a corporation across many possible jobs during their careers. The level of specificity of a measure should be carefully considered and matched to the job performance criteria developed in the preceding stage of the personnel selection process.

Job specific measures The most specific level for measuring relevant KSAOs for a job is performance on that specific job or the tasks of that job. For example, a probationary period of working on a job may be used as a selection instrument. If the conditions of the probationary period matched the conditions of normal employment, performance on the job during a probationary period should be able to predict performance for at least the immediate subsequent period on the job. In some sense, this is the purest case of using the principle that the best predictor of future behavior is past behavior.

However, this method of prediction does not have more general information value for the SIS. This selection method gives no further insight or understanding into the relevant KSAOs for the job or other jobs in the organization. Further, the results of the probationary job performance must still be fairly and systematically used for selection. That is, the multiple facets of performance must be synthesized into a selection score or used as a performance profile for selection. If performance facets are synthesized, condensing the multiple aspects of candidate's probationary job performance into one selection score can be a problem. If a performance profile is used, cut-points for successful performance on each critical aspect have to be established.

Without scientific validation several issues about using a selection method such as probationary performance remain. It may not be possible, for example, to evaluate the susceptibility of probationary job performance evaluations to bias or compounds that unfairly impact certain classes of job candidates. Similarly, the long-term predictive utility of probationary performance may be open to question. That is, the fact that the candidate can perform entry-level tasks under one supervisor during the probationary period may not generalize to performing other sets of tasks under other supervisors later on.

An alternative but closely related approach that would yield systematic information for the SIS is to use a representative work sample of the job for candidate assessment. One disadvantage of a representative work sample is the time spent to select or construct the tasks that represent the tasks on the job. Further, the sensitivity, reliability, and validity of the work sample measure would have to be established in the same manner as other critical SIS measures (Chapter 3). The advantage of administering a work sample under controlled conditions is that the performance on each of the work sample tasks can be scored separately and compared fairly across candidates. In one sense, the separate scores are an estimate of the capacity of the candidate for performing the tasks required by the job. Since a work sample is typically given in an artificial situation, the measured result may represent maximal performance rather than typical performance. Different aspects of the profile of work sample task performance can be related to other variables in the SIS to help establish sensitivity, reliability, and validity. Therefore, these scores give more detailed information about the candidate's performance and provide more information for the SIS.

Job cluster or family measures Job clusters or families are sets of closely related jobs. The sets can be determined by the judgments of subject matter experts or by empirical methods such as cluster analysis based on data in the SIS. Once these job clusters have been established, they can serve as the target for measuring relevant KSAOs. Having this more general target may broaden the set of relevant KSAOs. If the broader KSAO set is successfully measured, it should predict job performance in both a specific job and in the related jobs in the cluster.

For example, in the aviation domain the jobs of Captain, First Officer, and Flight Engineer are distinct but related. All three jobs require a high degree of aeronautical knowledge as well as monitoring and situation awareness skills. However, each cockpit position also has very distinct job responsibilities and related skills. If an airline needed to select flight engineers, for example, the KSAOs for the Flight Engineer position could be used to develop a selection battery for that specific position.

Unfortunately, this selection battery may not predict performance in either the First Officer or Captain positions very well because the distinct KSAOs for those positions are not measured. Therefore, treating the cockpit positions as a cluster of closely related jobs might be a more effective target for the selection battery. Selecting on this broader set of KSAOs for all three positions should allow for better prediction of performance of pilots as they make normal career transitions.

General career measures The most general and inclusive way to focus the instruments in a selection battery is to consider the likely career path of the employee. In most organizations, employee will follow one of a small set of possible career paths. In general, each segment of a career may begin with an appropriate training period, proceed with a supervised probationary period on the job, and end with an extended period of job performance under normal conditions.

Predictable career paths of multiple segments may occur in regular patterns that depend on opportunities afforded by the organization. Two common patterns could be a "technical" career path in which the individual proceeds to higher levels of technical expertise and responsibilities and a "managerial" career path in which the individual proceeds into higher and higher levels of management. In the aviation domain, the technical career path could have a pilot proceeding through the positions of Flight Engineer, First Officer, and Captain, for one type of aircraft followed by a similar progression for other types of aircraft. For some pilots, this typical progression may be interrupted by a career segment as an expert trainer or evaluator. The managerial career path, on the other hand, could have a pilot proceeding from the Captain position in the aircraft into the ranks of management in such positions as fleet or domicile Captain, training section Captain, and other high-ranking managerial positions.

Each distinct segment in the predictable career paths of employees may have a distinct set of relevant KSAOs. If the goal of selection is to predict the long-term job performance of an employee for this entire career path, then the relevant KSAOs for later segments should also be considered for selection. Unless these KSAOs depend solely on experiences or training in the early segments of the career path, they should be represented by measures in the selection battery. Clearly there is a cost effectiveness trade-off for this more extensive evaluation of an employee. If very few employees proceed on the possible career paths, than the cost of evaluation may not be justified by the benefit to the organization of better-qualified employees at the later career stages. Similarly, if the organization selects managers mainly from outside the organization, the cost-effectiveness may be low. If, however, management ranks are largely staffed by employees recruited from within and many employees make this transition, then selecting employees with the ultimate criteria of managerial performance in mind may be wise for the organization's future growth and development.

At the very least, the KSAOs for both training, supervised job performance, and normal job performance segments should be part of the selection battery. Even though all three of the stages refer to the same job, the KSAOs might be quite different. For example, in the initial training to be a commercial pilot both academic skills and psychomotor skills might be quite relevant. The candidate pilot must learn a great deal of technical information in a classroom setting and learn to use psychomotor skills to control complex, dynamic aircraft systems. The actual job performance as Flight Engineer or First Officer, however, requires teamwork skills such as coordinating with other crewmembers in addition to the basic piloting skills. Finally, in the role of Captain the pilot must exhibit team management and leadership skills as well as the ability to make final decisions for conducting each flight. Therefore, a relevant selection battery for commercial pilots could be comprised of academic, psychomotor, teamwork, leadership and decision making skills.

Choosing an appropriate selection battery

Once the content and level of specificity for the selection constructs have been determined, appropriate measures can be either selected or constructed. Selecting off-the-shelf measures has several advantages. First, it is the quickest method for obtaining measures. Further, information on the reliability and validity of an off-the-shelf measure may be available in reports of previous research. Finally, the testing corporation selling the measure will often score the responses and provide the results for a small fee.

However, an off-the-shelf measure will often have disadvantages. One disadvantage may be cost for those measures that are copyrighted and distributed by testing corporations. A second disadvantage for commercially vended measures is that the scoring key may be under the control of the testing corporation. This precludes using the item-level responses to code for the measurement of any other constructs. It also precludes re-analyzing item responses to ensure reliability and validity of the instrument. Since reliability and validity may alter for different populations or over time, the ability to continuously check on the scientific quality of the measure in the SIS is quite important and this is a serious disadvantage. Third, the content of this measure may not be quite right for the intended selection construct. That is, the measure may suffer from construct deficiency or bias. If the scoring and analysis are controlled by the testing corporation, it may be difficult or impossible to correctly adapt the measure to accurately reflect the intended construct.

Whether off-the-shelf measurements are used or not, the main principle is that the measure must be appropriate for the intended construct. If no appropriate measure can be found for critical KSAO, a measure will have to be either developed from an existing measure or constructed from scratch. Constructing a measure from scratch is very time-consuming and resource intensive. Therefore, this should be considered as a last alternative when no related measures can be found among published, non-copyrighted research measures.

If a measure used in a published research study measures a related KSAO and is not copyrighted, it may be possible to alter the content in such a way as to measure the intended construct. Starting from a set of items which has been successfully used to measure a related construct increases the likelihood that the altered items will successfully measure the intended construct. The scientific quality of such an adapted measure must, of course, be examined in the SIS in the same way as any other measure of a construct. The initial construction time and cost, however, should be greatly reduced for an adapted measure compared to one that is constructed from scratch.

Minimal spanning set The measures chosen for the selection battery should cover the job or career domain as efficiently as possible. That is, the measures should capture all relevant variance in the domain, but not overlap or be redundant. This is the principle of obtaining a minimal spanning set of measures for a domain. If the minimal spanning set of measures can be found, any other measures for this domain are redundant and could be expressed as a function of the minimal spanning set.

Consider a software-development company that wants to select managers, marketing staff, clerks, and programmers. Suppose that two relevant skills, social skills and data-management skills, distinguished these four jobs as described in Figure 9.6. In this simple example, each type of job is distinguished by a unique profile of the two skills. The figure indicates that these two skills are necessary and sufficient to distinguish personnel appropriate for the four jobs.

Both measures are necessary because if either measure is omitted, candidates could not be correctly assigned to jobs. For example, if the data skills measure is omitted, candidates scoring high on the social skills measure could not be correctly distinguished into Marketers and Managers. Similarly, if the social skills measure is omitted, candidates scoring high on the data skills measure could not be correctly distinguished into Managers and Programmers. Thus, both measures are necessary for assigning candidates into the four jobs.

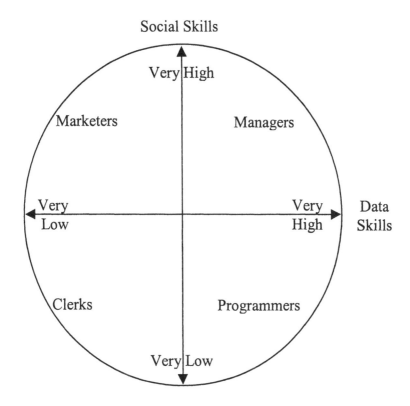

Figure 9.6 A minimal spanning set of two skill measures for selecting candidates into four jobs

The set of both measures is also sufficient for personnel selection. Given the information in the two measures, candidates can be uniquely and correctly assigned to all four jobs as shown in the figure. If the selection battery had direct measures of these two skills, the two measures would be a minimal spanning set for the selection into these four jobs.

A minimal spanning set of measures is not necessarily unique. An alternate set of measures that would be equally effective in this situation is presented in Figure 9.7. In this case, personnel appropriate for the same four jobs can be selected by two measures. One measure is a "manager vs. clerk" skills measure while the second measure is the "marketer vs. programmer" skills measure. Using these measures, a programmer would be selected by a low score on the "sales vs. programmer" measure and a medium score on the "manager vs. clerk" measure. Selection for each of the other three jobs would be similarly determined by a unique combination of scores on these two measures. Thus, these two measures are an alternative minimal spanning set for the selection battery. The important point is that the selection battery should be empirically checked for redundancy and completeness for adequate selection.

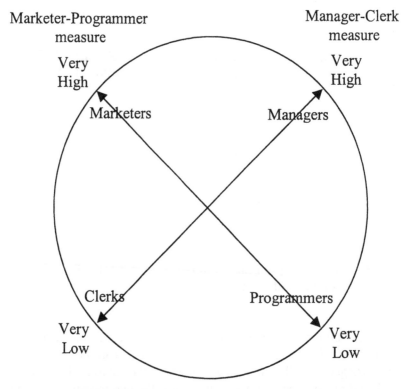

Figure 9.7 An alternative spanning set of two measures for selecting candidates into four jobs

Given the presence of measurement error and the initially uncertain knowledge of sensitivity, reliability, and validity, it may be difficult to determine at the outset that a set of selection measures forms a minimal spanning set for given domain. However, this is a long-term goal. That is, the development of the selection battery in the SIS should aim at both removing redundant measures and augmenting the selection set with measures that provide significant incremental validity. In this way, a minimal spanning set of selection measures that is both efficient and complete may be constructed in the long run.

Appropriate difficulty level   If possible, the potential measures should be pretested on an appropriate sample of job candidates to ensure that the distribution of scores will be acceptable. In particular, the measure should be neither too difficult nor too easy. A measure that is too difficult will lead to a "floor effect" of all the scores being grouped at the bottom. This type of test may also have a negative side effect of scaring off or discouraging well-qualified candidates. Conversely, a measure that is too easy will lead to a "ceiling effect" where all the scores are grouped together at the top. Even if floor and ceiling effects are avoided, however, the distribution of scores from the sample should be inspected to ensure that adequate variability exists because other biases, such as a central tendency judgment bias, can reduce the variability of a measure. Generally, reduced variance will make a measure less useful for selection and prediction of performance.

Relatedness to job performance   The directness of relationship of a measure to job performance should also be considered in battery selection. More direct measures such as work samples may have the advantage that they strongly and precisely predict performance in the job from which the work sample was taken. However, this precision may be at the expense of generality of prediction. That is, measures such as work samples that accurately predict performance in a specific job may not predict performance in other jobs. Conversely, more general measures such as abilities or aptitude measures may be better at predicting performance across a range of jobs, but suffer in their ability to strongly predict performance for any particular job. There may often be a trade-off between measures that strongly predict performance in a specific job vs. measures that moderately predict performance across a variety of jobs.

Therefore, the final set of measures included in the selection battery must be chosen with the explicit consideration of the goals of the selection. If the goals of the selection are performance in a specific job only, then the specific measures might be more appropriate. However, if the goal of the selection is performance across a variety of jobs or across a person's career with the organization, then the more general measures might be more appropriate.

Validation of personnel selection

The overall personnel selection theory consists of the integration of the theory of job performance with the theory concerning the individual differences of the persons performing the job. The personnel selection theory should be subject to the same confirmation and disconfirmation efforts as all the constituent theories in the SIS as discussed in Chapter 4. The process of validating personnel selection instruments, therefore, is essentially the process of validating the underlying theory. However, content validation, criterion validation, and construct validation have unique aspects when they are applied to the selection process.

Content validation

The idea of content validation is that the content of the items or scales of a measure should fairly represent the intended domain. For personnel selection, there are two relevant domains: the domain of the job tasks and the domain of the relevant KSAOs. To establish content validity for the performance criteria measures, these measures must be shown to be a fair and representative sample of the tasks in the task analysis for the job. Similarly, to establish content validity for the selection battery measures, these measures must be shown to be a fair and representative sample of the KSAOs that are necessary for job performance. Establishing content validity might be feasible for simple jobs with clearly enumerable tasks represented by a clearly defined set of KSAOs. However, establishing content validity for the measures in either the task domain or the KSAO domain might be quite difficult for complex jobs where this enumeration is difficult or impossible. Since most SIS development will involve selection for complex jobs, content validation is not sufficient.

Criterion validation

In the simplest possible case of criterion validation, the measure of a single KSAO would be correlated with a single performance criterion. The size of the correlation would index the degree of criterion validity. Although charmingly simple, this situation will almost never occur for complex jobs.

Generally, multiple KSAOs will be relevant for complex job performance. Therefore, criterion validation will have to take into account multiple predictors. If there is a single criterion of job performance (which could be either a simple criterion or a linear composite criterion), criterion validity can be evaluated using multiple regression. The multiple R indexes the overall prediction of the criterion. The significance of each KSAO measure in predicting that criterion is indexed by the regression weights.

In this case, the refinement of the selection battery to form the minimal spanning set discussed earlier is straightforward. If the measure of a KSAO is a non-significant predictor of the criterion, it can be eliminated from selection battery. Conversely, the incremental validity for other potential measures of relevant KSAOs can be examined by adding them to the regression in a process called hierarchical regression. If the measure for a particular KSAO significantly increases the prediction of the criterion of job performance, this measure can be added to the selection battery. Of course, the amount of incremental validity must be judged against the increased cost of assessment in order to make a cost-effective decision about adding the measure.

If, however, job performance is represented by a profile of performance measures or multiple performance measures across different stages of employment, a more complex analysis must be used. One approach to analyzing a profile of performance measures would be to take each performance measure separately as a criterion, and use the multiple regression approach with the set of KSAO measures as predictors. This approach may work in certain situations for which there are a small number of relatively independent or uncorrelated performance measures. For example, if the quality and quantity of the number of widgets produced were relatively uncorrelated performance measures, separate multiple regressions using the quality measure as the criterion and then the quantity measure as the criterion could give the necessary information for evaluating the selection battery.

Most often, however, the profile of performance measures will have more than two or three measures, and these measures may be interrelated. In such a case the extension of multiple regression that will give the necessary information to evaluate the selection battery will be a canonical correlation. The canonical correlation will relate the set of predictors to the set of criteria in the most efficient way possible. Since there are multiple criteria, the predictors may be related to the performance profile in multiple distinct ways. Therefore, the canonical correlation will show each possible way the set of predictors correlates with some aspect or facet of the criterion set.

By examining how different subsets of the predictors are related to different facets of the set of criteria, the selection measures can be evaluated. For example, a selection measure that has very low weights for predicting any distinct aspect of performance can be eliminated from the selection battery. Similarly, a predictor with high weights across all facets would be kept. The difficulty in this process occurs for predictors that predict one or two facets well, but not others. In this case, the decision must be made on the criticality of the job performance facets that are predicted well by the measure.

Construct validation

The critical advantage to embedding the selection measures and performance measures within a SIS is that the other information in the SIS will potentially allow the evaluation of construct validity in addition to content or criterion validity. Ultimately, the validation will depend on the validity of the criteria as much as the validity of the predictors. The constructs underlying the set of KSAO measures form one set of constructs that must be validated. The constructs underlying the performance criteria form the second set of constructs that must be validated. Finally, the critical part of the validation should be that the KSAO constructs strongly and significantly predict the performance constructs. This validation can proceed in distinct steps.

In the first validation step, the construct validity of the KSAO measures and the performance measures are separately evaluated using the other SIS information. First, the KSAO measures should have a factor structure that is predictable from the measure selection process. Further, the set of factor scores that represent constructs underlying these KSAOs should have theoretically interpretable relationships to other information in the SIS. A measure of job knowledge, for example, should have some positive relationship to the degree of job experience recorded in the personal database for each candidate. Similarly, a measure of job motivation should be positively correlated with a personality measure of conscientiousness. The pattern of interrelationships of the constructs underlying the KSAO measures with other SIS variables should confirm or disconfirm the validity of measuring these constructs.

The constructs underlying the measures of job performance should be validated in a similar fashion. The first step might be to factor analyze these measures and confirm that the expected set of performance dimensions is present. The next step would be to obtain factor scores that estimate performance on each of these constructs and correlate those scores with other relevant information in the SIS. A factor score representing the quantity of production could, for example, be correlated with job attendance records or time on task worksheet records for external validation.

The final step in the construct validation of the selection battery would be to interrelate the KSAO constructs and the job performance constructs. This interrelated evaluation of construct validity could be done using the method of structural equation modeling (SEM). In structural equation modeling, a set of measured predictor variables can be reduced to its underlying constructs and used to predict a set of constructs underlying the measured performance variables. In essence, SEM relates the constructs underlying the KSAO predictor variables to the constructs underlying the job performance variables. The initial bottom line is that the KSAO constructs should strongly relate to the import job performance constructs.

If the selection and performance constructs are not related as expected, the adequacy of the measurement instruments should be re-examined. A surface-level selection measure may be eliminated because it does not correctly correlate with the underlying selection construct. Similarly, a surface-level performance measure may be eliminated because it does not correctly correlate with the underlying performance construct. Further, if one of the KSAO constructs does not significantly predict any aspect of performance, all the measures for that KSAO construct can be eliminated from the battery.

Continuing evaluation of validity One advantage of the SIS approach is that it enables and encourages an ongoing evaluation of the validity of the selection battery and performance indicators. Although the development and validation of a selection battery can be conducted on a one-time basis, more information is derived from this effort if the theories and data are integrated into the SIS. At the least, this integration allows a long-term follow-up of the selected candidates. Long-term results may give a more complete picture of selection validity.

This SIS integration can have more long-term benefits. When integrated, the theory underlying the selection process becomes a part of the overall theoretical structure for the SIS and will be continuously evaluated. This ongoing evaluation is very valuable because it will help detect shifts in selection validities over time when they occur. Validities may shift due to the change in the job context, corporate climate or culture, or shift in the nature of the population being selected. No matter what the reason for the shift, it is very important to detect when a selection battery is losing its effectiveness for valid selection.

This ongoing evaluation also refines the theory underlying the KSAO measures and the performance indicators. The ongoing refinement of these theories is necessary to obtain a picture of the target system that is as complete and current as possible. Ultimately, this ongoing analysis may point to the necessity of revising the basic job analysis that was the beginning of the selection development process. That is, subsequent analyses may indicate that either the basic task structure for the job or the performance standards for the tasks have changed or shifted. Similarly, this ongoing analysis may indicate that appropriate measures for the KSAOs have changed or even that the relationship among the KSAOs has altered, which could shift the best measures for a minimal spanning set. Most importantly, these re-analyses could indicate when the relevance of particular KSAO constructs to critical aspects of job performance has significantly changed. The change in the structure of basic relationships would, of course, require a re-tuning of the selection battery.

In essence, the ongoing evaluation of the theory and data relevant to the task analysis and KSAOs represented in SIS databases will feedback and potentially alter the theories underlying these two facets of the job analysis. In many if not most domains of implementation for a SIS, the theoretical basis stemming from the job analysis will change over time based on the accumulation of empirical results. The job analysis in a SIS becomes a dynamic theoretical structure with ongoing validation rather than a static, one-time project for developing a selection procedure.

Other personnel selection issues

This section will cover other relevant issues personnel selection. The brief coverage of each issue is intended to raise awareness of the issues rather than an in-depth coverage of the topic. Personnel selection or organizational staffing texts (e.g. Schmitt, Borman, & Associates, 1993) contain more details on these issues.

Selection ratio and utility

The utility of the selection process depends in part on the percentage of the pool of candidates that must be selected to fill the available job positions. If the job candidate pool is small and the number of available positions that must be filled is quite large, the selection process at best can only weed out the very worst candidates. Many of the positions will perforce be filled by mediocre candidates. Conversely, if the job candidate pool is large and the number of available positions is small, the selection process when it is effective can result in only the the very best candidates being hired. Clearly the utility of an optimal selection process in the latter situation is higher.

From a short-term, cost benefit viewpoint, the justification for an extensive job analysis and selection process is weak when there are many jobs and few candidates. However, from the viewpoint of the development of the SIS, the utility of this process might still be quite high. In fact, the selection of a high percentage of the available candidate pool will insure that there is adequate variability in both the selection measures and job performance criterion measures to perform the empirical validation. When selection ratios are very low, in contrast, the validation may be difficult due to a restricted range in either the KSAO measures or the job performance measures. The advantage of the extra variability in KSAO and job performance measures will continue to help ongoing validation and theoretical refinement in the SIS over time. For example, the additional variability in the KSAO measures may help predict future performance criteria such as promotions or performance in later career segments. Therefore, although the immediate utility of the selection process for a high selection ratio situation is less, the long-term information value of measuring the KSAOs and job performance criteria may be higher.

Type I and II selection errors

Similar to the Type I and II errors in theory testing discussed in Chapter 4, there are also Type I and II errors in personnel selection. The final outcome of a selection process is that a job candidate is either selected or not. In one sense, the selection is a prediction of the future job success of the candidate. Similarly, the actual job performance of the worker can be classified as satisfactory or unsatisfactory in overall sense. When the predictions from selection process are contrasted with the satisfactory or unsatisfactory performance outcomes, the occurrence of a Type I or Type II error is clear as shown in Table 9.1.

If the candidate is selected and performs in a satisfactory manner, the selection process has functioned correctly. Similarly, if the candidate is not selected and would have performed unsatisfactorily, the selection process has functioned correctly. If a candidate is selected but does not perform satisfactorily on the job, the selection process has made a false positive error (Type I error). Conversely, if a candidate is rejected but would have performed satisfactorily if he or she had been hired, the selection process has made a false negative error (Type II error). Each of these types of error can have a distinct set of costs for the organization.

Table 9.1 Illustration of Type I and Type II selection errors

		Job performance result	
		Satisfactory	**Unsatisfactory**
Selection result	Candidate is **selected**	correct selection	Type I error
	Candidate is **not selected**	Type II error	correct rejection

The selection ratio discussed above will impact on the relative likelihood of Type I vs. Type II errors. If the selection ratio is small and only a few candidates are selected from a large pool, the selection process may be able to isolate "stars" who will indeed perform well on the job. In such case, Type I errors will be low. However, it is quite likely that many of the rejected candidates would have also performed satisfactorily on the job. That is, Type II selection errors will be high. Conversely, if the selection ratio is high and almost all of the job candidates are selected, the likelihood of Type II error is decreased but the likelihood that some of the candidates selected will not perform satisfactorily is higher (that is, Type I error is higher). The selection ratios and errors are particularly important if they adversely impact on subgroups of the population.

Adverse impact

One heuristic for judging adverse impact on the hiring of specific subgroups is the "four-fifths" rule. An initial pool of candidates for a job will have a known percentage of each subgroup. A selection process will have adverse impact on subgroup if the rate of hiring from the subgroup is less than four-fifths of the rate of hiring for the majority group. Put another way, the selection ratio for the subgroup should be at least 80% of the selection ratio for the majority group.

Adverse impact in a selection system may have undesirable consequences for the organization. Legal challenges to such a selection system could force the organization to show proof that the selection measures were directly related to critical job performance. Although legal issues are beyond the scope of this book, the legal complications of a selection process with adverse impact could be costly to the organization.

From the point of view of the SIS, however, the issue of adverse impact is a theoretical and empirical issue. If, for example, adverse impact is actually due to a large number of Type II errors for the subgroup, then the selection procedure is preventing the organization from hiring job candidates who would perform well. Clearly it is in the best interest of the organization to understand and rectify the aspects of a selection procedure that would result in these errors.

The theories and information in the SIS can be used to investigate group differences and try to understand the origins of adverse impact. If sufficient information exists about the performance of employees from a protected group, the same validation analyses that were conducted for the selection battery can be performed for this group. The comparison of the results for the subgroup to the overall validation results will give valuable information.

Take, for example, a selection battery validated by multiple regression with a composite job performance criterion. The overall R and the regression weights can be the focus of comparison between the protected group analysis and original validation analysis. The overall R for the subgroup could be either the same or different from the original validation analysis. If the overall R for the protected group is significantly lower than for the original validation analysis, the set of predictors is not working as well for this group as for employees in general.

The origins of this overall difference in predictability of job performance should be further investigated. Very possibly either the job performance criterion or the set of predictors does not have the same sensitivity, reliability, or validity for the subgroup. The sensitivity, reliability, and validity issues should be addressed by a broad set of analyses using the SIS data. Alternatively, it is possible that the set of KSAOs upon which the selection measures are based may not completely or accurately apply to members of the subgroup. Some empirical evidence of whether or not this is the case can be gained from a careful inspection of the regression weights in the multiple regression.

The regression weights for each predictor should be compared for the protected group vs. the original validation analysis. The pattern of significant weights may be quite different even if the overall multiple R is the same. Any significant difference in one or more of these weights implies a difference in the basic connection between the KSAOs and the job performance criteria for members of the subgroup. This would, for example, be the case if the members of this group approach the job in a qualitatively distinct manner.

Finding this type of important difference could suggest either a new job analysis that focuses on the job performance of members of the subgroup, a search for more appropriate KSAO measures for this group, or other possibilities such as distinct training for members of this group. In essence, a theory of why and how the job performance of members of the subgroup is different from other groups would be constructed, validated, and used. Potentially, the theory and related empirical results could be used for both the selection process and the training process, which is covered in the next chapter.

Chapter summary

This chapter has covered some of the basic issues of personnel selection and evaluation with respect to a Scientific Information System. The information in a SIS can be used to help job analysis, establishment of performance criteria and corresponding measures, and the definition and measurement of personal qualities for personnel selection. Conversely, the inclusion of the theories and data developed for these steps in the personnel selection process can be a valuable addition to the information in the SIS. Embedding the theories and data in the SIS greatly expands the scope of both short-term and long-term validation of selection procedures. Continuing to collect and embed such data in the SIS also allows for a continuing evaluation of theoretical validity and the detection of shifts in validity due to changing circumstances. Thus, a SIS augments the information available for the scientific development of personnel selection instruments, and derives more potential utility from the information that flows from the personnel selection process.

10 Personnel Training with a SIS

Overview

Although selection can help provide qualified employees, for many if not most complex jobs training will be required for optimal performance. In the aviation domain, for example, extensive training is provided for each unique aircraft and for each duty position in the crew (Captain, First Officer, etc.). For integration into a SIS, the training of employees should be systematically linked to the selection process discussed in the last chapter and to the precise model of job performance that was discussed in preceding chapters.

Training and selection

There are several distinct ways in which selection and training can work together to optimize system performance in an organizational context. Most commonly, selection may be used to ensure a base level of performance which training then augments (Figure 10.1). That is, the selection process weeds out candidates who cannot perform at some basic, threshold level. Subsequent training raises the level of performance to that desired on the job.

A second distinct way in which selection connects to training happens when the selection is for the capability or ability to learn aspects of job performance rather than a base level of performance (Figure 10.2). In this case, the selection process is aimed at the qualities of the candidate that will facilitate more efficient or higher quality learning of job-related skills. Candidates who score higher on the selection process should show an increased rate of learning or better learning outcomes at the end of training and achieve the desired level of performance (group "A" in Figure 10.2). With more capable trainees, training may be correspondingly reduced which would result in a saving of training resources.

Selection can also be designed to replace all or a part of the required training for a job (Figure 10.3). That is, if candidates have sufficient proficiency on components of the job, the training for those job components can be omitted. This approach to selection can save even more training resources then selecting for base level performance or trainee ability because entire segments of the training curriculum may be omitted.

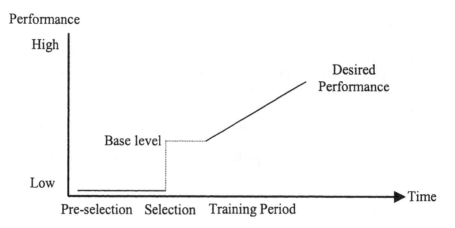

Figure 10.1 Augmentation model of selection and training

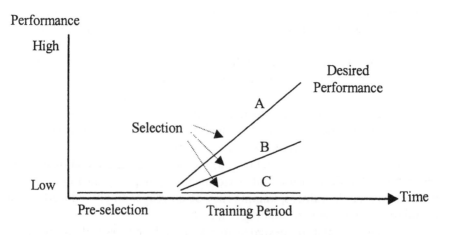

Figure 10.2 Facilitation model of selection and training

However, this approach assumes that the training for job performance is decomposable at some level.   If training for job performance is decomposable, the candidates can be trained on only those aspects of the job for which they do not have the required proficiency. This approach also requires a job pool of sufficiently qualified candidates and strong evidence for the validity of the selection process for predicting performance on specific components of the job. If these requirements are not met, the replacement approach may not be feasible.

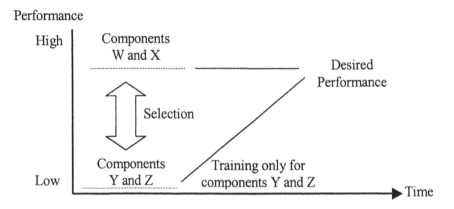

Figure 10.3 Replacement model of selection and training

Training and job performance

When training is used to increase performance from threshold to job-satisfactory levels, the training outcomes must connect to the satisfactory levels of performance defined by the theory of job performance in the SIS. In particular, when a "train to proficiency" approach is used, the level of proficiency that defines satisfactory performance must be clearly and objectively defined to guide the training process. To the extent that the theory of job performance has been empirically validated, the precise measures that mark satisfactory performance and the required level of performance on these measures can be objectively specified. These performance benchmarks can be empirically established using the SIS job performance data.

When the training focus is learning aspects of job performance, training needs should be directly connected to job-relevant knowledge, skills, abilities, and other characteristics (KSAOs) defined by the theory of job performance. The job analysis discussed in a previous chapter will specify the job-relevant KSAOs and be validated using appropriate SIS data. Typically, each component of training should be designed to specifically address one or more of the validated KSAOs. These KSAOs also guide the evaluation of training since the effectiveness of the training in increasing the designated KSAOs should be empirically evaluated.

When the job is decomposable and partial training of job components is the approach used, the job components chosen for training should be guided by the theory of job performance. The theory of job performance in the SIS must support the assumption that the designated job performance components can be separately trained and effectively integrated with other components. In addition, the job performance theory and relevant SIS data should specify the appropriate measures of performance for each job component and define satisfactory performance levels for the component. Partial training should be carefully designed using a well-validated job performance theory wherever possible.

For any of these approaches to selection and training, the training component must be connected to the theoretical principles underlying the selection process, on the one hand, and the theory of job performance, on the other hand. In addition, the training component may also be a separate focus of theoretical development, particularly if training is a major component of the preparation for the job. The training given to astronauts prior to a space shuttle mission, for example, is both extensive and intensive. Months of training are required for days of mission performance.

When the theory of job performance is initially developed, it may specify job-relevant KSAOs but not give precise information on how to *train* these KSAOs. Since training KSAOs may be quite distinct from job performance, the training aspect may require its own theoretical development. In this case, the training theory becomes a part of the theoretical structure of the SIS and a focus for ongoing empirical evaluation just like all the other theoretical components of the SIS.

Similar to the other theories in the SIS, the training theory may undergo revisions from a relatively simple initial form to more complex versions as the data in the SIS are used to refine it. The simplest form of this theory might be a list of the basic facets that influence the training outcomes, while more complex versions of this theory may track the instructional sequence or process in a more precise manner. Examples of both simple and more complex training theories along with their implications for the SIS will be covered in the sections below. References such as Jonassen and Brabowski (1993) give more detail on learning and training in general and texts such as Tefler (1993) or Tefler & Moore (1997) apply to aviation training.

Facet theories of training

Constructing a facet theory

The simplest approach to forming a theory of training is to designate the major facets that influence the outcomes of the training, and then elaborate those facets. One simple facet theory of training is that three major facets jointly influence training outcomes: the students or trainees, the teacher or trainer, and the training program itself. This generic theory is illustrated in Figure 10.4.

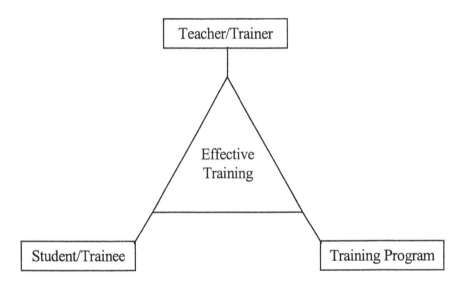

Figure 10.4 Three major facets of effective training

Each of these facets must be appropriately elaborated to make this a testable theory of training. Initially, the components of each facet that have the strongest effect on training should be listed as exhaustively as possible. These components should then be translated into the relevant constructs or variables that are expected to critically influence training effectiveness. For actual inclusion in the SIS databases, of course, each construct or variable must be associated with an appropriate measure or set of measures.

To illustrate this process, take the facet of the student or trainee. Clearly, differences among students or trainees may influence the effectiveness of training. To elaborate this facet, the differences among students that would have the greatest impact on training outcomes should be listed. For example, relevant student variables that could strongly influence training effectiveness are student motivation, knowledge, previous experience, learning ability, preferred learning style, and so forth. To be included in the SIS, each of these variables would have to be associated with a specific measure or set of measures and stored in an appropriate student or trainee database. In the aviation domain, for example, relevant information about the experience of each pilot would include the total number of flying hours, number of hours as pilot-in-command, qualifications for different types of aircraft, results of previous training, and so forth.

The other two facets would be elaborated in a similar manner. Relevant teacher or trainer variables that could influence training effectiveness are his or her motivation, knowledge, experience, teaching ability, teaching style, and so forth. Just as for students, the appropriate measures for each of these constructs would have to be designated and the information stored in a teacher or trainer database. An example of this in the aviation domain is the information about instructors or evaluators that is gathered as part of the FAA's advanced qualification program (AQP). This information can include, for example, teaching experience, classroom evaluations by students, other forms of instructor effectiveness information such as the percentage of trainees passing exams, and evaluation information such as the leniency-harshness of each instructor.

Relevant variables of the training program that could influence training effectiveness are training content, organization, timing, and the method or process of training each component. These aspects of training are more usable and integratable in the SIS when they are represented in a coherent theory of training. This theory of training should include key instructional principles related to the effectiveness of training. In the aviation domain, for example, this type of instructional information is represented in the program audit database (PADB) for fleets that are participating in the FAA's AQP program. In general, appropriate data is available from course syllabi, training materials, training program footprints, and so forth. However, this basic data may have to be re-coded to correctly reflect relevant instructional principles so that they can be systematically analysed in the SIS.

Take, for instance the principle of "just in time" instruction, which proposes that instruction should only be delivered directly before actual practice on a procedure or skill. Each module or segment of the training curriculum would have to be examined to evaluate how quickly the information was put into practice after initial training. The metric could, for example, be the number of days that intervene between the conceptual introduction of a skill and actual practice in the training program. If the principle of "just in time" instruction is correct, instructional modules that have no delay between the introduction and applied use of information should be more effective than instructional modules that have one or more days of delay.

A second example would be the principle of active participation of the students in instruction. Again, each module or segment of the training curriculum would have to be evaluated for the degree that active student participation was required. Modules requiring role-playing or re-enactment of the training, such as simulator training sessions, would be very high in active student participation. Conversely, instructional modules using classroom lecture may be relatively low on active student participation. If the principle of active participation is correct, the instructional modules higher on active student participation should be found to be more effective. The critical point is to empirically evaluate the principles and theory of training.

Evaluation of facet theories

Facet theories suggest that the components of each facet affect major outcomes of training, either singly or in combination. Since the exact process is not a target of facet theories, the training outcomes are the appropriate focal point of evaluation. The most basic type of analysis that would show the simple or joint affects of different facets on training outcomes is an analysis of variance. For these analyses, the training outcomes are the dependent variable(s) while the different values of critical facets are the independent variables. The purpose of the analysis is to evaluate either the effect of a single critical variable, or the joint effects of two or more variables on instructional outcomes.

Effects of a single critical variable If a facet of training such as "just in time" instruction or active participation is expected to have a unique, separate impact on training outcomes, then a simple one-way analysis of variance is appropriate. Suppose the important training outcome is the average number of attempts required to learn a particular skill or procedure. The number of attempts required for proficient performance would be the focal dependent variable for the analysis. The independent variable for "just in time" instruction would be the number of days of delay between the introduction of a skill and practice of the skill or procedure. This analysis is illustrated in Table 10.1.

Table 10.1 One-way analysis of variance for "just in time" instruction

	Number of days between conceptual introduction and practice:			
Independent Variable:	Training segments with no delay (Skill set A)	Training segments with 1 day delay (Skill set B)	Training segments with 2 days delay (Skill set C)	Training segments with 3 or more days (Skill set D)
Dependent Measure:	Average # of attempts for set A skills	Average # of attempts for set B skills	Average # of attempts for set C skills	Average # of attempts for set D skills

If the principle of just in time instruction is correct, the set A skills taught in training segments with no delay should have the lowest average number of attempts for acquiring proficient performance. Set B skills taught in training segments which impose a one-day delay between conceptual introduction and skill practice should have a higher average number of attempts for proficient performance. Similarly, skill sets C and D taught in training segments that have a 2, 3 or more days delay should have an even higher number of required attempts. If the principle of "just in time" is applicable to this particular training, the observed results should follow this pattern.

The presence of the expected pattern of results can be checked by simply examining the average number of required attempts for the skill sets with no delay, one-day delay, two-day delay, and three or more days delay. In some cases, the appearance of the expected pattern may be so clearly present (or clearly absent) that no statistical analysis is necessary. But in most cases, the analysis of variance adds two types of important information to a simple check of the averages. The first piece of information is whether the obtained pattern of means is different from what would be expected by chance fluctuations or not. The second piece of information is the strength or size of the differences in means relative to an objective baseline.

These pieces of information should be considered in order. If the differences in the means are not noticeably different from what would arise merely from chance fluctuations, then there are no systematic differences as predicted by the principle or theory. If, for example, the mean number of tries in Table 10.1 were essentially equal in all conditions, the principle of "just in time" training would not have the expected affect and would be disconfirmed. In this case, there is no sense in estimating how strong a difference the number of days delay makes on skill acquisition since there is no detectable effect of it in the first place.

However, if the analysis of variance shows that the differences in the means are large enough to be significantly different from chance, then there is systematic variability due to the number of days delay and this should be further analyzed and understood. First, of course, the pattern of mean differences must be checked to make sure that the means are in the expected order. In this case, the " no delay" training segments should have the lowest number of skill acquisition attempts, and the 1, 2, and 3 -day delay training segments should have successively higher averages for skill acquisition attempts. Clearly, if the means are not in this order, the principle of "just in time" training is disconfirmed.

If the means are in this order, the strength of the effect of delay on skill acquisition can be estimated from the information in the analysis of variance. For example, Hay's (1981) omega-squared strength of effect measure can be used to index how strongly delay of practice impacts on skill acquisition. This baseline for the index is total variability. The size of the effect is represented as a proportion of total variability. Facet variables that strongly influence training outcomes are practically as well as theoretically important. Variables with strong impact on training outcomes can be used to optimize the effectiveness of the training and guide the further development of the training program.

Examining facet variables in isolation, however, is only a part of the picture. Even such a basic principle as "just in time" instruction might interact with other training facets such as the nature of the content of the training module. That is, suppose some types of skills or procedures need to be practiced immediately after the initial conceptual training, while the practice for other types of skills can be delayed from the initial training without ill effects. Checking for these possible interactions among the facet variables is very important for developing a more thorough and accurate theory of training.

Joint affects of critical variables The simple facet theory presented above suggests the possibility of joint or interactive affects of different facet variables. A possible interaction between the student facet and the teacher facet is that certain types of instructors or styles of instruction may more effectively train certain types of students. For example, ex-military students might get along better with an authoritarian style of instruction while civilian students might prefer a more egalitarian or laissez-faire instructor.

An example of interaction between the teacher/trainer facet and the training program facet is that some instructors may be more effective at training modules with specific content in which they specialize. If so, instructors assigned to training modules that represent their specialty areas should be quite effective while instructors assigned to modules outside their area of expertise would be less effective. A slightly more complex extension of the analysis of variance can be used to investigate the possible joint affects or interaction of facets of training. Although more complex, these analyses will give a correspondingly greater amount of information.

Suppose, for example, that a training program had four modules taught by four instructors, and these instructors cycled through the training of each module. In such a case, the data on the average performance of the students for each module when taught by each instructor could be put in a table such as Table 10.2. In this case, the dependent variable is the average performance of each class taught on each module--for example, the average percent of correct answers on a quiz or exam. There are two independent variables, one that represents the trainer facet and one that represents the training module or segment. The appropriate analysis is a two-way analysis of variance. This analysis would give information on whether or not trainers make a difference, whether or not modules make a difference, and whether certain trainers are better for certain modules. If real differences exist, answering each of these questions can be important for evaluating and developing a successful training program.

Table 10.2 Training outcomes indicating differences among the modules

		Trainer			
		T1	T2	T3	T4
	A	High	High	High	High
Training	B	High	High	High	High
Module	C	Low	Low	Low	Low
	D	High	High	High	High

The pattern of average performance outcomes in Table 10.2 indicates a difference among the training modules. In particular, module C has consistently lower performance scores no matter which trainer is doing the instruction. In the analysis of variance, this pattern of results would be called a "main effect" of the training modules. If a main effect of training modules is found, the averages for each training module can be compared to determine which training modules are significantly lower or higher scores. In this example module C has significantly lower-than-average scores and modules A, B, and D may have significantly higher-than-average scores.

If significant, the pattern of lower scores for a module should be further analyzed to ascertain the cause of the lower scores. The content of the module might be more difficult, the instructional method for that module may be ineffective, the motivation of the students for studying that content might be lower, and so forth. Each of these potential explanations should be analyzed and empirically checked using other SIS information.

Sometimes this auxiliary information will already be available in the SIS, while in other cases new information will have to be added to the SIS. An example of the former would be the case where skill sets A, B, C, and D are also evaluated regularly for current job incumbents. This would happen in the FAA's AQP, for example, where a wide range of pilot skills is assessed on an ongoing basis for all pilots in a fleet. The relative performance of incumbents on skill sets A, B, C, and D could be checked to see whether the same pattern of lower performance on skill set C occurs. If so, the observed differences in training may be due to an inherent quality of the skills such as their relative difficulty.

An example where information would have to be added to the SIS would be evaluating the contribution of student motivation to this pattern of results. For example, questionnaires could be added in which students were asked about their motivation for learning each module. This information could be integrated into the SIS and analyzed to see whether the lower scores for module C would be due to lower student motivation.

A different pattern of training outcomes is indicated by the average performance results presented in Table 10.3 below. In this pattern of results, the low scores seem to be attributable to trainer T2. Trainer T2 has classes that have uniformly lower performance scores than the other trainers, and this pattern is consistent across the training modules. In the analysis of variance, this would be on "main effect" of trainers. As before, if a main effect of trainers were found, post-hoc tests would be used to determine which trainers had significantly lower or significantly higher overall results.

Trainers that are significantly more effective than the average could be used as role models for the other trainers. Trainers that are significantly less effective than the average may indicate some problem with the trainer. In this example some problem with trainer T2 is indicated. This problem should be further diagnosed using SIS information in a similar manner to that discussed above for instructional modules. Different aspects of the motivation, experience, and so forth of the trainers should be analyzed and empirically tested to see if they account for the observed differences in training effectiveness. Possible remedial action for trainer selection or instruction would, of course, depend on the underlying cause determined for the poor effectiveness.

Table 10.3 Training outcomes indicating differences among trainers

		Trainer			
		T1	**T2**	**T3**	**T4**
	A	High	Low	High	High
Training	**B**	High	Low	High	High
Module	**C**	High	Low	High	High
	D	High	Low	High	High

However, the effectiveness of training may depend both on the trainer *and* the material trained. This type of joint effects of two different facets would produce a different pattern of outcomes which is often called an "interaction effect". One example of a pattern of results indicating such an interaction is presented in Table 10.4 below. In this pattern, trainer T1 is effective for module B and C, but not for modules A and D. Trainer T2, in contrast, is only effective for module A. Conversely, trainer T3 is only effective for module D. Finally, trainer T4 is effective for modules A and B, but not modules C and D. In summary, the pattern of results shows that each trainer is effective on some modules but not others.

The analysis of variance would detect the significant unique effects of the combinations of trainers and training modules as significant interactions. If such a significant interaction effect is detected, post-hoc tests can be used to determine the combinations of trainer and module that result in significantly higher or lower performance. Determining the exact nature of the interaction effect has important practical implications.

In this case, for example, the judicious assignment of trainers to modules will result in much better overall student performance. That is, module A can be assigned to trainer T2, module B can be assigned to trainer T4, module C can be assigned to trainer T1, and module D can be assigned to trainer T3. Using this assignment, the average performance of all student classes should be high (and higher than any other possible assignment scheme). Conversely, the assignment of trainer T1 to module A, trainer T2 to module B, trainer T3 to module C, and trainer T4 to module D must be avoided as it would result in a disastrously low level of outcomes.

Table 10.4 Training outcomes indicating a trainer-content interaction

		Trainer			
		T1	**T2**	**T3**	**T4**
Training Module	**A**	Low	High	Low	High
	B	High	Low	Low	High
	C	High	Low	Low	Low
	D	Low	Low	High	Low

In addition to the information on the interaction pattern, the source of these differences among the trainers and training modules could be further analyzed and investigated by checking both the aspects of the trainers and aspects of the training modules that seem to cause either good or bad training results. Using the data in the SIS to further analyze and explicate the interaction effect is important for developing the training theory. The training theory is developed as explanations for the patterns of observed results are either confirmed or disconfirmed. In general, however, a facet theory can only be developed so far because it takes a broad, simple view of the relevant factors in training.

Further development of the training theory may require a theory that takes into account details such as the stages or phases of training for the job. Such theories are more complex to construct and validate than a simple facet theory, but they offer correspondingly more precise information about the optimal training sequence for the SIS. Pragmatically, more precise training information in the SIS together with the more precise theory potentially affords more ways to increase the effectiveness of training.

Stages and phase theories of training

Constructing a stage or phase theory of training

Stages In some domains, training occurs in a natural set of stages. In basic education, for example, most training occurs in stages patterned on the academic year and arranged in grades or classes. In aviation, training for a newly hired pilot often occurs in the basic stages of indoctrination training, training of fleet-common KSAOs, and then training of fleet-specific KSAOs. In addition, subsequent stages of training include recurrent training to maintain proficiency and upgrade training to transition from First Officer to Captain. Each of the stages represents a distinct training event in the career of a pilot, and each training stage can be further subdivided into phases of training. Viewing training as a set of stages and phases gives a more detailed picture of training than just considering basic training facets. However, this comes at the expense of more complexity in the theory and corresponding SIS data.

Phases Kirkpatrick (1976) hypothesized a common, generic set of phases for a training sequence. An adaptation of Kirkpatrick's theory of training phases is given in Figure 10.5. The first phase of training begins with the initial reactions of the trainees to the trainer, the training program, the training facilities and so forth. The quality and strength of these initial reactions help determine the effectiveness of the actual learning of the desired KSAOs in the second phase of training. The extent to which desired KSAOs are learned in the training environment sets a limit on the extent to which job performance can be altered in the transfer phase. The transfer phase of training requires the employee to shift the KSAOs learned in the training environment to the job environment. Finally, the results phase of training is the long-term collective effect of any changes in job KSAOs that do occur.

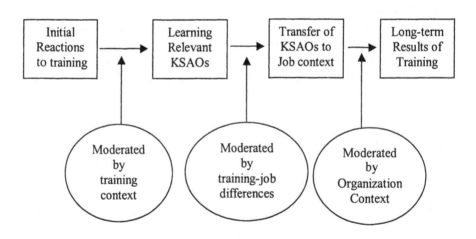

Figure 10.5 Adaptation of Kirkpatrick's (1976) phases of training

In the aviation domain, for example, the focus of training is often on highly skilled cognitive or psychomotor performance. In the case of psychomotor performance, one moderator of how well the initial reactions to training will translate into learning of the relevant KSAOs could be the quality of the simulators in which the skilled performance is learned. Aircraft simulators may vary from low-fidelity desktop simulators to extremely high-fidelity simulators such as those used to train commercial pilots. Even with the best initial reactions and motivation to learn, the learning of difficult psychomotor skills such as a landing procedure would be moderated by the quality of the training simulator.

Once the relevant KSAOs are learned in the training situation, the transfer of these KSAOs to the job is also moderated by differences in the training and job contexts. One very general principle of transfer is that the more similar the training and job context, the greater the extent to which the learned KSAOs are likely to transfer to the job. Conversely, strong or noticeable differences between the job context and training context could lead to poor or incomplete transfer of the KSAOs to the job. In the worst possible case, training creates KSAOs that would actually *decrease* job performance below no-training conditions (negative transfer).

In the aviation domain, the FAA's AQP has encouraged the development of training and evaluation methods that are very similar to the job context. Such methods as line-oriented flight training (LOFT) and line-oriented evaluations (LOE) use realistic flight scenarios enacted by normal crews in high fidelity simulators. These methods encourage transfer of relevant KSAOs between the training and job contexts.

Satisfactory performance on the job after training does not, however, guarantee that this performance will be maintained in the long run. As the time after training increases from months to years, many other factors can potentially moderate the long-term effect of training. In particular, steps taken by the organization that affect the safety and efficiency of the workplace as well as the motivation of the employees may become quite important. Pay systems, for example, that are perceived as unfair or discriminatory may result in erosion of motivation to perform well on the job. Similarly, a failure of the organization to invest in modern, up-to-date equipment and the support infrastructure necessary for good job performance may result in a performance decline.

As these examples illustrate, the advantage of a more thorough theory of training such as this phase-based theory is that a more detailed and useful picture of the entire training process can be developed. The development of the theory should be guided by the previous analyses of information in the SIS and relevant instructional literature, but it must also be tested wherever possible. This extensive empirical evaluation of the training theory should be based on the SIS information; therefore, all of the major variables for each phase of training and the major moderating variables should be included in the SIS measures. If this SIS information is available, the theory can be evaluated by performing the appropriate analyses to check on the causal flow implied by the phase transition diagram.

Evaluation of phase theories

Just like the analyses of variance for facet-based theories of training, there are both simple and complex ways of analyzing the information for this type of phase-based theory of training. The transition from one phase to the next is the simplest possible focus for the analysis and is covered in the next section on moderated regression. The evaluation of the entire phase-transition theory along with all of its moderators is naturally more complex and is covered in the subsequent section.

Analyzing a single-phase transition   Analyzing the transition from one phase to the next narrows the focus of the data analysis to an initial phase and the subsequent phase. The critical variables in the initial phase should be examined to see if they change the critical variables in the subsequent phase. In addition, the role of the potential moderators such as context in influencing this relationship should also be empirically assessed. One efficient way of doing this is using moderated regression.

Essentially, the idea of a moderator is that the relationship between variable X and variable Y depends on the value of the moderator. For some values of the moderator this relationship may be stronger, while for other values it may be weaker. The most common form of moderation is that the relationship is stronger for higher (or lower) values of the moderator, but progressively weaker as the moderator decreases (increases) in value.

For an example, take the very first two phases in the training model for basic aviation skills. The critical variable of the initial motivation of the student in phase one should predict how well the student will learn a difficult skill such as landing an aircraft in phase two. However, the extent to which the motivation will cause better learning in phase two will be moderated by the quality of the simulator used for training (Figure 10.6). For a high-quality simulator, differences in student motivation should strongly predict differences in subsequent learning of the landing skill. For a low quality simulator, however, differences in student motivation may only weakly predict differences in subsequent learning, or not at all.

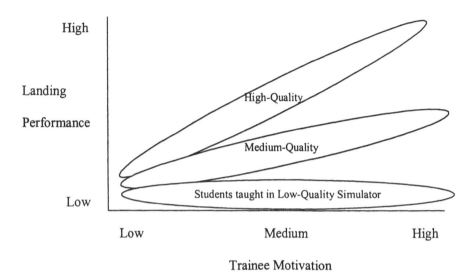

Figure 10.6   Quality of simulator moderating the relationship between student motivation and learning a landing skill

This form of moderation can be empirically analyzed by creating a new variable that is the product of the original predictor and the moderator. Suppose that different groups of students used a high-quality, medium-quality, and low-quality flight simulator. To test moderation, a new variable is created by multiplying the student motivation variable by the simulator quality variable (appropriately scored). For example, simulator quality could be coded as 3, 2, 1 for high, medium, and low fidelity, respectively, and multiplied by the motivation score for each student.

The moderated regression focuses on the learning score from phase two as the dependent variable. The analysis has three distinct levels as shown in Figure 10.7. At Level 1, student motivation score is entered as the predictor. Student motivation should have at least some positive, significant prediction of learning the landing skill. At Level 2, the simulator quality variable is entered as an additional predictor of skill learning. If the simulator quality has a direct effect on learning the landing skill, this predictor will also be significant.

At Level 3, the product variable of student motivation by simulator quality is entered as a predictor. This is the critical test of moderation. If the pattern of results in Figure 10.6 is obtained, this product variable will also predict learning of the landing skill. Such a result would confirm that the quality of the simulator moderates or changes the contribution of the student motivation to skill learning. Regression analyses can evaluate the contribution of critical variables of each phase to subsequent phases and can assess the impact of potential moderator variables.

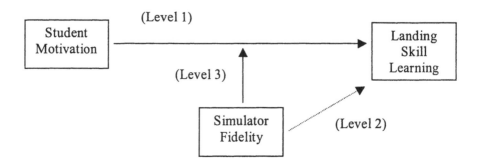

Figure 10.7 Example of regression steps to test the moderating effect of simulator fidelity on the relationship of motivation to skill acquisition

Although simple, this regression method can be used for evaluating a wide range of the hypothesized relationships among the training phases. One advantage of having this training information embedded in the SIS is that the job performance information and long-term results can also be tied to initial reactions and training outcomes. The complete set of information would allow all phases of Kirkpatrick's model to be empirically evaluated after a training intervention.

However, with such a large number of variables, the repeated use of the moderated regression analysis across the phases becomes cumbersome. Integrating the results from many analyses into a complete picture of the training process and its results may be quite difficult. Therefore, a more global, connected evaluation of the entire phase framework is desirable and can be provided by a structural equation modeling approach.

Analyzing multiple training phases   Analyzing multiple training phases requires first that the critical variables of each phase be identified and measured, along with the important moderators that could influence the relationships among these variables. The focus of the analysis is on the entire pattern of covariation among the set of variables. A hypothesized structure of relationships among variables must be specified. The analysis process evaluates the extent to which the observed pattern of covariation fits this hypothesized structure. The results of the analysis are both global indices of overall fit of the hypothesized and observed pattern of relationships, as well as the more specific evaluations of each hypothesized relationship.

An example of the type of hypothesized structure that can be analyzed using structural equation modeling is given in Figure 10.8. This example from the aviation domain concerns learning the skill of using aircraft automation such as the flight management computer (FMC). This example specifies possible KSAOs as well as potentially important moderator variables for automation training and use.

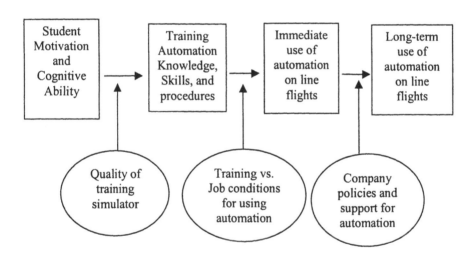

Figure 10.8 Example of training phases for aircraft automation skills

The one requirement for empirically analyzing this set of relationships is that the measures can be correlated with each other in a sensible fashion. In this example, the measures from all four phases (initial student reactions, automation training, immediate automation use, long-term automation use) are potentially measured on the same basis, namely, the individual pilot. If each KSAO or performance measure in these four phases can be assessed for an individual pilot, the data can be correlated across phases. This set of correlated data forms the basis for the analysis. The pieces of information in the SIS must be tied together in a sensible fashion for the analysis of this type of causal model.

The hypotheses of this causal model can, of course, be tested separately using methods such as moderated regression. Figure 10.9 gives a concrete example of how these causal hypotheses would be tested separately. Using the example of training pilots to use aircraft automation, possible moderating variables have been specified for each stage. The middle panels show expected results if the moderators affect the initial training and transfer phases.

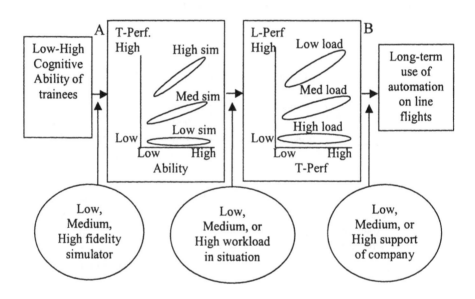

Figure 10.9   Example of training phases for training automation skills
T-perf indicates performance in training
L-perf indicates performance in line operations

In this figure, the transition from the cognitive ability of the trainees to training performance is moderated by the quality of the simulator used for training automation.   A high fidelity simulator facilitates learning, particularly for high ability trainees. Conversely, a low fidelity simulator inhibits learning for all trainees. The moderation is shown by the different relationships of ability to training performance for high, medium, and low simulator fidelity in panel "A".

Panel "B" shows a moderator affect of changes in the situation for the transition between training performance and line performance. The automation skills transfer strongly from training to line situations under low workload. However, under line situations with high workload (overload), such as departures or landings at extremely congested airports, the newly trained pilots have trouble performing the automation skills. That is, under high workload the performance of all of the newly trained pilots is low regardless of how well they did in training.

The tests for panel "A" and panel "B" can, of course, be done separately by separate moderated regressions. The advantage of the structural equation modeling approach is that these separate analyses are combined into one overall analysis and evaluation of the causal model. While more complex, the combined analysis offers much more potential information.

The correlation or covariation among all the measures is the basic input to the structural equation modeling (SEM) program. One of the basic requirements for this kind of approach is that the sample size is sufficient to uniquely estimate all of the input correlations. This is particularly important if there are a large number of measures and possible links among the variables in the SEM analysis, which could occur for a complex causal model. For further details on the execution of SEM, see more specialized references such as Tabachnik and Fidell (1996).

Consideration of the phases of training can be made more specific than this generic model. For example, Telfer and Biggs (1988) subdivided the instructional cycle into phases for 1) preparing the student, 2) gaining the student's attention, 3) communicating the lesson 4) interpreting the trainee response, 5) providing feedback, 6) facilitating understanding, retention, and transfer to new situations. This general procedure of decomposing phases into sub phases can be carried out at ever-finer levels of detail. When the phases and sub phases are very finely elaborated in this manner, the transfer effects and effects of moderators may imply specific processes that could be used in a more detailed process model of training.

If phases and subphases are converted to more basic processes, the theory becomes a dynamic, process-oriented model for training. This type of detailed, process-oriented theory is more complex than a simple facet theory or a phase theory of training. Consequently, a process theory will correspondingly require more SIS data and sophisticated analytical methods. However, this type of approach gives corresponding advantages in distinguishing different types of training and organizing a training program more effectively.

Process theories of training

Constructing a process theory of training

A training process theory will typically include processes for describing the learning of job-relevant KSAOs on the part of the trainee. Additionally, this theory might include processes for describing crew or group-level learning of KSAOs for jobs that emphasize the performance of tightly-linked crews, teams, task forces, etc. Adding this additional level would increase the complexity of the training theory, but give a potentially better account of all relevant training processes for domains with interdependent multi-person systems. Here we will focus on the training processes for a single person or trainee that would be relevant in most, if not all, training situations.

Forming a process theory of training necessarily involves implicit or explicit theoretical assumptions about the nature of learning. The appropriate theory of learning may depend on the type of individual being trained, the type of content being trained, and the instructional context. Since it is not possible to cover multiple theories of learning, the process approach will be illustrated with a simple generic learning framework for an individual trainee. The processes for this generic framework are given in Figure 10.10.

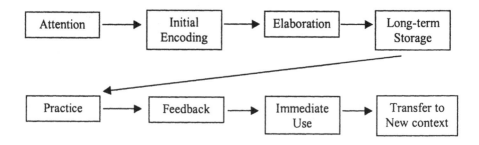

Figure 10.10 Chain of basic learning processes for training

In this figure, the hypothesized learning processes start with attention to the material to be learned. The processes continue with the encoding, elaboration, storage, practice, feedback, immediate use, and finally the transfer to new situations of the learned material. For clarity, only the major transitions from one process to another are shown; the many other potential links among the processes are omitted. Some of these processes can, of course, be further decomposed into component subprocesses. This decomposition procedure is similar to the decomposition of phases discussed earlier, but involves subprocesses rather than subphases. The difference is that the subprocesses specify precise dynamic sequences of changes that occur in a particular time frame and are linked to other processes or sub processes in an explicit way. Therefore, the decomposition procedure for subprocesses will involve an elaboration of more details of each one and a tighter integration among them.

Typically, the theory of learning will also specify principles that either facilitate or inhibit each process. A facilitating principle for the elaboration process could be that elaborating the meaning of new material in terms of previously learned or well-known material increases comprehension and retention. A facilitating principle for the practice process might be the superiority of distributed practice vs. massed practice for acquiring relevant skills. These principles are important for structuring the training for maximum effectiveness, but are also part of the theoretical structure for training that must be empirically evaluated using the SIS information.

Different processes may be emphasized for different types of learning. To illustrate this, consider a possible training cycle that would first assess relevant KSAOs and then assign trainees to different types of training based on those results. An example of this type of training cycle is graphically represented in Figure 10.11.

In this figure, the initial assessment of KSAOs is very important to determine to what extent the trainee already has the correct KSAOs for job performance and, conversely, the extent to which "incorrect" KSAOs may be inhibiting either performance or the learning process. The different outcomes of this initial assessment can lead to qualitatively distinct types of training. These types of training may have correspondingly distinct goals and processes. A succinct summary of the different goals is given in Table 10.5.

For job components already above criterion levels, training is not necessary. Generally, however, some training will still be necessary for the other job components where performance is below criterion. Partial training would typically emphasize the remaining KSAOs necessary for complete and adequate job performance. This training should include the integration of all necessary KSAOs into total job performance.

If the initial performance is essentially at a zero baseline, then the training is *ab initio* or initial training. This training must start with the premise that the necessary KSAOs do not exist and will have to be constructed from the ground up. The methods and content of training will focus on the construction and consolidation of new KSAOs. Initial training may, for example, more strongly emphasize the conceptual basis for the initial learning of skills than would augmentation training.

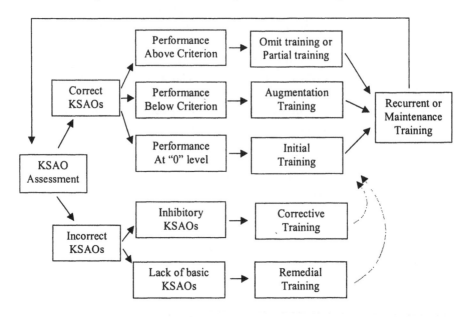

Figure 10.11 Initial KSAO assessment leading to different types of training in a training cycle

Table 10.5 Different possible types of training and goals of training

Type of Training	Goal and content emphasized by the training
Partial	Train subset of KSAOs for job components that are not at performance standard
Augmentation	Strengthen a correct but weak KSAO until performance levels are satisfactory
Initial	Create a new set of KSAOs required for satisfactory performance
Corrective	Overlay or de-emphasize an incorrect set of KSAOs with emphasis on a correct set of KSAOs
Remedial	Learn KSAO set required for effective training of the job-related KSAOs
Recurrent	Reinforce and maintain correct KSAO set and performance levels

Incorrect KSAOs are associated with two distinct types of training, depending on whether the KSAOs are inhibiting performance or the learning of KSAOs necessary for performance. Any KSAOs that directly inhibit, interfere with, or degrade job performance are incorrect. This inhibition may occur, for example, when incorrect KSAOs are accessed instead of the correct KSAOs during job performance. When this occurs, corrective training that is focused on decreasing the influence of the incorrect KSAOs relative to the correct KSAOs is appropriate. This training may include critiques of why the incorrect KSAOs are wrong as well as justification of the correct KSAOs. Corrective training could also be a prelude to subsequent retraining emphasizing and reinforcing the correct KSAOs. The first curved, dotted arrow in the figure indicates this subsequent transition.

A lack of basic KSAOs required for learning is typically associated with remedial training. The remedial training is focused on supplying the necessary preliminary KSAOs so that the trainee can benefit from normal training. Candidates who successfully complete remedial training will subsequently transition into the normal training sequence. The second curved, dotted arrow in the figure indicates this subsequent transition.

Finally, some types of training are focused at maintaining the appropriate levels of job performance and associated KSAOs on the job over a long time span. In the aviation domain, this is called recurrent training. Recurrent or maintenance training takes place after an initial level of performance and competence have been demonstrated by the trainee, but some additional time has elapsed. The basis for this type of training is the fact that the performance level and KSAOs have been satisfactory at some point in the past; therefore, the trainee "knows" the relevant KSAOs and correct standards for job performance. However, knowledge may fade, skills may degrade, and motivation to perform may slip. This training focuses on reinforcing that which is already known or performed in order to reach acceptable standards. The specific training focus may be on refreshing knowledge, sharpening skills, increasing motivation, or supporting any other relevant KSAOs.

Clearly, the different types of training that could occur with this simple training cycle have distinct goals for training. Types of training with different goals suggest that the training processes and methods might also have to be different. Table 10.6 illustrates the impact of different types of training on the learning processes discussed earlier.

The type of training that is emphasized in many contexts such as education is the initial training of new KSAOs. This training assumes that the student is a "blank slate" upon which the new training will instill the requisite KSAOs. For example, in kindergarten the skills of word recognition and reading may be assumed to not be present for all children. Therefore, the emphasis for this training is on new (to the trainee) KSAOs that are related to satisfactory levels of performance.

Table 10.6 Effect of types of training on learning processes related to training

Type of Training	Learning Process						
	Attend/ encode	Elabor- ate	Store	Practice	Feed- back	Immed- iate use	Trans- fers to
Initial	new KSAOs	new KSAOs	new. KSAOs	rehearse new KSAOs	perf. criteria	KSAOs in training	KSAOs on the job
Partial	KSAO subset	KSAO subset	KSAO subset	integrate KSAO subset	perf. criteria	KSAOs in training	KSAOs on the job
Augmen- tation	→		activate weak KSAOs	rehearse weak KSAOs	perf. criteria	KSAOs in training	KSAOs on the job
Correct- ive	shift to correct KSAOs	correct KSAOs	correct KSAOs	rehearse correct KSAOs	correct KSAOs	avoid incorrect KSAOs	KSAOs on the job
Remed- ial	basic KSAOs	basic KSAOs	basic KSAOs	rehearse basic KSAOs	basic KSAOs	KSAOs in training	KSAOs in training
Recur- rent	→		refresh old KSAOs	rehearse old KSAOs	perf. criteria	KSAOs on the job	Long term job KSAOs

For initial training, the new KSAOs serve as the focal point for the attention, encoding, elaboration, storage, and practice processes. Further, the trainer uses the performance criteria associated with these KSAOs to give the trainee feedback. The immediate criterion use of the training is that the trainee's immediate job performance meets the required standard of performance. In the aviation domain, for example, the pilot in training is formally evaluated both at the end of training for a specific aircraft and also subsequently on the job by a different instructor/evaluator pilot who flies with the trainee and insures that the appropriate performance standards are being met.

Partial training differs from initial training in focusing only on the subset of KSAOs and components of performance that are not at the required level. This subset is the focus of the attention, encoding, elaboration, storage, and practice processes. However, to facilitate integrated recall and practice, the practice process should connect the newly learned KSAOs with the rest of the KSAOs required for effective performance.

In contrast, augmentation training relies on the fact that the KSAOs are present in the trainees, even if not in the strength required for criterion job performance. Therefore, the attention, encoding, and elaboration processes can be de-emphasized since the KSAOs have already been encoded and elaborated, at least to some extent. Rather, the focus for augmentation training is to activate the relevant KSAOs and rehearse or reinforce them until they are sufficiently strong to generate satisfactory performance.

For augmentation training as well as initial or partial training, the ultimate focus is on job performance. Therefore, the criteria used for training feedback are the job's performance criteria. The training focus shifts, however, for corrective or remedial training. This shift in focus necessitates the use of different criteria for training feedback.

Corrective training, for example, is more focused on avoiding the activation of inhibitory KSAOs. Since incorrect KSAOs exist in this case, the attention process must be shifted from the incorrect to the correct KSAOs for job performance. The elaboration process would focus more on the correct KSAOs, but may also contrast these with the incorrect KSAOs. The storage process would focus on strengthening the cues for correct KSAOs so that the likelihood of retrieval of the correct KSAOs is greater than the incorrect KSAOs. Congruently, practice would be aimed at rehearsing correct but not incorrect KSAOs and feedback would be similarly directed at reinforcing correct but not incorrect KSAOs. The performance standard for corrective training might give strong emphasis to not using the incorrect KSAOs, in addition to emphasizing the use of correct KSAOs.

Remedial training is focused on the basic KSAOs required for the training processes rather than the KSAOs directly required for job performance. Although the sets of KSAOs may overlap, they may also be quite distinct. The training set of KSAOs might include basic knowledge that is necessary to understand more advanced knowledge, basic reading or education skills that are necessary for forming more advanced skills, and basic abilities that are necessary components of more complex abilities. In addition, basic KSAOs may include knowledge and skills necessary for the training method but not directly related to job KSAOs. An example is the knowledge and skills required to operate a computer in computer-based training.

For remedial training, all training-relevant KSAOs are the focus for the attention, encoding, elaboration, storage, and practice processes. The feedback for the trainee is based upon whether the basic KSAOs are sufficiently strong for effective learning in a subsequent training session rather than directly referencing job performance. Similarly, the immediate criterion of use of these KSAOs is whether they meet the needs of the training program, and the desired transfer of these KSAOs is to the training context rather than to the job. Remedial training will, of course, often be a preliminary part of the training process that is followed by a normal training sequence such as initial training.

In contrast, recurrent training will typically occur long after the successful completion of a normal training sequence. Therefore, the KSAOs and performance were satisfactory at some time in the past for each trainee, and the focus is more on refreshing or reactivating those KSAOs than on new or corrective learning. As in augmentation training, this reactivation can de-emphasize the attention, encoding, and elaboration processes. Rather, the focus for recurrent training is on reactivating the KSAOs and performance levels that have presumably been stored as a result of previous training and evaluation. These old KSAOs are the focus for rehearsal that is designed to bring performance up to the level of the performance criteria.

However, the goal for recurrent training is often maintaining the level of performance on the job over long term. This different emphasis may result in several qualitative differences in the training. For example, inducing durable motivation of the trainees to enact the KSAOs may be a high training priority. Congruently, the focus for training performance standards may be to have the trainee internalize and use the performance standards. For example, the trainees could be encouraged to use their internal standards for judging their job performance on a day-to-day basis.

Practically, the different focus and relative emphasis on these processes for different types of training would serve to help design more efficient and effective programs for each type of training. The appropriate data in the SIS should include measures of the relevant KSAOs as well as measures of the training processes themselves. This information would be used both to evaluate the results of training for the trainees and to evaluate the theory of learning upon which the training processes are based. In the short run, the evaluation of the results of training for the trainees is very important for employee development and for efficient organizational functioning. In the long run, the evaluation of the theory of learning that underlies the training processes will help shape a more correct theory which gives more effective and efficient training for the job.

The processes specified in the above table may also, of course, depend on pre-existing qualities of the trainee. Different types of trainees may have different inherent facility for the attention, encoding, elaboration, storage, or practice processes for a given set of KSAOs in a particular domain. Some trainees would, therefore, require a different emphasis on these processes during training. For example, trainees who have trouble consolidating KSAOs into long-term memory might require more emphasis on elaboration, storage, and practice processes in the training sequence. If confirmed, these individual differences should form a part of the selection battery so that each trainee can have optimized forms of training. Therefore, the evaluation of process-oriented theories of training should include a consideration of possible individual differences in the trainees that would affect training processes and outcomes.

Evaluation of process theories of training

A process theory of training will generally account for changes in KSAOs or performance over time. The relevant time span for the theory may include the training sequence, transfer to the job, and long-term trends in KSAOs on the job. The SIS theory should link training, learning, or decay processes to KSAOs and job performance. Although the example emphasized cognitive learning processes, other processes such as forgetting or interference processes could be relevant as well. The exact nature of the processes underlying training may be different for different domains (e.g. psychomotor skill vs. cognitive domains), but the specified processes should always explain the acquisition of relevant KSAOs over the training sequence. Clearly, a process-oriented theory of training will generally be a dynamic theory or model of changes in the trainees over time. Like any key part of the SIS, this dynamic model must be evaluated; however, in certain cases this evaluation process can be simplified.

Simple evaluation Evaluating a complex theory in a simple manner must be done very carefully. In a simple evaluation, many aspects of the complex theory will be ignored. To the extent that the ignored aspects would change the empirical results, a simple empirical evaluation may be quite misleading. If, however, the ignored aspects would only make minor changes or no changes in the empirical results, then a simple empirical evaluation may be essentially accurate.

The crux of evaluating a complex dynamic theory in a simple manner is to decompose the complex theory into essentially independent modules wherever possible, and to pick for each module the most important or dominant input, process, and output measures. Decomposition of a complex theory into quasi-independent process modules is possible were the processes can be grouped or linked in some manner such that the processes within a group are tightly interrelated, but the linkages between the groups are fewer and more easy to specify. In the case of the generic processes for training presented earlier, the attention and initial encoding processes might form one such group, the elaboration and storage processes might form a second group, and the practice and feedback processes might form a third group (Figure 10.12).

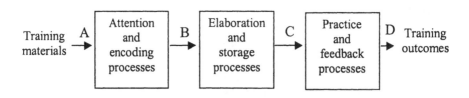

Figure 10.12 Simplified version of a process model for training

This simplified model of the learning processes underlying training has three basic processes. The training materials are input for the attention-encoding processes (arrow "A"). The attention-encoding processes produce a meaningful result in the mind of the learner (arrow "B"). This meaningful result is also the input for the elaboration and storage processes that form long-term memory. The elaboration and storage processes form a long-term memory trace (arrow "C"). This long-term memory trace acts as input for the practice and feedback processes that consolidate the overt aspects of the learning. The final result of this learning process is both cognitive and behavioral change of the trainee as a training outcome (arrow "D"). This simple example illustrates how the outcomes of each process in the sequence can be the inputs for the succeeding or subsequent processes.

To do this type of simplification, the most important or dominant inputs, processes, and outputs should be chosen for the simplified model. The goal of finding dominant inputs, for example, is to determine a subset of all possible inputs that controls most of the important variation in subprocesses and outcomes. Similarly, the goal of choosing critical processes or clusters of processes is to identify processes that strongly link to inputs and outcomes while ignoring processes with weak links. Critical outcomes include the outcomes required as important inputs for all critical processes in the dynamic chain as well as the critical final training outcomes.

In essence, a simplified evaluation treats the process model as a linked set of input-process-output modules. The actual evaluation would be empirically checking the links between inputs and relevant subprocesses, and the net effect of both inputs and processes on outcomes for each module. For this evaluation, simple empirical methods such as multiple regression or other forms of multivariate analysis can be used. Thus, if the theoretical structure of the processes can be appropriately simplified, simple forms of empirical evaluation may suffice.

Complex evaluation If the theoretical structure cannot be simplified, the interlinked processes must be evaluated with dynamic modeling methods. Completely evaluating a process theory of training would be essentially similar to evaluating other dynamic theories in the SIS. As discussed in Chapters 7 and 8, the evaluation process requires appropriately detailed and time-stamped measures of process variables as well as outcomes. This extensive base of information in the SIS serves as the foundation for constructing detailed process models and validating them. The only difference is that, in this case, the process models concern training; therefore, the measures and information for the SIS will primarily derive from the training sequence rather than on-the-job performance.

However, the extensive auxiliary information available in the SIS may also greatly enhance the evaluation of a process theory of training by giving more information about training inputs, training outcomes, or training moderators. One class of relevant training inputs are individual differences in the trainees. Examining the SIS information for systematic individual differences that affect training outcomes may point to differences in the underlying processes or outcomes. Developing the process model to account for these differences should broaden the scope of trainees to which the model applies and should increase the effectiveness of training based on the model. Therefore, the selection information for the trainees available in the SIS should be analyzed for any systematic relationships with either indicators of training processes or training outcomes.

In addition, the long-term accumulation of outcome information about post-training job performance in the SIS can also be used to refine a training process model. For the current example, the transfer of training to the job context is an explicit process step. Therefore, the exact relationship between KSAOs and level of performance at the end of training and all aspects of later performance on the job should be carefully examined to either confirm or disconfirm transfer of training. If certain KSAOs or certain performance components are found to transfer while others do not, the theory of training must be modified to account for these findings.

Further, the process markers for evaluating the theory can also be validated with job performance information where possible. In the previous example, the storage process could be marked by the rate of learning. If the rate of learning was recorded during training as a process marker, this value could be validated against the extent of later on the job learning. It would be very important to know if the trainees who exhibited higher rates of learning during training would continue to show higher rates of learning from experience while on the job. If that were so, the rate of learning could become an important goal for both the selection and training of an employee.

In the aviation domain, for example, aircraft avionics and operating details may be extensively updated over time. In such a case, it is very important to predict which pilots will spontaneously track these changes and learn appropriate operational behavior and which pilots will not. Training resources can be conserved and more efficiently used by targeting the pilots who are unlikely to spontaneously learn about and correctly use avionics innovations.

The complete evaluation of complex process theories of training is, of course, more difficult than facet or phase-based theories of training. However, the process theories offer the most precise guidance for effective training at different points of the training cycle and for different types of trainees. The processes should also point to more effective methods of training and training contexts. Therefore, the net gain in training effectiveness in the long run should be weighed against the increased cost and time required to develop and validate a process theory of training in the SIS.

Chapter summary

This chapter has reviewed the role of a SIS in developing and validating a theory of training. This theory of training can be as simple as a facet based model for training or as complex as an interlinked process model of training. In general, the initial theory for training when the SIS is first constructed would be a simpler theory such as a facet-based model. As the SIS is developed, a more complex theory of training such as a stage or phase theory may become necessary. To obtain the highest levels of explanation, prediction, and control of the training process, however, a more complex and detailed process model may have to be developed.

The sequential development and refinement of theory will generally be true for any major aspect of the SIS. To the extent that more complete, precise, and valid information is required for any aspect of the SIS, the theories underlying that section of the SIS may have to be further developed and refined. In general, the development and refinement will lead to more complex but also more accurate theories about aspects of the target system. Evaluating these more complex theories requires more effort and appropriate information, but offers scope for a wider variety of practical applications as well as more effective and efficient interventions. The cross-linking and empirical evaluation of the extended theories helps validate the entire theoretical structure of the SIS.

11 Analyzing the Job Context

Overview

Both selection and training prepare employees for the job. Although the tasks and demands of the job are taken into account in establishing, for example, required knowledge, skills, ability, and other qualifications (KSAOs), the focus is on the individual and selecting or developing an individual for the work environment. An alternative approach to focusing on the individual is to focus more on the job context and the interaction of the individual with this context as the job is being performed. This broadens the theoretical coverage of the SIS to include a wider variety of potentially important influences on work performance. These contextual influences may occur at different levels. For example, the job context may include the devices or artifacts with which a person must work, the physical and social context of the work, and the broader organizational climate or culture of the workplace.

In essence, broadening the theory to include the job context is elaborating a theory of the situation and how the situation affects all aspects of job performance. The effects of context on job performance may be elaborated at the individual level; team, crew or small group level; or some larger social, political, or functional unit of the organization. This broadening of the theory is graphically depicted in Figure 11.1.

In Figure 11.1, the concentric circles represent the different levels of both the social and physical context of the job. Typically, each level would have some degree of interdependence with the levels above or below. For example, the team, crew, or small group level typically depends on inputs from individual workers at the lower level and works to satisfy upper-level goals, directives, or projects stemming from some larger social, political, or functional unit of the organization. Conversely, the mid-level team, crew, or group produces a product, report, or result of some type that is input or feedback for the larger organizational unit and affects the functioning of that unit.

Although it may not be as apparent, there is potentially the same differentiation of levels and interdependence of levels for the physical job context. That is, an individual worker will typically deal with a set of devices, tools, or artifacts of some type on the job. The nature and arrangement of these things both constrains, and is constrained by, the overall arrangement of the total set of physical things in the group workspace. The size, shape, or functionality of the individual's work devices may limit the possible arrangements of physical things at the group level. An obvious example is that the physical size and required working positions for a device such as a lathe will preclude the location of other working devices around it. Conversely, limitations of such physical aspects of the group workspace as the total floor space or working room will limit the number and arrangement of individual workstations within that space.

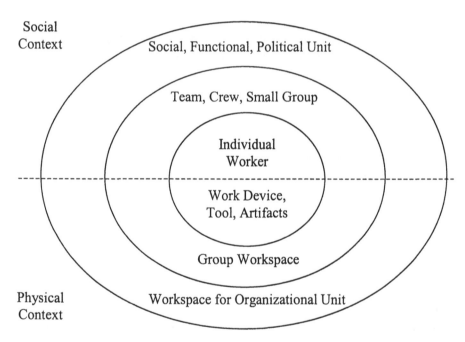

Figure 11.1 Overview of social and physical context at different levels

This interdependence also occurs between the group workspace level and a workspace defined for the entire organizational unit. If an organizational unit is housed in one building or structure, the architectural constraints of that structure will constrain the possible arrangements of workgroups and their associated workspaces. Conversely, the set of workgroups that comprises an organizational unit will put a limit on the possible physical spaces and architectural arrangements that would be feasible if they are all housed in a single building.

An organizational unit may, of course, be physically dispersed across different buildings or other structures. If so, a different set of constraints and influences will occur due to this separation. For example, any required intercommunication of these workgroups may be accommodated by electronic rather than face-to-face communication. That type of shift could, of course, have predictable effects on the amount and quality of communication and aspects of inter-group cooperation and productivity that depend on communication. These effects would, therefore, be part of the theoretical elaboration of the dispersed or distributed work context and its effects on performance of the unit.

The theoretical elaboration of the multiple levels of both social and physical effects can take advantage of the accumulated results of different areas of psychological research on these issues. Two major areas of psychology that have focused on these issues are human factors (HF) research and industrial organizational (I/O) psychology. Although these fields overlap in examining aspects of job performance, they tend to emphasize different levels of analysis and different theoretical viewpoints.

Industrial organizational psychology includes both researchers and practitioners who focus at the organizational level as well as those who focus on smaller organizational units, such as individuals. In general, I/O psychologists emphasize the human aspects or components of a system rather than the physical or technical aspects of the system. Typically, organizational psychologists would emphasize the outer level of social job context. Their research could illustrate how outer-level variables such as organizational climate and culture would affect other aspects of the system such as worker morale, productivity, and so forth.

Some I/O researchers also emphasize the team, crew, or small group level. Results from this body of research could illustrate how mid-level variables such as leadership, morale, group communication and cooperation are interrelated or related to lower-level or higher-level variables. Leaders may, for example, influence the job performance of individual workers by increasing levels of motivation, setting appropriate goals, structuring the job performance more efficiently, etc. Also, leaders typically respond to the upper-level goals or directives from the organization's management structure.

Finally, industrial psychologists emphasize the individual level of analysis, typically focusing on selection and training issues. Results from this body of research could be very relevant to either the selection or training aspects of the SIS. Other I/O research at the individual level may focus on the direct antecedents or consequences of better job performance. In field research of this type, job performance is typically measured at the level of an entire job for a relatively large timeframe such as a week, month, or year. In laboratory research of this type, performance may be indexed over one or more tasks with a timeframe of one to several hours.

However, neither field nor laboratory I/O research would typically focus on a very detailed, moment-by-moment explication of the dynamic thoughts and actions of the worker doing a specific task. In contrast, human factors research focuses on the detailed interactive nature of the worker in the work context as represented by the tools, devices, or artifacts required to do the job. Human factors theories focus as strongly on the physical and technical components of the job context as on the individual and social components.

Historically, the field of human factors originated with a concern for a detailed analysis of the individual working with some type of tool, device, or artifact to do a designated task (see Meister, 1999 for a comprehensive coverage of this history). Although deriving much of its original methods and theoretical viewpoint from the parent discipline of psychology, the field of human factors developed unique theoretical viewpoints and a large degree of interdisciplinary interaction with fields such as engineering that focus on the physical context. The most basic theoretical viewpoint for human factors was the systems viewpoint that emphasizes the interaction of the human and physical/technical aspects of a situation. Typically, this theoretical viewpoint was developed for the detailed processes underlying specific task performance. Developing this type of detailed, combined system encouraged the use of information from other disciplines.

Understanding the technical and physical aspects of the system fostered the inclusion of the engineering viewpoint in HF (Meister, 1999). The engineering viewpoint was further encouraged by the use of HF theories and information in design of new technology by HF practitioners and others. In a similar fashion, the detailed examination of human-machine interaction over extended periods of time and for a variety of different tasks also encouraged input from such fields as physiology. Such topics as the effects of fatigue or aging on human and system performance have become integral parts of the field.

More recently, the field of human factors has emphasized the cognitive as well as physical interactions of the person with these devices. This development was congruent with the cognitive revolution in the rest of psychology. Initially, cognition was integrated in research on such topics as control of dynamic systems or human-computer interface issues, where cognition clearly plays an important role. The cognitive aspect of interaction may be generally important for many if not most complex jobs and tasks that are required in technological societies. In aviation, for example, the pilots must program the flight management and navigation computer with the computer interface provided in the cockpit.

Current HF research has also broadened the focal point of the field by focusing on the mid- and upper-levels of the social aspect of the job context. That is, some human factors research targets teams or crews. In the aviation domain, for example, crew resource management (CRM) research focuses on crew-level as well as individual pilot performance. Further, the macro ergonomics branch of human factors focuses on the broader social contextual issues such as the organization levels of the job context. For more details on HF in aviation, Garland, Wise, and Hopkin's (1999) handbook gives a good overview.

Therefore, there is some overlap between the I/O and HF fields at all levels of the job context. However, even when the fields overlap, the focus of the HF research tends to be more task-specific and detailed in terms of the system processes and outcomes. In constructing a relevant theory of the situation for the SIS, therefore, results from I/O and HF psychology may complement each other at each level of analysis. The development of an integrated theory of the job context for the SIS may proceed in steps ranging from a simple theory of important facets of the target system to a complete and fully detailed dynamic model of the system processes. These three steps are similar to the development of a theory of training as discussed in the Chapter 10. Each step will be illustrated in one of the following sections.

Facets of the system

Synthesizing multiple levels of interacting social and physical systems creates a potentially very complex theory. Condensing this complexity into a set of basic facets may be over-simplifying the target system. Nevertheless, this extreme simplification may be necessary to start the development of the theory in the SIS and to generate an initial set of relevant system measures. The process of developing a facet theory is to first establish the most basic facets or components of the system, then to subdivide these facets or components into smaller subfacets, and ultimately to specify relevant measurable variables.

Meister (1999) proposes three major facets for human factors systems: general system variables, system structural variables, and general behavioral variables. These variables can be reorganized into the basic facets of the human/social system, the physical/machine system, and the combined or overall system (Figure 11.2). The human system includes the individual or social variables that affect processes and outcomes in the target system at any level: individuals; teams, crews or small groups; and larger groups or organizational units.   The physical system includes the electrical, mechanical, chemical, or electronic variables that affect the processes and outcomes of physical or virtual machines, devices, or tools. If these things are interdependent or coupled in any way, the physical system also includes the interdependent sets or networks of the machines, devices, or tools. The combined or overall system includes general influences on both the human and physical systems as well as the interface or interaction aspects concerning the links between the human and physical systems.

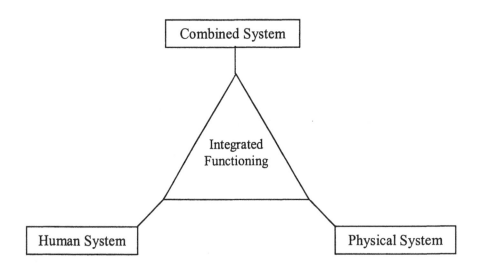

Figure 11.2 Three major facets of a performance system

Each of these facets must be elaborated into sub-facets and ultimately into constructs and specific variables that will be measured and tested in the SIS. Table 11.1 presents a possible set of subfacets, many of which are taken from Meister (1999). Clearly the variables in this table are examples that could be relevant for a wide range of performance systems. For any particular performance system, however, this list could be incomplete and may contain irrelevant variables. For other possible SIS domains, the entries in this table would have to be changed.

Table 11.1 Example of variables for a facet theory of a target system

Human system facets	Physical system facets	Combined system facets
Number of humans	Number of physical systems	Overall system complexity
Organization of humans: teams, crews, or groups	Organization of physical subsystems	Organization & balance of human and physical system
Human system goals	Physical system goals	Overall system goals
Human capabilities, KSAOs	Designed capabilities of the physical system	Overall system resources and capabilities
Human interdependencies	Physical interdependencies	Nature of human-physical system interdependencies
Physiological and psychological constraints	Electrical, chemical, and mechanical constraints	Overall system constraints such as time, money, laws
Marker variables for human system processes	Marker variables for physical system processes	Marker variables for inter-linked system processes
Human system operating environment	Physical system operating environment	Overall political, economic, legal context/environment

Although some simple facets may be directly reducible to a single construct and a corresponding measure or set of measures, most of these facets will be represented by more than one construct or measure. For example, the number of humans in the social system would, in some cases, be a simple count of assigned workers. However, in many if not most complex systems there are many types of employees that must be distinguished when counting the workers. Part and full-time employees would typically be distinguished so that the part-time employees are properly counted in full-time equivalents. Partial assignment of each employee to different tasks within the system should also be counted appropriately in the SIS measures. Finally, any qualitative distinctions among the employees that are relevant to the functioning of the target system, such as permanent employees vs. contract employees, should also be considered for SIS variables.

For the human system facets, Table 11.2 gives some examples of typical constructs and possible variables. In any given application, of course, only the variables deemed most important to the functioning of the target system would be included as measured variables in the SIS. Initially developing a fairly complete list of relevant variables, however, is useful both to avoid overlooking important factors and to facilitate SIS modifications during the ongoing SIS evaluation process.

Table 11.2   Examples of relevant constructs and variables for the human/social system facets

Human system facets	Measurable constructs or variables
Number of humans	Number and type of humans assigned to each job task or process
Organization of humans: teams, crews, or groups	Number of organization levels Distributed vs. centralized structure Permanent vs. temporary teams, groups
Human system goals	Human goal and subgoal structure Difficulty, specificity, goal path clarity
Human capability KSAOs	Profile of KSAOs for each worker
Human interdependencies	Power, communication, or management links among humans
Physiological and psychological constraints	Fatigue, stress, and workload Required reading level, social skills, etc
Marker variables for human system processes	Sensation, perception, cognition, decision, communication, and action processes
Human system operating environment	Morale, job satisfaction, organizational culture and climate variables

When SIS information is analyzed, variables that are shown to be ineffective in predicting system performance can be eliminated while new variables can be added. This modified list of influential variables helps the subsequent development and elaboration of the theory of the job context. A list of key variables is particularly important in elaborating the facet theory into a more detailed stage/phase theory or process theory as discussed in subsequent sections.

In a similar manner, the facets of the machine/physical system can also be elaborated into relevant constructs and measurable variables. Sample constructs and variables for facets of the physical system are given in Table 11.3. Some of these measured variables may also directly affect the human facet of the target system. For example, high temperatures in the physical environment may affect human processes and actions (e.g. aggression) as well as affecting a machine such as a computer (e.g. heat-related central processor failures).

Table 11.3   Examples of relevant construct and variables for the machine/physical system facets

Physical system facets:	Measurable constructs or variables:
Number of physical systems	Number and type of machines assigned to each job task or process
Organization of physical subsystems	Arrangement and interconnection of the machines assigned to job tasks/processes
Physical system goals	Designed or operational goals for each machine or physical subsystem
Designed capabilities of the physical system	Rates for input, throughput, and output for each machine or physical system
Machine/workstation interdependencies	Physical, logical, or computer links among machines or work stations
Electrical, chemical, and mechanical constraints	Electromechanical or chemical operating limits for each machine or subsystem
Marker variables for physical system processes	Critical input-output relationships for each machine or subsystem
Physical system operating environment	Temperature, pressure, vibration, or other relevant environmental variables

The interconnections between relevant variables of the human/social facet and variables of the machine/physical facet are clarified in the listing for the combined system. That is, listing of variables for the overall system should focus on the interface between the human/social system and the machine/physical system. The variables that are particularly important in that interface should be listed for the overall system facet. This is illustrated in the variables listed for the combined or overall system in Table 11.4.

Evaluation of a facet system   The evaluation of the facet system involves a check of the separate as well as combined effects of the measured variables on aspects of the target system functioning or output.   Each function or output measure serves as the dependent variable in the analysis.   The influence of the facet on this dependent variable can be analyzed by standard techniques as discussed in Chapters 9 and 10. If the facet involves qualitatively different conditions (e.g. qualitatively distinct working conditions), some version of the analysis of variance could be appropriate. If the facet variable is continuous (e.g. incentive pay levels for successful performance), multiple regression would be reasonable.

Table 11.4   Examples of relevant constructs and variables for the overall system (interface)

Combined system facets:	Measurable constructs or variables:
Overall system complexity	Number and type of human-machine units assigned to each job task or process
Organization & balance of human and physical system	Arrangement and interconnection of humans and machines, tools, or devices
Overall system goals	Overall short-term and long-term goals for human-machine system
Overall system resources and capabilities	Resources and capabilities for the entire human-machine system
Nature of human-physical system interdependencies	Human-computer or human-machine interface links
Overall system constraints	General system limits of, time, money, personnel, space, legal issues, etc.
Combined, inter-linked overall system processes	Critical input, throughput, output measures for combined human-machine processes
Overall system context and environment	Political, social, technical, legal, and physical operating environment

When all variables for the analysis come from the same level, the analysis is straightforward. For example, if all predictor variables concern a single individual working with a single machine and the dependent measure is the output of that worker-machine combination, the predictor variables and the criterion or dependent variable are on the same level. An analysis of variance or multiple regression analysis would then be appropriate for analyzing the total, separate, and joint affect of the major facet variables on outcomes.

However, when one or more of the independent variables is at either a higher or lower level than the dependent variable, the levels should be carefully considered and recognized in the analysis. Suppose, for example, the productivity of working team was the dependent variable. Predictors concerning each individual's motivation, KSAOs, and so forth would be measured at the lower level of a single person while the criterion is measured at the team level. Since each working team is comprised of two or more individuals, each of these individuals would have separate scores for the particular KSAO measures. To correctly analyze this type of data, the lower level information should be aggregated into group level measures and then analyzed. One approach, is to average individual KSAOs in each team for use as a predictor variable. Alternatively, a form of hierarchical analysis can be used that first partials out the effects of the lower level variables and then completes the rest of the analysis.

Predictors measured at a higher level than the dependent variable must also be carefully considered. For example, measures of the organization's culture or climate would be at a higher level than the working team. There must be some variation in this measure for it to be analyzable. Suppose, for example, that all work teams in a sample were taken from the same organizational division. If climate was homogeneous in the division, a measure of organizational climate would be a constant for all the cases and therefore have zero variance and not be analyzable. Instead, this kind of measure would have to be considered as a context-setting variable for all the results of the analysis.

However, suppose working teams were sampled from different organizational divisions with different organizational climates. In this case the organizational climate predictor variable would have variability and its effects could be analyzed using standard statistical methods. mentioned above. Using hierarchical methods, team performance could be predicted from individual-level KSAs, team-level variables, and the upper-level organizational climate measure. The information from these analyses could be used as a basis for building more detailed stage or phase theories of the functioning of the total system.

Stage-phase theories

For some target systems, it may be worthwhile to construct a stage/phase theory of the processes and output from all levels of the system. This type of theory is useful, for example, if the system has functional periods that are qualitatively distinct, clearly identifiable, and stable over some time interval. The qualitatively distinct stages or phases may, for example, be caused by changes or constraints in the broader context of the organization. The end of a government budgetary cycle, for instance, may predictably cause a phase of spending all the remaining money in the budget so that the budget is not subsequently decreased. During such a phase, the qualitative functioning of the unit might be organized around the selection and acquisition of all necessary material for the next work cycle. After the spending deadline, however, the qualitative functioning of the unit may return to the usual set of task goals for that organization.

Alternatively, phases may also be identified by a careful examination of the processes of different levels of the social and physical systems over time. That is, the activity of the entire system may at some point be marked by a process at the lowest level of the man-machine combination while at other points the activity is focused on the mid-or upper-levels of the social-physical system. One simple way to try to identify this type of phase is to plot the variability of the critical indicator variables at each level over time. If the variability at different levels systematically changes over time, these changes may indicate system phases. An example of this is given in Figure 11.3.

In this figure, the initial period of variability up to time "A" concerns an individual worker and his or her workstation. This person may, for example be working on a computer to calculate a spreadsheet concerning monthly expenditures. This interface activity bridges the human and machine because the human is working with the computer as an integrated system to obtain the spreadsheet projections.

Suppose the worker finds a projected budget deficit, and subsequently thinks about the reasons for being over budget and possible solutions. This individual cognitive activity is represented by the variability leading up to time "B". The individual then reports the budget situation to his or her team leader, and budget issues are discussed at the next team meeting (represented by the variability at the team level ending at time "C"). The team decides to call an executive meeting of their division so that possible solutions for the budgetary problem (such as restructuring of work duties) can be considered and implemented (variability ending at time "D"). This information may be distributed to all divisions using electronic networks, a conferencing telephone and distance-based data display system, or other communication devices that are part of the workspace for the organizational units. Therefore, the final cycle of activity (after time "D") is again an interface phase in which the organizational unit uses its workspace tools and devices to carry out upper-level communication on budget issues.

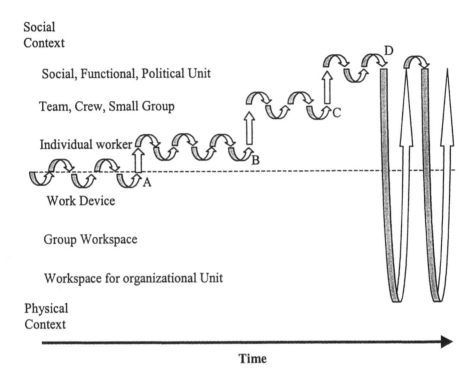

Figure 11.3 Pattern of variability over time that may indicate system phases

An example from the aviation domain would be the phases of activities prior to, during, and after the normal departure of a typical flight (Figure 11.4). The airline's dispatch system and the national aerospace computer system are linked together to constrain the possible routes and times for a given flight. In the U.S., for example, a landing slot at the destination airport must be available before a commercial flight is allowed to depart. Typically, the Captain accesses the information from these systems and interacts with other computer system such as the weather computer to plan the flight (point A). The Captain will then brief the First Officer on the intended path of the flight as well as pertinent weather information and other operationally relevant conditions. This crew interaction for communicating and setting the goals and schedules for the flight is represented by the time prior to point B. The takeoff phase involves an intensive interaction between the designated pilot flying and the aircraft as it accelerates down the runway and ascends (point C). During the later phase of the ascent or at the beginning of cruise, the aircraft is typically set into auto pilot mode. During this phase, the auto pilot is conducting the moment-by-moment control of the aircraft while the crew monitors and alters the inputs for the auto pilot as necessary (point D).

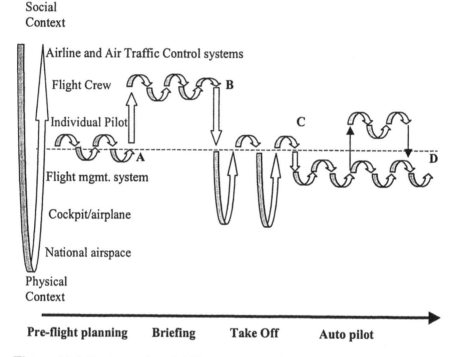

Figure 11.4 Pattern of variability over time for part of a commercial flight

These simple examples illustrate how tracking the values of key variables at each level over time may help indicate qualitatively distinct stages or phases. If this type of pattern is repeatedly and reliably found, the stage/phase theory can be organized around these qualitatively distinct phases. Noticeable variability in key input, process, or outcome variables at particular levels of the social/physical system would mark each distinct phase. For levels of social/physical system *not* involved in a phase, the values of key variables may be marked by a coherent and relatively stable or non-varying pattern of system processes and outcomes. In the aviation example presented previously, the takeoff phase may be marked by expected variability in the cockpit control inputs made by the pilot flying. In contrast, however, relatively little or no variability would be expected during that period at the levels of the ATC systems, the flight management system, and so forth.

In elaborating this type of theory, three types of variables are critically important. First, the key variables that mark each phase in the process must be identified. Second, the trigger variables that cause the transitions from one phase to another must be specified. That is, events or values of variables that cause the system to shift into another type of functional mode must be carefully defined. Third, the moderator variables that strongly influence the processes and outcomes of each phase must also be defined and the nature of the moderation specified. These three types of variables are illustrated in Figure 11.5.

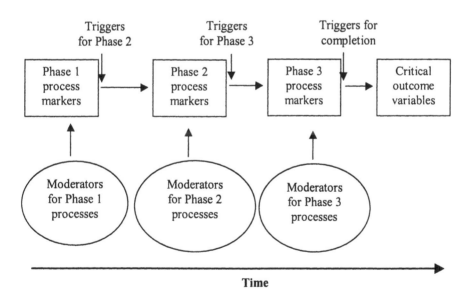

Figure 11.5 Illustration of marker, moderator, and trigger variables for a phase theory of a target system

A concrete example of aircraft departure using the above framework is given in Figure 11.6. In this figure, the phase 1 process markers are the preflight planning and completion of the takeoff checklist. The moderators for phase 1 are heavy or light loading of the aircraft, weather conditions such as fog or icing, and the departure conditions at the airport such as congestion and expected delays. The trigger for the transition from phase 1 to phase 2 is the ATC clearance for takeoff (dashed arrow). Plausible variables for phase process markers, moderators, and triggers have been inserted for the subsequent phases and will be used for discussing validation.

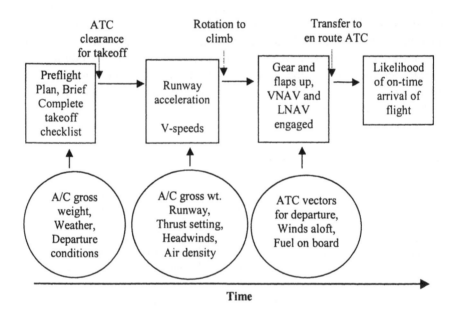

Figure 11.6 Illustration of marker, moderator, and trigger variables for a phase theory of a typical aircraft departure

In validating a phase theory, the measurements of the key variables must be appropriately tagged by time. That is, trigger, process marker, and moderator variables must be measured with sufficient time fidelity that they can be used in the SIS for the empirical evaluation of the phase theory. The hypothesized trigger variables must occur prior to the subsequent phase. In this example, the trigger variables for phase 2 must occur prior to the phase 2 process marker variables. That is, the ATC takeoff clearance must occur prior to the acceleration of the aircraft down the runway. Similarly, the trigger variables for phase 3 must occur before the phase 3 marker variables. In this case, the rotation to climb must occur before the retraction of a landing gear. In general, the trigger variables must be measured at some point prior to the measurement of the marker variables for a phase, and the marker variables for a phase should be measured before the outcome variables for that phase.

Unless moderators are presumed to be absolutely stable, the measurement of the important moderator variables must also be timed well enough that they can be unambiguously associated with the appropriate phases. In this example, the value of the hypothesized moderator variables for the takeoff roll such as headwinds and air density should be measured concurrently with the process markers for takeoff. That is, the headwinds and air density should be measured exactly at the time of take off to precisely predict the length and time of the takeoff roll. Similarly, the hypothesized moderators for the other phases should also be measured concurrently with the process marker variables for those phases. When these variables have time marks of sufficient accuracy that the time course of the hypothesized phases can be clearly established, the empirical evaluation of the phase theory can proceed.

Evaluation of the system phases A good phase-based theory should be able to account for the basic outcomes in each phase and transitions from one phase to another. Therefore, the set of predictors that are measured in each phase should be able to predict the variables marking important outcomes from that phase. Further, the set of theoretically specified predictors should also predict the triggering events or variables that cause a transition to the subsequent phase.

Since the events and processes in a particular phase may depend on the events and processes of previous phases, the information from prior phases may serve as predictors for subsequent phases. In the above example, some aircraft processes that are involved in takeoff and climb are similar. Based on the similarly, the theoretical assumption could be that the system should exhibit these processes in a stable manner. In this case, if normal takeoff power was available during the takeoff phase, adequate power should also be available for the climb phase. In general, if similar processes occur in prior and subsequent phases, the process marker variables from earlier phases might also be used for predicting processes in later phases.

If predictors and criteria have a simple linear hypothesized relationship, the empirical evaluation would typically use some form of simple data analysis such as multiple regression analysis. In particular, if moderated relationships are predicted, one appropriate analysis would be a moderated regression analysis. The essential criteria for the empirical evaluation would be the accuracy of prediction of outcomes for each phase and the accuracy of predicting the timing and nature of the transition from each phase to the next phase. Once a stage/phase theory is validated, it can serve as a framework for constructing a more detailed process theory of the target system.

Process theories

When it is necessary to have a very precise theoretical account of the target system, an initial facet or phase theory of the target system may be elaborated into a process theory. This development would be justified when the benefits of the precise theory would outweigh the development costs or when development is necessary for legal, regulatory, or political reasons. One example might be the causal processes underlying catastrophic accidents that jeopardize the survival of the organization.

The development of a process theory should take advantage of the information contained in the preliminary facet or phase theory as well as information derived from exploratory analyses of the SIS data. Typically, the processes indexed by summary constructs in a facet theory or by marker variables in a phase theory will be elaborated into more complete theoretical accounts of each process and subprocess of the target system. The exploratory analyses can either refine the account of the previously expected processes or suggest new processes for the developing theory.

This conceptual elaboration of processes should be both qualitative and quantitative. Wherever possible, the qualitative and quantitative aspects should be combined into a corresponding computational model that becomes the focal point for empirical validation. If the target system includes multiple interlocking levels, the elaboration must take these levels into account.

Elaboration across levels can either be top-down or bottom-up. A top-down elaboration starts at the more general levels of the social/physical system and successively develops the more detailed levels. Conversely, a bottom-up approach first elaborates the most basic level of the human-machine system. Subsequently, the upper levels of the social-physical system are elaborated and appropriately connected to the basic level. Bottom-up conceptual elaboration process is illustrated in the following sections for a generic complex task performed in a multi-level environment.

Elaborating basic human-machine processes A simple, integrated process account of the human-machine level is illustrated in Figure 11.7 for a generic task with significant cognitive demands. In this figure, the central cycle of worker-device interaction processes is in the center of the ellipse. In this cycle, human information processing leads to a decision that results in a response or controlling action of some kind. This control action is sensed by the device as a control input and processed to yield a change of display or physical output for the user. The worker/user senses this output or result, and the interpretation of the result will typically feed directly into further information processing, which begins the cycle anew.

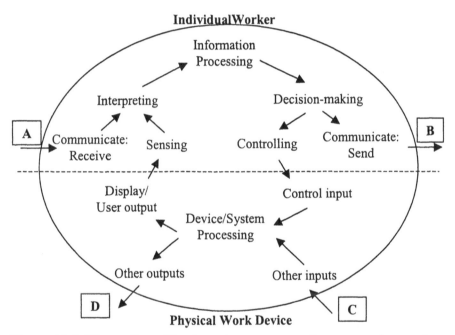

Figure 11.7 Illustration of basic system processes for one worker and device

To obtain a computational model, both the processes and the links in this diagram must be represented in a precise form such as a mathematical or computer model. In a computer model, the processes would be represented by programs, algorithms, or subroutines that would take the available inputs, execute the hypothesized process, and produce the appropriate output. The links would represent the activation of a subsequent process by a prior or preceding process, which could be modeled as subroutine calls, parameter transfers etc. These activation links would also carry the necessary values of inputs for the subsequent processes and outcomes from prior processes. The processes at the basic human-machine level can also connect to processes at the upper levels of the system as denoted by the labeled arrows A, B, C, and D in the figure.

The labeled arrows A, B, C, and D illustrate the links of this cycle to the outer levels of both the social and physical contexts. Arrow A represents the reception of communication from some other person. Most often this communication will be from a crewmember, co-worker, supervisor, or subordinate. This communication must be encoded and interpreted for meaning. Indirect forms of communication may require the use of a device to receive the communication (e.g. voicemail, email), but for simplicity this complication is not presented in the figure. Once the communication is encoded and interpreted, its meaning is input for information processing in the basic cognitive cycle.

Arrow B represents sending communication to some other person in the social context. Again, the means of communication may be either direct or indirect (e.g. spoken vs. written or electronic), but only direct communication is diagramed. Indirect communication may require the use of another device or artifact such as a computer terminal, but the goal of communication in either case is informing another person. Therefore, this arrow functionally represents either a direct or indirect link to the other levels of the social context.

Arrows C and D represent corresponding connections of the physical device to other levels of the physical context. For simple, isolated physical machines, these connections may be lacking or only rarely occur. For example, the "typing pool" in the days of the typewriter was composed of isolated physical machines that had essentially no interdependence.

Currently, however, word processing may be carried out on a set of networked computers. This network allows the output of one machine to become the input for another, and vice versa. Due to this interdependence, the activities of a worker using one machine may impact other machines. An extreme example is issuing an illegal command and crashing the network. A more common example is being logged onto an application or file that only allows one user, thereby preventing the access by other users.

An important instance of links C and D is a worker's use of a device to seek necessary information. An example would be using an information database in a library to find appropriate information. In this case, the worker uses the interface of his or her terminal to send queries out to the appropriate database servers (arrow D). The database servers would represent the next level of the physical system. The servers process the queries and send the results back to the worker's machine where it is displayed, which is represented by arrow C in the diagram.

Particularly for complex systems the interdependencies of a single worker's device with other physical devices may be both frequent and important. In aviation, for example, each pilot can use an interface to access the flight management computer system. The inputs made by one pilot to this system directly affects the use of the system by the other pilot. The basic interaction of a single pilot and the flight management system is illustrated in Figure 11.8 for the task of setting a higher altitude.

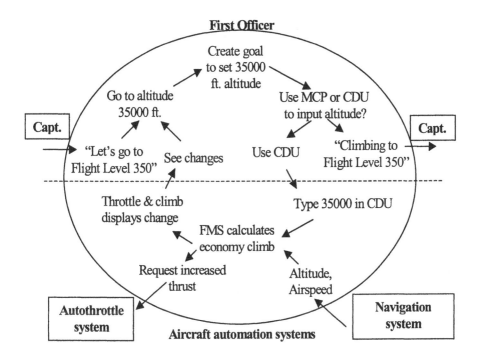

Figure 11.8 Illustration of basic system processes for a pilot setting an altitude

As illustrated in this figure, the basic cycle begins with the Captain requesting a higher altitude. The First Officer interprets this communication and sets the goal of changing the aircraft altitude to 35,000 feet. The First Officer must decide whether to accomplish this task using the mode control panel (MCP) or using the control display unit (CDU) to input the new altitude. The First Officer confirms to the Captain that he is changing the altitude and decides to use the CDU to accomplish this. The actual control input is typing 35,000 into the CDU and entering it in the appropriate field.

A similar chain of events occurs on the physical side among the aircraft automation systems. The flight management system (FMS) takes altitude and airspeed inputs from the navigation system and calculates an optimal economy rate of climb that achieves the goal while saving fuel. The FMS requests exactly the required amount of thrust to achieve this climb from the autothrottle system. Assuming these commands have the desired effect, the increased throttle settings and rate of climb are displayed on cockpit instruments. Closing the loop, the First Officer monitors these changes in the displays to ensure that the intended altitude increase occurs as expected.

Understanding performance for this type of interlinked process system requires a careful explication of all significant links between processes. The links among processes within one level of abstraction must be carefully specified. In addition, the links to processes in systems at other levels of the physical and social context must also be carefully specified. For low-level processes, only the links to the upper-level social and physical context are necessary. When the mid-level processes are elaborated, however, both the potential links to lower-level as well as upper-level levels must be specified if they are relevant.

Mid-level process elaboration   Once the basic level processes and interdependencies have been specified, the conceptual elaboration can proceed to the next higher level. On the social side, the next level would involve human teams, task forces, or working groups. On the physical side, the next level would involve a set or network of machines or machine-based processes that are interdependent in some respects. This next level is illustrated in Figure 11.9.

On the social side, the individual worker is commonly a part of an interdependent crew, team, or workgroup of some type. The communications and actions of the individual worker act as input to the group's communication structure and processes. The communication structure and processes may link with the team's social structure and more general team processes. Conversely, the communications and actions of the other team members can directly impact the individual worker. Team communications or actions can give required information for the individual worker or affect teamwork or task performance of the individual in a variety of possible ways.

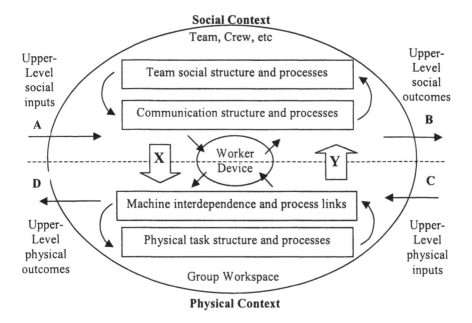

Figure 11.9 Process elaboration at the team, crew, and small group level

On the physical side, the device used by an individual worker may be part of a network or interdependent set of machines or physical processes. Most commonly, the machine or processes interdependence will be based on the physical task structure and required processes for the job. An example of direct interdependence of physical machines would be an assembly line. An assembly line is an efficient interconnection of machines for doing complex, multi-stage assembly.

The assembly line forces a strong interdependence among the machines. In an assembly line, the first machine on the assembly line must complete its task before the object is released to be worked on by a second machine, and so forth down the line. This machine interdependence constrains the possible processes that can be performed on any particular machine at a given time.

The machines of the physical environment can, however, be virtual as well as physical machines. For example, a main computer can be emulating separate workstations for a set of workers. Each of these virtual workstations could be connected via a server. In such a case, the workers could exchange work products across the network. If a worker had to wait for products from another worker to complete his or her task, the constraint would be very similar to the constraints of a physical assembly line. However, more subtle interdependencies are also possible in this case. The overall processing rate of the server may depend on the combined load of all workstations. In such a case, a single worker doing a high-demand task could slow down the response times or system effectiveness for all workers in the network.

For clarity, the potential interface links between the team or crew social level and the physical aspect of the group workspace are summarized and represented by arrows X and Y. These links may, of course, be as important as the human-machine interface links at the lower level. An example where the interface at the group level would be important is the physical devices and processes used for team communication amongst a dispersed team. In this case, the capabilities of the communication devices to send and receive documents, spreadsheets, images, audio recordings, or other forms of information amongst the team would determine the limits of team communication (arrow Y). That is, the interface and capabilities of the group workspace system would constrain team communication, coordination, etc.

The interface constraints can, however, also flow in the opposite direction. That is, aspects of the team social or communication structures may limit the use of the communication devices (arrow X). For example, even though it may be technically possible for a low-level employee to send an email message directly to the Chief Executive Officer of the organization, it may rarely be done. Communication restraints due to social power or status differences may therefore limit the actual use of communication devices or a communication network and consequently limit group information interchange.

As previously, the A, B, C, and D arrows in Figure 11.7 indicate the links of the middle-level social and physical system with the upper levels. The team, crew, task force, or small group takes social input from the upper levels of the organization. This input may consist of projects, goals, budgets, or other parameters of the work situation that are set by the upper level of the organization. Conversely, the output or products of the team or workgroup, such as team reports, typically are communicated back up the social levels of the organization.

Similarly, the mid-level physical system may interact with the larger physical context. If the system is organized around the output of a physical product, the product produced by a work team would typically be input for other types of physical processing. For example, the products produced by a workgroup may be transferred to a storage area and processed for distribution by other units of the organization. Similarly, the raw materials for a physical production task are generally provided by other units or processes in the organization.

If the system is organized around an information-based product rather than a physical product, the nature of the links to the upper levels could be similar although the products would be more conceptual than physical. That is, raw information inputs could be provided by data-searching devices or other organization units designed to obtain this information. Similarly, the processed information output from the work team could be sent out for further processing or distribution by other units. If the product is a service, the qualitative nature of links to upper levels of the social/physical system might be very different, but the mid-level processes might still be strongly linked to, and very dependent on, the upper level processes.

In the aviation example, several mid-level processes are relevant to the First Officer setting a new altitude. The mid-level of the context is represented by the air traffic control (ATC) system. This context has both a social aspect in the air traffic controller and a physical aspect in the machines that he or she is working on in the air traffic control centers (Figure 11.10).

The first step in the processes at the mid-level would be the crew's discussion of the situation. This discussion process may involve the value the crew places on fuel efficiency and the knowledge that fuel efficiency increases with altitude. This discussion could result in a decision to request a higher altitude from ATC. The request for higher altitude is processed by a controller, who represents the social aspect of the upper level ATC system. Assuming the crew is cleared to the higher altitude, the Captain would give the order to increase altitude to the First Officer as discussed above.

The First Officer's entry of the new altitude using the CDU is represented by the innermost ellipse and was discussed previously. In addition to this human-machine interaction, both the Captain and the First Officer are using other aspects of the aircraft automation systems, represented by arrow "X". The Captain may, for example, be adjusting thrust settings, transferring fuel, or checking the navigation display, weather radar, and so forth. The effects of these actions change the status of the aircraft and the corresponding displays, which is the feedback to the crew and is represented by arrow "Y". Together, arrows "X" and "Y" represent the interface between the crew system and physical aircraft system at this middle level.

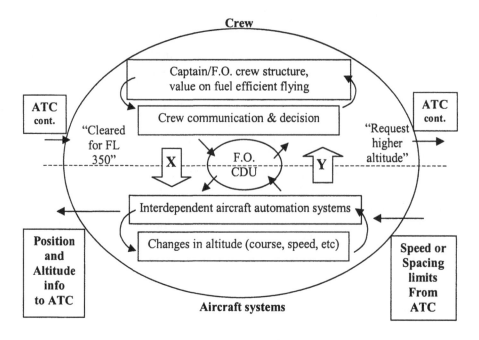

Figure 11.10 Process elaboration at the crew level for increasing altitude

Inputs and outcomes from the upper-level ATC system may also affect the physical aspects of aircraft system. The potential inputs and outcomes are illustrated in the bottom half of the figure. For inputs, the speed or spacing limits imposed by the upper-level of the air traffic control system are constraints on the physical operation of aircraft. These inputs set the boundaries or limiting values for the operation of the physical aspects of the aircraft system. These limits may represent time, position, speed, or altitude requirements for the aircraft.

Conversely changes in these key parameters of the aircraft are outcomes of the mid-level processes that act as inputs for the upper-level ATC system. For example, as the aircraft ascends to the new altitude, it changes its position and altitude, which are input to the upper-level ATC system. To complete modeling of a multi-level system such as this one, the appropriate processes of the upper-level systems must also be elaborated and modeled.

Elaborating upper-level processes To complete a multi-level process theory using a bottom-up approach, processes at the upper levels of both the social and physical context must be specified. To do this, the processes that are relevant at the upper level must be elaborated and connected to the processes occurring at lower levels. An example of this interconnection is given in Figure 11.11.

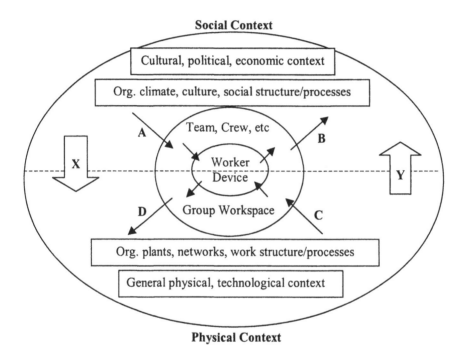

Figure 11.11 Process elaboration at upper social and physical level of a system

The specific processes would, of course, depend on the specific target system being modeled. However, for many jobs the general social, political, and economic context of the organization as well as its climate, culture, and social structure and processes would have an impact on the teams or working groups (arrow A). Similarly, the general physical and technological context for doing the work as well as the way the organization has structured its plants, networks, and work structures and processes would impact on the devices, artifacts, and physical aspects of the group workspace (arrow C). In elaborating this level of the model, the problem may be to prioritize among a welter of possible influences and to obtain reasonable measures for the upper level processes that are included in the model.

Arrows X and Y represent the potential interface between the social and physical systems at the upper level. For example, the development of the Internet technology in the physical context has changed some aspects of commerce and the economic context (arrow Y). Similarly, the open political and regulatory structures of the United States culture may have facilitated the development, spread, and use of Internet technology (arrow X).

To complete the aviation example, the upper level of system processes is depicted in Figure 11.12. For this example, the upper level embedding system of the air traffic control network was chosen. An additional upper-level system could be the corporate structure for the airline.

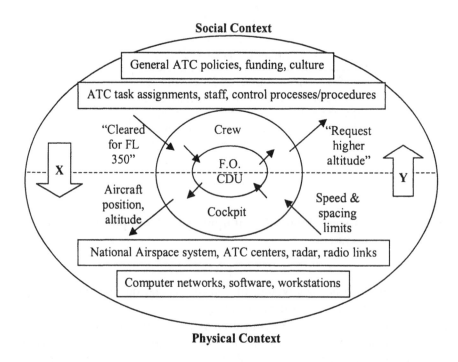

Figure 11.12 Process elaboration at the upper level Air Traffic Control system

At the social level, the essential link between a crew and the ATC system with current technology is a two-way voice communication channel. Crew requests are input to a controller acting as part of an ATC team. The controller uses ATC policies and procedures, input from other controllers, and his or her perception of the situation to respond to crew requests or give control directives. The social interaction between the crew and ATC is dynamic as the crew can request a change in a directive or refuse to comply.

At the physical level, the essential link between the aircraft and ATC system is the radar detection system and associated display terminals and workstations. The position, call sign, and altitude of each aircraft are displayed. The controller's perception of this set of displayed information plus information from other devices constitutes one aspect of the social/physical interface at the upper level, depicted by arrow "Y".

The controller must use this display and other information to enforce appropriate speed and spacing limits. The controllers use of his or her workstation plus other information channels is represented by arrow "X". One aspect of controller operations is to appropriately position and space the aircraft, limit their speed, and so forth. Since the actions of any single controller can directly affect the situation for other aircraft being controlled by other controllers in adjacent sectors, this upper level system is an interdependent social and physical system.

In achieving a satisfactory multi-level model, it may not be necessary to fully elaborate all levels. In many if not most situations, the focal point for the modeling will be on one level. In the aviation domain, for example, the focal point may be on the crew level of interaction with the aircraft systems. In such a case, the processes and linkages for that focal level would be fully elaborated, but the processes and linkages on the other levels could be simplified. For example, the processes and linkages at the upper ATC level and at the lower level of the First Officer interacting with the CDU could be greatly simplified.

The simplification must be done carefully. The simplified elaboration of the other levels must still be sufficiently complete to accurately respond to inputs from the focal level and give appropriate outcomes to that level. In this example, the upper-level ATC processes should be simulated accurately enough to mimic the likelihood that request for higher altitude is either granted or denied. Similarly, the lower-level interaction of a single pilot with the CDU must be sufficiently accurate to simulate the time required for the task and the likelihood of errors. As long as the simplification does not remove or distort key processes, the evaluation and potential use of the multi-level model will not be affected.

Evaluation of the system processes Whether constructed bottom-up or top-down, a good process-based theory of a target system should be able to account for the details of the dynamic changes in the system over time. The time scale that is analyzed may differ for the different levels of analysis in the model. At the human-machine level, for example, the time span would be governed by a single task or a coherent set of tasks for the job. At the team or workgroup level, a team project or product might govern the appropriate time span for the model. At the upper level of the model, the time span for modeling processes in the social context might depend on a political, managerial, or budgetary cycle. The time span for modeling changes in the general physical context might depend on the pace of technological development, innovation, or the replacement of work machines in the production cycle.

The empirical evaluation process relies on the computational model of the system structure and processes. If structural aspects of a system can be examined independently of the processes, techniques such as factor analysis or multi-dimensional scaling can be used as covered in Chapter 6. Typically, evaluating static structural aspects of the theory is much simpler than evaluating the dynamic aspects, but it is also much more limited in its results and implications.

Evaluating the multi-level dynamic processes of a target system involves capturing the expected dynamic changes in the target system over a suitable time span. Evaluating multi-level processes requires appropriate amounts of time-marked data at each of the designated levels. The techniques discussed in Chapters 7 and 8 on dynamic modeling can be extended to include the multiple levels of dynamic processes for this type of model. The connections among processes within a level (including the interface links) must be specified as well as the process linkages across the levels. As the theory becomes more complex, the complete evaluation of the theory requires correspondingly more time and resources. If evaluation resources are limited, the evaluation can be prioritized and focused on limited areas of the theory.

When prioritizing the use of evaluation resources, the evaluation can be focused on aspects of the theory which are conceptually important, problematic, or practically useful. For example, the aspects of the job analysis that justify the worker selection methods currently in use or the training given to workers might conceptually be more important than other parts of the theory. If so, these aspects of the model would be one of the first targets for empirical validation.

Alternatively, the focus may be on problematic aspects of the system. For example, either accidents or serious incidents happening in the workplace might indicate problems that should be a focal point for evaluation. In such a case, the parts of the theory most closely related to the genesis of such accidents or incidents would be given higher priority in the empirical evaluation process.

Finally, the focus may be on the most pragmatic, cost-effective evaluations of the processes. Pragmatically, the key processes at each level that have the strongest influence on final system outcomes may be prioritized for empirical evaluation. Validation of these key processes would give the greatest potential bottom-line benefit for the evaluation costs.

The feasibility of partial evaluation of areas of the theory should be guided by the SIS theoretical structure and, if available, the computational model. That is, the model can be used to quickly simulate and understand how the area to be evaluated is influenced by other linked processes. If the degree of influence or interference is low, the specific area of the theory can be independently evaluated. In aviation, for example, if crew processes were the area of interest and the model showed that crew processes in the cockpit were relatively independent of larger organizational/social factors, then the crew process area of the theory could be separately evaluated. Thus, the empirical evaluation process may work from "centers of theoretical interest" out toward the full theoretical version. In this manner the evaluation task can be handled by fewer people and fewer resources, although the evaluation may be stretched out over time.

Chapter summary

This chapter has presented an integrated social/physical systems viewpoint for including the job context in the job performance theories in a SIS. In other domains, the contextual elements may be different but would be elaborated and included in the SIS in a similar fashion. The theory and data about the context of the target system can be initially developed as a facet theory, later elaborated into a stage or phase-based theory, and finally elaborated as a multi-level system of interconnected processes. The elaboration of such a system may be top-down or bottom-up. The bottom-up development process for a generic performance situation was illustrated with aviation examples. Other hyper-complex multi-level domains may be theoretically modeled and empirically evaluated in a SIS by using comprehensive system viewpoints and methods discussed in this chapter.

12 Information Management

Overview

Once a SIS is started, the theories and data in a SIS are a potentially rich source of information for management. This information can be evaluated for understanding the past history, current state, or the likely future states of a target system. Understanding the current state and past history of a system helps identify problem areas that should be targeted by management. Being able to accurately predict the future effect of changes in the system helps management design appropriate interventions to either solve problems or optimize overall system performance. The use of a SIS to gain information is covered in the first section of this chapter.

However, this information must be appropriately disseminated not only to all appropriate levels of management but also to other information stakeholders. An information stakeholder is any relevant party whose actions or inaction can effect the operation of the target system. Stakeholders can be at all different levels of management or even outside of management. An example of a non-management stakeholder, for instance, could be an employee union. The information needs of the different stakeholders must be carefully assessed since they might be quite distinct. Based on the information needs, appropriate SIS information should be analyzed and feedback designed and disseminated in a timely and effective fashion to all relevant parties. However, privacy concerns and other legal issues should be taken into consideration when releasing data from the SIS. The identification of stakeholders, assessment of their information needs, and designing appropriate feedback is covered in the second section of this chapter.

The pragmatic test of the information from a SIS is the use of this information by managers or other responsible parties to make appropriate changes in the target system. Changes in employee selection, training, or redesign of the job or job context are examples of potential system changes. Any change should be carefully constructed and implemented based on all relevant information in the SIS. After implementation, the effects of the change must be evaluated to see if they are congruent with the intended effects. Creating and evaluating change are the final topics discussed in this chapter.

461

Analyzing and forecasting performance

Analyzing and predicting performance trends

One of the major jobs of a manager is to understand the current and past state of a target system. Optimal managerial decisions require a complete and accurate understanding of the system's current status and functioning. The history of the system can aid this understanding. In particular, the trends of performance from the past to the present may be important information. For example, if the performance of an important target system were gradually declining, the early detection of such a decline and prompt remediation of system problems by management could be quite important.

Trend analysis The simplest way to analyze the history of a system is to plot the changes in key variables or processes over time. Trend analysis is typically used to track the changes in the average or mean value of measured variables across time intervals. Appropriate time intervals must be selected that are neither too large nor too small. If too large a time interval is chosen, the graph will not be sufficiently detailed to show interesting fine-grained trends. If too small a time interval is chosen, the data points will typically be based on less information and be more unstable, and this instability may also obscure important trends. Therefore, selecting an appropriate time interval is important.

The amount and reliability of the information in the SIS will play a role in how fine a time interval can be used for the trend analysis. The more data that is available and the more reliable that data, the finer the possible time intervals for the trend analysis. Information on the reliability of the data should be available from the basic scientific analysis of the measures in the SIS as discussed in Chapter 3. If the data have high reliability, less data will be required to estimate system performance for a specific time interval. Conversely, if the data are somewhat unreliable, stability can only be obtained by increasing the effective sample size. If the data are uniformly spread over time, increasing the size of the time interval (e.g. changing from weekly to monthly averages) can increase the effective sample size, but this will be at the cost of losing fine-grained trend information. Alternatively, more data must be obtained for each time interval.

To obtain more data for a given time interval, aggregation over sets of similar systems should be considered. The theoretical and empirical information in the SIS should be used to justify the aggregation. This may involve consideration of two essential questions. First, are the systems being aggregated theoretically similar in their basic processes and critical variables? For an aviation example, the performance of crews from different fleets could be sensibly aggregated if the job performance models in the SIS for the different fleets were very similar models and emphasized the same key variables.

The second consideration for aggregation is empirical similarity. That is, are there any important empirical differences among the sets of systems being aggregated? To be combined, the groups of cases should have very similar averages, variability, and patterns of relationships among key variables, each of which can be empirically checked. Checking for mean differences between groups of cases could use, for example, a simple multivariate technique such as Hotelling's T^2 test to check on mean differences for a set of variables. The equivalence of variances and covariances could be checked with similar suitable methods. If no significant differences are found, then the groups of cases would be similar enough for a joint analysis.

Once the time interval is chosen and cases aggregated (if appropriate), trend analysis can be performed using a simple version of a multiple regression. In this regression, the time interval is used as the predictor variable while a key system performance variable is the criterion. To check for non-linear trends over time, the squared, cubed, quartic, and so forth powers of the time variable can also be used as predictors. Each predictor is added in turn using a hierarchical regression approach, and checked for significant predictive power. The interpretation of the results should focus on understanding the significant trend components.

For forecasting, the equation that best fits the past trends is extrapolated into the future. That is, the significant linear and non-linear regression weights are used with the values of future time intervals to project a predicted line of future performance. This process is illustrated in Figure 12.1. The process of extrapolation is a very simple assumption that the current trends will continue in exactly the same manner into the future. In Figure 12.1, the slope of the trend from point A to point B is assumed to continue into the future (from point B to point C).

The advantage of simple regression-based trend analysis is that it is completely empirical. That is, no theory is required to perform this type of analysis. Therefore, this type of trend analysis would be particularly appropriate in the early stages of the construction of a SIS when the SIS theories have not been thoroughly empirically validated and there is no other basis upon which to make an extrapolation.

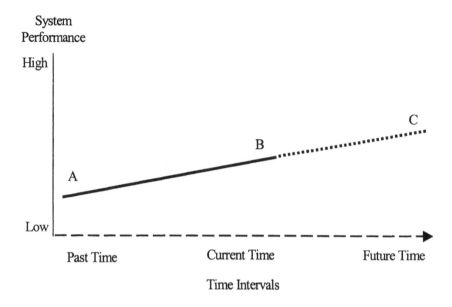

Figure 12.1 Example of a linear trend line and extrapolation to the future

Unfortunately, the simple approach will not always work. In Figure 12.2 for example, the past performance trend is complex and it is not clear exactly which trend will continue into the future. If the most recent slightly negative performance trend (A) continues into the future, performance will gradually decline. If, on the other hand, the steep negative descent of performance at the recent point B is resumed in the future, performance will sharply decline. Finally, if the trend at the earlier time C resumes, performance will actually improve in the future. Without a deeper theoretical and empirical understanding of the system, extrapolations can be dangerous.

A further disadvantage of completely empirical trend analysis is that it cannot sensibly predict the effects of any particular *changes* in the target system. In particular, this atheoretical approach cannot predict the effects of specific changes in selection, training, job design or job context because the observed trends are not tied to a theoretical understanding of the system.

To make changes on the basis of an empirical trend analysis, the data analyst or the user must supply an understanding of what might be causing the observed trends and use this understanding as the basis for the changes.

The results of making changes based on this intuitive process are, of course, apt to be no better than the user's understanding of the target system. Therefore, the understanding of the target system and predictions for possible changes that are based on a well-validated theory are much more apt to be correct.

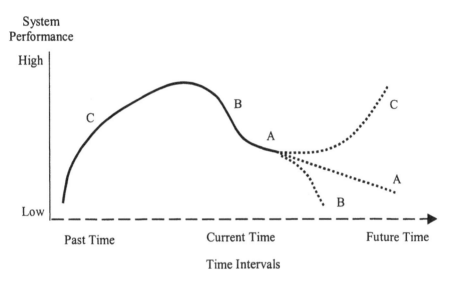

Figure 12.2 Example of a trend line and possible extrapolations to future times

Theoretically predicting effects of changes

To accurately predict all important effects of changes, the theoretical information in the SIS should be integrated. The links between the core theory of job performance and the theories underlying job selection, training, and job context should be clearly and precisely delineated in a combined, augmented model of job performance (Chapters 9, 10, and 11). Constructing the augmented model may already give sufficient understanding of the performance implications for the proposed change that no further analysis is necessary. For example, the augmented model may predict uniformly positive effects for the changes. This would support implementation of the change. Conversely, if the model predict uniformly negative effects, the proposed change could be eliminated from further consideration

For complex systems with many important links between the proposed change and the final effects on performance, the anticipatable effects of the proposed change may be difficult to discern or be a mixture of positive or negative effects. Suppose, for example, that a proposed change is to automate a part of a complex human-machine job. Automation may contribute positively to performance by reducing human fatigue and increasing the reliability of critical machine processes. Automation may also negatively contribute to performance by decreasing the human involvement in the system and task motivation, with consequent decreases in monitoring efficiency and error detection as well as possible increases in undesirable variables such as complacency or apathy. In such a case, the effects of the proposed change would have to be very carefully analyzed to determine if the ultimate performance effects would be positive, negative, or some mixture of both.

Systems that have very complex, interlinked processes may be difficult to analyze in an intuitive fashion. Instead, a more systematic and detailed form of analysis would be necessary. One approach to this more detailed form of analysis would be constructing a computational version of the augmented model of job performance and the relevant subsystems such as selection, training, or job context. If the augmented model can be developed into a computational version, the results of the possible changes can be simulated and data collected from the simulations. The results from the multiple simulations will be a distribution of values for the key performance variables for each time interval, as discussed in Chapters 7 and 8. These trends are the expected effects of the proposed change.

Although the initial construction of the computational version of an augmented model requires effort, the simulated results can be broadly useful. In particular, "what if" projections of the effects of the proposed change in slightly different simulated conditions can be quickly and easily explored. For example, a change in the basic conditions of the target system could be represented by a change in the input variables for the simulation. The results of the simulation with the changed input variables could be compared to the original trends for expected job performance outcomes. In this manner, the effect of the proposed change under different conditions as well as the effect of possible alternative changes can be extrapolated into the future and compared.

Where significant differences are detected, the causal origins of these differences can be further explored using the computational model. That is, the computational model and log of each event or interval in the simulation execution can be used to unambiguously determine the cause of the observed differences in the outcome variables. Suppose, for example, that a planned change such as the introduction of automation to a human-machine system was found by simulation to have significant negative long-term effects on certain aspects of performance such as a decrease in error detection. In this case, the simulations in which an undetected error occurred could be further analyzed for the timing of the error together with the causal variables and processes 'leading up to the error. This more detailed information could guide the design of alternate versions of the automation implementation that might have fewer undesirable side effects. In particular, an implementation that included a simulated change in the presentation of relevant error-detection information to the human operator might reduce this side effect. Using the computational model in this manner, the design of the planned intervention could be both analyzed in detail and optimized. Extending this approach, a computational model can also be used to optimize the system in general.

Optimizing target system performance

Overall optimization of the performance of a target system requires that all important aspects of the relevant subsystems should be included in a combined or integrated theory. For a theory of job performance, the critical aspects of employee selection, training, and job context could be combined with the core theory of job performance to form an integrated theory. This integration must be done carefully and give due consideration to links among these theoretical subsystems.

All the relevant subsystems can potentially interact with each other and with job performance (Figure 12.3 gives an overview of possible relevant subsystems). Certain aspects of employee selection, for example, may only be important for job outcomes if certain aspects of the job context have a specific state or value. As an illustration, selecting for academic intelligence may only cause large differences in job performance if the job-context information for job instructions and job execution is presented at a college-reading level. In this case, the requirement of a high reading level in the job context makes the selection for academic intelligence become relevant to job performance. This and other possible interactions among the theoretical subsystems must be taken into account when constructing an integrated theoretical viewpoint.

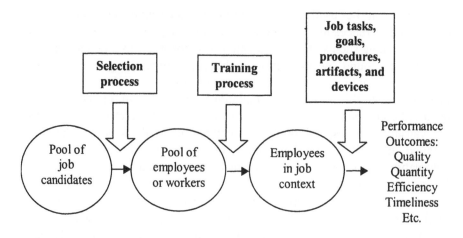

Figure 12.3 High-level overview of processes for an integrated job performance theory

Although an intuitive inspection of the integrated theoretical viewpoint may be sufficient to point out possible optimal configurations in certain clear-cut cases, in general an integrated theory will be too complex for easy intuitive use. Instead, a computational version of the integrated theory will be necessary for optimization. The integrated computational theory should be explored using computer simulations similar to those discussed for performance trends in the section above. The goal of the simulations is the same as for predicting the effects of system changes, only broader in scope. For the integrated theory, potential configurations of the selection process, training process, job context, and job design should be explored for their effects on system outcomes.

For optimizing a system, the key aspects of system performance should be combined into an aggregated performance index. That is, different aspects of system performance such as the quality, quantity, efficiency, and timeliness of performance need to be combined into a single summary index. This combined summary performance index is used to evaluate the multifaceted system outcomes so that they can be combined into an overall better or worse evaluation of the outcome profile.

Since the evaluation of possible system configurations and ultimate optimization of the target system depends on the accuracy and validity of this performance index, it is very important to get it right. The aggregation of different aspects of performance outcomes into an overall index should receive the same intense scrutiny that other aspects of the theories in the SIS have been given in the development process. Once the performance index is designed and validated, the optimization of the system can proceed by the empirical simulation process described above.

In most cases, however, the integrated theory will be quite complex. Therefore, it will not be possible to explore all possible combinations and values of all possible variables for all the theoretical subsystems. Instead, a systematic approach to finding optimal values that will be feasible with limited time and resources must be used. One such systematic approach would be a "Hill-climbing" approach that proceeds by finding incremental improvements in system performance from a starting configuration.

One problem with a hill climbing approach is that it might find a local optimum rather than a global optimum. That is, the optimization process finds a better situation than the current one, but does not find the truly "best" solution. This problem is more likely when the analysis explores only a limited range of simulated situations. One way to minimize this possibility is to do a broad-based search across a wide range of starting configurations. A possible approach to do this is to take the most critical parameters from the integrated theory and examine the system performance for a broad subset of those values.

The simulation could be run for a broad variety of starting situations, that is, each having a unique set of values. From this broad set of results, the most promising subset of two or three initial starting configurations can be selected. Each of those starting configurations can then be optimized further using a hill climbing technique to make fine-grained adjustments to the system parameters and processes that gives the best results. In this way, the globally best solution is more likely to be found.

One important result from analyzing the optimal configurations of the target system may be that there is more than one way to reach optimal performance. For example, a very selective initial selection process with minimal subsequent training might be one way to reach optimal performance, while a weak initial selection process combined with very intensive subsequent training would be another way. The trade-offs among these alternate possible configurations for the total system are important information for managerial planning and decision making. The results of possible optimal systems as well as information about the predicted effects of certain planned changes are relevant information from the SIS that must be disseminated to the appropriate parties to have the desired impact.

Disseminating information

Identifying information stakeholders

An information stakeholder is any person or party that has a legitimate need to know about some aspect of the information in a SIS. In the aviation domain, for example, information stakeholders for a job performance SIS could be the pilots themselves, the pilot union, airline management, training and safety departments, local and national representatives of the Federal Aviation Administration, and the flying public. The legitimacy of an information stakeholder can be established because of a legal contract (pilot union), legitimate organizational functions (management, training, and safety departments), federal, state, or local regulations (FAA representatives), or more general social obligations (the flying public).

For disseminating information, appropriate representatives of each stakeholder group must be designated. For a group such as the federal, state, or local regulators of an industry, the designated stakeholder is often quite clear. In this case, the designated representative will be legally appointed and empowered by the appropriate agency. In other cases, however, the designated representative may not be clear. For a union representative, for example, there may be ambiguity as to which person or persons in the union would best represent the information needs of the union. In the case of the flying public in aviation, for example, the designation of a stakeholder representative might be exceedingly difficult or even impossible due to different potential representatives from public interest groups with conflicting agendas. After information stakeholders are designated, their specific information needs can be carefully assessed in preparation for designing and disseminating appropriate feedback.

Assessing information needs of stakeholders

In general, the different goals and agendas of the stakeholder groups will lead to quite different information needs and limits on what can be disseminated. An example of information goals and needs for stakeholder groups in the aviation domain is given in Table 12.1. In some cases, the contractual or regulatory duties of the information stakeholders will be very explicit and can be used to develop information needs. In other cases, however, the stakeholder representatives may only have a vague idea of what they need to know.

Table 12.1 Example of different goals and information needs of stakeholder groups

Stakeholder	Goals	Examples of information needs
Pilots	Professional pilot performance	specific strengths and weaknesses of performance
Union	Evaluation fairness for union members	reliability and validity of pilot/crew assessments
Management	Running a profitable airline	efficiency aspects of performance
Training center	Effective pilot and crew training	detailed evaluation of training and selection effectiveness
Airline safety officer	Safety for each airline fleet	detailed evaluation of unsafe or near-unsafe performance
Local FAA	Safety for an entire airline	aggregate performance trends for each fleet
National FAA	Safety of national airspace system	aggregate performance trends across fleets, airlines, and areas

Each of these groups also may impose limitations on the data that can be entered into the SIS or the information that can be disseminated from the SIS. A union, for example, may prefer that the information in the SIS is stored anonymously rather than being attached to an employee name or number and that the information not include a pilot's union status. Pilots may wish to be reassured that no personal information will be stored in or released from the SIS. Similarly, company management may wish to limit the dissemination of any information that may reveal company secrets, trends in company performance, or any form of specific performance information about employees.

One approach to clarifying the information needs or limitations of the stakeholder representatives is to interview them in order to understand better their information requirements. The interview approach is particularly feasible when the stakeholder group is small and homogeneous. In this case, the answers to an interview or survey would consistently indicate an information agenda. When a stakeholder group is broad and diverse, a small sample of representatives may be brought together in a focus group that discusses their information needs and create an information agenda.

In both the personal interview and focus group methods for establishing information needs, problems may occur with representatives who are not articulate or simply do not have a clear idea about what they need to know. Often such persons will not be able to directly express their information needs, but will still be able to react to concrete examples of information. That is, they cannot clearly express what they need to know, but if they have a specific example in hand, they can evaluate whether or not they need that piece of information.

In this case, using a "rapid prototyping" approach of developing sample reports and having the representatives evaluate these reports may be the most feasible method for establishing information needs. A rapid prototyping approach will typically occur as a series of cycles where the reaction to initial prototypes is used to modify and revise successive versions of the information prototypes. Therefore, assessing and meeting stakeholder information needs may require a cyclical, incremental process of evaluating the needs and designing forms for appropriate information feedback as discussed below.

Designing information feedback for stakeholders

Designing information feedback for stakeholders should take into account both the information needs assessed in the previous step and the preferences of the stakeholder for dealing with information. That is, the information feedback should be as natural and easy to use as possible without losing necessary information. There are two types of essential information in the SIS -- conceptual and empirical information. Conceptual information is represented by the current theoretical or conceptual structures that underlie the SIS. The empirical information is represented by the accumulated data in the databases and all qualitative and quantitative analyses that have been performed on that data.

These two facets of SIS information are not totally distinct as the theory guides the measurement and use of the empirical data, and the results of empirical data analyses will change the theories over time. Nevertheless, the facets are somewhat distinct in that the theoretical structure is conceptually based and should change rather slowly over time while the empirical data are accumulating steadily and may change rather rapidly. Most routine reports and regular use of the SIS information will emphasize the empirical data. Therefore it is particularly important that the empirical data reports meet the information needs of the stakeholders. Specifically, the information presentation should match the preference of each stakeholder for modality of presentation, information metaphors, and processing style.

Modality preference Different types of people have a preference for information presented in different ways. Pilots, for example, due to job self-selection for visual acuity and extensive training and experience, will often prefer visual information. An example of visually oriented information would be graphs, charts, or figures. Conversely, people who by inclination or training are very verbal may prefer verbal information. An example of verbally oriented information would be concise verbal summaries, outlines, or bulleted points that contain the relevant information. Some types of people, such as statisticians or accountants, are used to dealing with numbers and would prefer a numerical summary of the information. An example of a numerical summary would be a tabulation of means and standard deviations for relevant variables. Whatever the modality preference for the stakeholder, the feedback should be designed to match that preference. In addition, the feedback should take advantage of specific information metaphors that might be used by each stakeholder or that are common in a culture.

Information metaphors An information metaphor is a precise symbol or signal that has an inherent, intuitive meaning for the person. This metaphor can be used to design an information interface that has inherent meaning for the stakeholders. For example, a stop sign in the U.S. is the color red and has a specific octagonal shape. An information metaphor for a system problem, therefore, could make use either of the color coding of red or the visual shape of the stop sign. In the aviation domain, pilots are used to seeing color-coded information that is red for undesirable conditions, yellow for potential problems, and green for normal functioning.

To take advantage of this color metaphor, a prototype report on maneuver performance for pilots at one carrier was designed to use a system of color-coded flags (Table 12.2). The information stakeholders were Fleet Captains who were pilots managing fleet operations. In this system, the mean performance on each maneuver was compared to the overall average performance across all maneuvers. If the mean was significantly higher or lower than the overall average performance, it was flagged by appropriate colors to bring it to the attention of the Fleet Captain.

Table 12.2 Using a color metaphor to present performance information

Color of performance information	Meaning for user
Red	Maneuver performance was very *significantly below* the overall average
Yellow	Maneuver performance was significantly *below* the overall average
Black (no flag)	Maneuvers NOT different from the overall average
Blue	Maneuver performance was significantly *above* the overall average
Green	Maneuver performance was very *significantly above* the overall average

Metaphors may also be symbolic. For example, the symbol of a pointing finger to indicate a particular point of interest is one symbol for directing attention to the relevant information in a report. Similarly the symbol of a raised open hand may indicate that the user should stop, look, and think about the information of a specific section because it is important. Both symbolic and color metaphors are, of course, culturally bound as well as potentially influenced by training and experience. Therefore, the use of these metaphors might have to be tailored to each group of information stakeholders.

If natural metaphors do not exist, a metaphor for the information interface can still be constructed by adopting a consistent set of presentation conventions. The convention should be designed to be as simple and natural as possible, and at the very least it should not contradict the use of symbols by the stakeholder group. Once learned by the stakeholders, these presentation conventions can become an efficient way to communicate information.

In the aviation domain, for example, one convention developed for visual presentation of information was to use a circle to represent the variance of a variable and labeled arrows to represent causal effects. This metaphor is illustrated in Figure 12.4 for hypothetical information from an analysis of crew performance. The circle for the criterion variable represents the partitioning of the variable's variance in a fashion similar to a pie chart. The arrows represent the causal impact of prior crew teamwork and technical performance evaluations on the final evaluation of the Captain and First Officer.

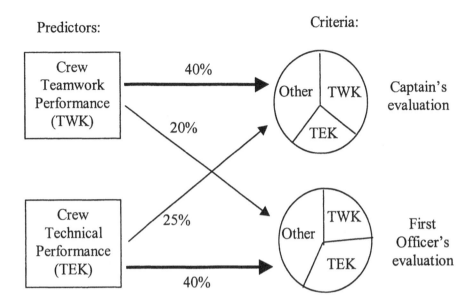

Figure 12.4 A convention for presenting crew evaluation information

In this figure, the size of the arrow represents the relative size of the causal impact of the predictor on each criterion. The label associated with each arrow is the numerical representation of how much variance the predictor uniquely accounts for in the criterion. The variance accounted for information is also visually represented in the corresponding segment of the pie chart for each criterion circle. That is, the size of the pie segments in each circle corresponds to the percentage of variance accounted for by each predictor. The essential information of this figure is that the Captain's evaluation depends more strongly on the teamwork of the crew, whereas the First Officer's evaluation depends more strongly on the crew's technical performance. This pattern of results could suggest different cockpit roles with the Captain as a team leader and the First Officer as the pilot executing many of the technical tasks.

This figure uses both visual and numerical modalities to communicate information. Where this is possible, users with either a visual modality preference or a numerical modality preference can both be accommodated. Further, the orientation of the causal arrows in this figure take advantage of a natural left to right scanning tendency in the U.S. culture. Other aspects of the stakeholder's processing style for dealing with information should be taken into account when designing the final form of the information feedback.

Information processing style People may differ in their preferred manner of processing information. Research such as that summarized by Jonassen and Grabowski (1993) indicates a broad range of individual differences in learning and information processing. Some people may prefer to burrow into the details of a certain part of a report first and then cover more general parts (information burrowers). In contrast, other people may prefer to scan across all the highlights of report and only process details when necessary (information browsers). Some people may wish to find how one piece of information is similar to, or can be lumped together with, another piece of information (levelers); others may wish to find how one-piece of information is different from or contrasts with other pieces of information (sharpeners).

If the information report prototype is a paper report, the presentation format must be fixed. That is, the set design for the report must try to accommodate the modality preference, information metaphors, and processing style for a particular stakeholder group. However, one interesting alternative is to have an interactive electronic report. Using appropriate hardware and software, it is possible to design an interface for a set of information that allows it to be displayed in different ways. Potentially, the software could show the same basic information in different modalities and using different metaphors. Further, the interactive display could do this in a user-controllable manner that would allow the user to process the information according to his or her preferred style, which could increase the motivation of the user to more thoroughly review and understand the information.

There are, of course, potential costs in learning to use such an interactive information interface as well as potential problems of improper use. However, where successfully implemented the flexibility of such an interactive information interface should increase the effectiveness of communication of the information to each group of stakeholders. Communicating this information effectively sets the stage for any efforts to create changes in the target system.

Creating and evaluating change

Creating and evaluating change based on the information SIS is a process that has several basic steps. The steps are illustrated in Figure 12.5. First, based on the information disseminated to the stakeholders from the SIS, the need for change must be determined. Second, action teams with appropriate representatives must be designated both to check on the need for change and to design appropriate possible changes using the problem solution cycle of Chapter 5. Once the changes are designed, they should be carefully implemented and recorded in the SIS databases. Finally, the net result or effect of implemented changes must be assessed and evaluated to see if the intended results are obtained. If this evaluation indicates the needs for change have not been successfully met, the cycle may have to be repeated as indicated by the dotted arrow.

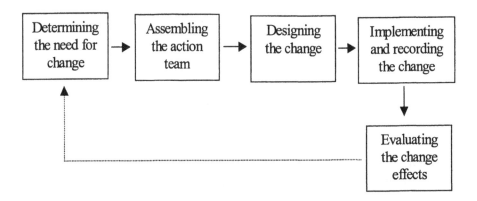

Figure 12.5 The change cycle

Determining the need for change

The need for change can be determined either from the theoretical or empirical information in the SIS, or it can come from an outside source or event such as a new law. On the theoretical side, the need for change can be determined from a normative or expected theory that is disconfirmed. These normative theories could concern the selection, training, job context, or the job performance theoretical subsystems. If, for example, a manual or formal job description states that a job should be performed in a certain prescribed fashion and the empirical evaluation indicates that it is not, this could become a focus for change in either job training or the job context.

Conversely, the need for change can also occur by confirmation of a theory that has undesirable or unwanted implications for the target system. For example, a theory of gender differences in the workplace might hypothesize that females are treated differently than males during certain critical job processes. If confirmed, this theory of gender discrimination might point to the need for change.

Most commonly, however, the need for change will stem from the SIS data concerning selection, training, job context, or job performance (Chapters 9, 10, and 11). Typically, the "bottom-line" system performance data will play an important role in determining the need for change. The bottom-line system performance would be composed of both the presence of positive aspects of system performance (e.g. quantity, quality, efficiency) and a lack of the negative aspects of performance (e.g. errors, incidents, accidents). Both positive and negative aspects of performance can be combined to indicate overall better or worse system performance.

Finding negative trends in overall system performance as discussed in the section on trend analysis above would be one signal of a possible need for change. However, other patterns of potential system problems should also be explored in the empirical data analysis. With qualitative performance data, the presence of serious incidents or accidents may indicate a need for change. In the aviation domain for example, the occurrence of unintentional incursions of aircraft into active runways might indicate a need for changes in the standard operating procedures (SOP) for taxiing or for training emphasizing adherence to current SOP.

With quantitative performance data, another pattern that should be tested is for a sudden spike downward in performance even though the overall trend is level. This is illustrated in Figure 12.6. The overall trend in performance up to the current time is level (line A). At the current time there is a downward spike in performance (line B). It is critical to distinguish whether B represents the start of a new downward trend in performance (line C) or is a chance fluctuation, in which case performance should return to the original baseline (line D). One way to empirically assess this is to check whether B is significantly different from the trend line of performance represented by line A.

With either qualitative or quantative data, the observed results may have to be compared to an appropriate benchmark to decide if the need for change is serious enough to start the change cycle. If an incident occurs, for example, the seriousness of the incident and its implications for the target system might have to be elaborated to evaluate the need for change. If a downward spike in performance occurs, the seriousness of the performance variable must be considered. A preliminary meeting of subject matter experts and involved information stakeholders could evaluate this information.

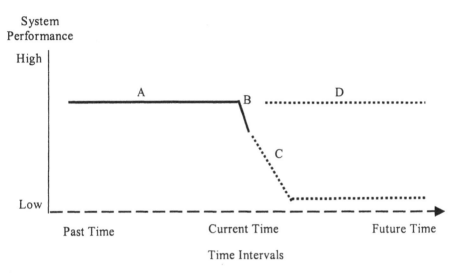

Figure 12.6 Example of a downward spike in performance

For quantitative data such as average system performance, this type of team could establish a value for key variables that indicates some unsafe or unsatisfactory level of system performance. In the aviation domain, for example, the acceptable and unacceptable benchmarks for system performance might be clearly defined by subject matter experts using methods of risk analysis. These benchmarks could be used for a comparison of the monthly performance data from the SIS to determine the need for change. This comparison could use the color-coded flag system discussed above. If, for example, system performance were to come close to the unacceptable benchmarks, the results could be "yellow flagged" in the report. Similarly, if the performance significantly exceeded the unacceptable benchmark, the results could be "red flagged" in the report. A red flag result would require the action of either a designated person or the assembly of an action team which is the next step in the change process.

Assembling the action team

The action team should include persons who can represent the viewpoint of the stakeholder groups that are affected by the low system performance or problems. In addition, the action team should have subject matter experts who will help the team in both the interpretation of the problem and the design of possible interventions. Finally, the action team should either have as a member, or work closely together with, a data analyst who is familiar with the theories and databases of the SIS. If each type of expertise resides in a different person, the smallest feasible action team would consist of one stakeholder representative, one subject matter expert, and the SIS data analyst. If one person has multiple sources of expertise (e.g. a data analysis who is also an SME), the team could be smaller.

The first goal of the action team is to focus on the problem indicated by the past empirical results (typically poor performance of the target system). However, the initial focus should be on *exploring the problem* and ensuring a correct and complete diagnosis of the ramifications of this problem in the target system, rather than focusing immediately on solutions. First, the action team should confirm the nature of the problem. That is, the past data (such as "red flag" results) that caused the action team to be assembled should be re-evaluated. Both the meaning and severity of the indicated problems within the target system should be reconfirmed by reinspection of the results, reinterpretation of the results, and reanalysis were necessary.

This reanalysis process will often involve "drill down" and "ripple out" analyses. A drill down analysis is a further elaboration of a basic result into more detailed components. For an aviation example, suppose the average performance of a fleet of aircraft on the yearly pilot examination had declined to the extent of causing a red flag alert. The drill down analyses would seek to pinpoint the exact locus of the problem. The performance indicators could be, for example, cross-classified and analyzed by relevant factors such as Captain vs. First Officer evaluations, pilot flying vs. pilot not flying evaluations, the evaluators giving the examination, and the results across different phases of flight in the examination. Further, the precise performance items that were used to compose the performance average could be decomposed and separately analyzed. In this process, the stakeholder representative has the ultimate need to know that motivates the analysis. The subject matter expert understands the basic meaning of the data and the data analyst performs the data analysis and reports back the meaning of the statistical results. Thus, the three types of action team members must work together to more deeply analyze the problems of the target system.

Not all problems will require further action. Suppose, for example, that drill down analyses reveal that the lower performance scores were due to the tests being given by more harsh or strict evaluators for a particular month. Correcting the monthly performance data for this artifact or bias in the evaluation process might eliminate the negative performance trend. In such a case, the original problem would be due to a confound or artifact of the evaluation process and would not require further action beyond, perhaps, calibration training for the evaluators.

In addition to the drill down analyses, the team should work to explore the collateral results and implications of the indicated problem (i.e. "ripple out" analyses). That is, the occurrence of the problem should be cross-analyzed with the extensive information in the SIS in order to see if there is any pattern of associated results. For example, if a crew performance problem was found to be more typical of First Officers than Captains, the causal impact of flying experience might be further explored using the SIS information. This exploration would take advantage of the associated experience information in the personal database for each pilot. The occurrence of the problem incidents could be cross-analyzed with the experience level of the pilots to see if a lack of pilot experience is really the causal factor.

Similarly, the occurrence of the focal problems during a certain phase of flight could be cross-analyzed with the crew's performance on earlier phases of flight in order to see if a diagnostic pattern or performance profile would emerge. In particular, if earlier problems in the flight cause an "error chain" to occur, detecting the initial cause and the steps in this process might be very important. Correcting the problem may require intervention at a much earlier phase of operations than indicated by the original analysis of the low performance data. In general, ripple out analyses evaluate the broader implications of the problem and can give valuable additional information both for understanding the problem and for designing an appropriate change.

Designing the change

Using the information from the drill down and ripple out analyses, the action team should then create and evaluate possible interventions designed to resolve the problem. In any complex system, however, making any significant change creates the possibility that unwanted side effects can occur. For evaluating the possible occurrence of side effects, the theoretical as well as empirical information in the SIS is invaluable. The design team should use the SIS theories to consider what the entire range of expected effects would be from a given intervention. If a relevant computational model has been previously created for predicting the effects of selection, training, or job context, this model can be used to extrapolate out the expected effects of these types of planned interventions.

If it has sufficient scope, a relevant computational model can also be used to evaluate the relative worth of alternative problem solutions. The model can be modified as appropriate for each solution and then used to simulate the entire range of the expected effects of the solution. Using this extensive and detailed profile of expected effects, the action team can more fully evaluate the anticipated consequences of each solution and choose the best one for implementation.

The process of evaluating the effects of change from the model results may be simple or complex. If target system performance can be accurately condensed into a single critical outcome variable or a weighted composite, evaluating the consequences of alternative changes or solutions becomes a simple comparison of values for that variable. However, if multiple aspects of a performance profile must be simultaneously considered, the evaluation process is more complex. In such a case, the evaluation may rely on methods such as evaluating performance profiles or using multiple cut-points for satisfactory performance levels across multiple criteria.

To the extent that results for a particular proposed change exceeds all these performance cut points, it is an acceptable solution. To the extent that the results for one proposed change equal or exceed the expected performance profile for all other alternatives, it is a "dominant" or better solution across the board. If a dominant solution can be found, it is a clear choice for implementation. More typically, each proposed solution may be superior to other solutions in some respects, but not others. In such a case, the action team must evaluate the relative pros and cons of the modeled results for each proposed solution in order to choose the one for implementation.

Once a satisfactory or optimal planned change has been selected, there are at least two qualitatively distinct strategies for implementing the change. The first strategy is a "go for broke" approach of immediately and totally implementing the solution across all system venues in its complete and final form. This approach is making the strong presumption that the intervention will solve the problem as predicted and that no significant negative side effects will occur. Clearly in the early stages of SIS development when the theories and data may be incomplete and imprecise, this presumption may be incorrect, and this approach is a risky strategy. When a SIS is firmly established with extensive empirical support, this strategy is much safer.

The second strategy is a more conservative "small group try out" approach in which the solution is implemented for a small subset of the target systems and its effects carefully monitored. Using this approach, the small group try out on systems that have the intervention can be compared with other systems that do not. Essentially this is a field experiment where the experimental group has the implemented solution and the other systems serve as a control group. This approach is less risky in that a smaller set of systems is affected by the proposed changes, and the effects of the changes can be fully evaluated before the final implementation. However, this more cautious approach also takes more time and may not be feasible when the nature of the problem requires an immediate solution (e.g. critical legal or safety problems).

Implementing the change

Implementing a change as either a small group try out or as a global and total change requires a plan for implementing the intervention and for gathering data to measure the effects of the intervention. In a job performance situation, interventions may involve employee selection, job training, the work context, or other sub-systems. This wide variety of possible interventions may correspondingly involve a wide variety of potential stakeholders in the implementation process. The implementation plan must be developed in conjunction with relevant stakeholders and appropriate regulatory agencies.

A key part of the implementation plan would be a timeline for phasing in the interventions. A timeline is particularly necessary when certain aspects of intervention must be implemented before other aspects of the intervention can be completed. In aviation, for example, FAA approval for a proposed change must be obtained and pilots must be trained in a new procedure or on the new equipment before a procedure can be officially implemented as standard operating procedure for a fleet. In our aviation research, obtaining official approval for a revised Quick Reference Handbook that covered operational emergencies required several months. In general, obtaining necessary agreements and permits may require a significant amount of time and potential modifications to the plan to satisfy different stakeholders.

The implementation plan should include the plan for data collection and evaluating the effects of the intervention. It is very important that both the analysis plan for the data is completely designed and that the appropriate data collection measures are in place prior to the implementation of intervention (Holt, Boehm-Davis, and Beaubien, 2001). Otherwise, there is a real danger that the intervention will be implemented in such a way that no data will be appropriate for evaluating its effects. An example of this mistake is to implement an intervention for all the target systems in the organization and not collect any pre-intervention baseline performance data. With only post-intervention performance data and no control or comparison groups, the effects of the intervention cannot be accurately empirically evaluated.

The plan for the data collection should include at the very minimum the theoretically expected positive and negative effects of the intervention. The relevant theories of the SIS should be examined to find potential positive and negative effects of the intervention, and data collection measures designed to capture these expected effects. However, based on the ripple out analyses discussed above, data should also be collected on the possible collateral or side effects of the intervention. The goal is to anticipate and evaluate the possible positive and negative effects of the intervention as completely as possible.

The implementation of the data collection process for evaluating an intervention may require a change in the content or structure of the databases in the SIS. Clearly, a total and global implementation that permanently changes the structure of either a selection system, training system, or the job context will change some relevant information in the SIS. At the very least, the change must be noted and archived along with the performance data in SIS so that the exact point of implementation is clear. This allows the pre-intervention data to be objectively distinguished from the post-intervention data, which is necessary for evaluation analyses. In most cases, a significant change in any of the subsystems will require new variables to be recorded in the SIS.

In contrast, the small group try out approach may use a separate, temporary database that would not necessarily be part of the SIS. Even in this case, the appropriate database keys should be included so that the information in the small group try out evaluation database can be connected with the more extensive SIS information for more complete data analysis such as drill down analyses of results. This connection will allow, for example, the more complete evaluation of the effect of individual differences among employees on system performance by connecting the individuals in the small group try out with the information in the SIS personal database.

Evaluating the effects of the change

For problems of system performance, evaluating the effects of a change or intervention will typically involve the evaluation of a change in performance. The change in system performance may, however, be a change in the mean level performance, a change in the variability of performance, or a change in the nature of the interrelationship of measured performance variables. The initial analyses should focus on confirming or disconfirming the expected effects of the intervention. The expected effects would include the direct effects of the intervention as well as any collateral or ripple out effects. The evaluation of the expected effects is the heart of the evaluation, but the information in the SIS can usefully extend these analyses.

Exploratory analyses should investigate unanticipated effects of the intervention. For a small group try out, these analyses would start with a very global comparison of the intervention group with the non-intervention group across a broad variety of performance measures. Based on the results of this global comparison, more detailed analyses could be done for specific aspects, processes, or components of performance. These exploratory analyses will be particularly important at the initial stages of the development of the SIS where the theories may not be a complete and accurate guide to the expected changes caused by the intervention.

The final summary report should be prepared for the relevant stakeholders that summarizes the results of both the planned and exploratory data analyses. This report should detail both the positive and negative consequences of the intervention. In particular, the comparison of the desired results to the original problem or need must be included in the report. If the original problem or need has not been completely resolved, the action team may have to go back to a reconsideration of the problem and the design of additional or alternative interventions. That is, they may have to re-do the problem solution cycle.

Finally, the evaluation information should also be used to refine the theoretical basis of the SIS. In particular, the confirmation of the theoretically expected results would support the theoretical aspects of the SIS from which the intervention was derived. Conversely, the disconfirmation of expected results would reduce the credibility of the corresponding theory. In this manner, the pragmatic information about the effects of change will also aid the scientific development of the SIS in the long run. In order to obtain the full range of long run benefits of the SIS, management must be concerned with the scientific development of the system as well as the use of the system. That is, the organization must become involved with the business of science.

The business of science

Business-oriented science may be in some ways similar and in other ways different from academically-oriented science. Similarities would exist in the common scientific approach to principled measurement, theory construction, and theory evaluation. More specifically, the similarities would include the basic concepts of sensitivity, reliability, and validity of measurement, a conceptual or abstract representation of the precepts of the theory, and some type of systematic approach to theory evaluation that would often include empirical methods.

There are also, however, noticeable differences. The goals of academic scientists often involve pursuing a personal research agenda, promotion and tenure at an academic institution, a large number of scientific publications, and establishing a scientific reputation in a traditional academic discipline. In contrast, the goals of a business scientist may involve pursuing a corporate research agenda, securing a good position and advancement in the organization, a large number of practical applications of the research, and a potential bottom-line impact on the profit and viability of the organization. Thus, business science is more likely than academic science to emphasize a theoretically complete and accurate account of system functioning that is sufficiently good to be used for practical interventions or changes that will have desired effects. That is, from a business perspective the theoretical account must be good enough to design changes in the human or physical aspects of complex systems that will have strong and predictable results that make some pragmatic difference.

The focus on a theoretical account that is complete and accurate enough for designing practical change has important ramifications for the scientific process as well as the outcomes of business science. First, the goal of a complete theoretical account makes the issues of defining and assessing the completeness of the SIS theoretical system become a very high priority. The definition and assessment of theoretical completeness was covered in more detail in Chapter 4 on theoretical evaluation. Second, the scientific development of a theoretically complete account may require a highly complex theory or set of theories, particularly if the target system is complex. One example would be the complex multi-level theories of human/physical systems such as those covered in Chapter 11. Third, the evaluation and pragmatic use of a highly complex theory or set of theories may require a very precise, rigorous, and systematic integration of the set of theoretical precepts into a tightly integrated theoretical account of the target system. This makes the mathematical or computational integration of the

SIS theories a high priority. These differences in emphasis could lead to a more complete and extensively integrated theory for business science.

The measures and data in the SIS should be used to evaluate the integrated theory. For this evaluation, the amount of data available for testing the theory should exceed the complexity of the theory. The SIS computer systems for storing data are essentially extensible without limit. Therefore, the data on the target systems can be extended and augmented until it is sufficient for theoretical evaluation. In general, the only limitation will be the cost of collecting, storing, and analyzing large amounts of data.

However, the amount of data could become a problem in a SIS that has data on relatively few target systems but hypothesizes an extremely complex integrated theory. One advantage of the SIS for this evaluation situation is that relevant data on the target systems will potentially be accumulated over a long period of time. This extensive warehouse of relevant data may ultimately have sufficient data to empirically evaluate the integrated theory even if the number of target systems is relatively small. In addition, the information in a SIS allows a better judgment about whether data collected on other target systems can be combined or aggregated in the empirical evaluation. Combining data from different subsamples (e.g. from a different SIS or from other research) may provide sufficient data to evaluate a complex integrated theory.

A further advantage of the SIS for this evaluation process is that the data may reflect the entire population rather than a sample. For example, an aviation SIS may include all information on the population of pilots for a fleet. If so, the statistical results from the data are population parameters rather than statistical estimates and are not subject to sampling error. Having data from the entire population rather than having to estimate the population variables from a sample is a significant advantage for empirical evaluation. Therefore, the availability of data using a SIS will often allow the empirical evaluation of integrated theories that have even a high degree of complexity, as would be expected in many human-machine systems.

In a similar manner, the precisely integrated theory in a SIS allows greater scope for the pragmatic use of the SIS information. The expected results of interventions such as changes in worker selection, changes in training, or changes in the job context can be clearly predicted even for complex systems. This allows management decisions to be made on a well validated, scientific basis. These decisions should have, therefore, a much higher likelihood of obtaining the desired organizational result. If the information from the SIS can be disseminated in appropriate academic publications, this information may also enrich the relevant scientific disciplines. This raises the issue of whether the same types of results will be found using a SIS as in traditional academic science in the same domain.

Business science vs. academic science

Scientific focus If it is true that "science is science" no matter how or by whom it is practiced, then the results of business science and academic science would be the same. If, however, business science and academic science have different goals, this may lead to the adoption of different methods and standards for executing the scientific process. Therefore, it is legitimate to ask if the scientific results will be the same.

Adopting a SIS is natural for business science because the focus is on a comprehensive and pragmatically useful theoretical and empirical account of the target systems within the organization. A manager needs to know what is true for his or her organization and how to change it. This requires a high degree of theoretical completeness that allows for accurate prediction and high levels of control. Most often, managers would be less concerned about how well their theories generalize to other organizations and more about how well they can be used in the home organization.

In contrast, academic research may emphasize a more general search for theoretical principles that can be validated across multiple organizations, multiple domains, and so forth. The academic research is usually oriented to showing that a theoretical statement is necessarily true in a broad range of organizations, but does not stress theoretical completeness in any one organization or setting. Academic research on complex human systems typically involves separate samples in separate studies across domains and organizations that focus on the theoretical issues considered important by academic researchers. Validating key theoretical concepts is important while the unique aspects of each organization or domain and the pragmatic utility and comprehensiveness of the account may be less important. Therefore, although the SIS approach can be used for academic research, there may be fewer incentives to do so given academic research goals. There may also be greater costs for a SIS approach due to the necessity for keeping track of data from the target systems on an ongoing basis and having ongoing cycles of validation and analysis.

Change of focus The focal point of SIS development may also shift in distinct ways for business science compared to academic science. For business science, a shift in focus may be caused by changes in legal, operational, contractual, or technological systems. Further, the business development of areas in a SIS may halt with a *satisfactory* account of key variables and processes. Measured data for these satisfactory areas may also be reduced to essential measures necessary for monitoring key aspects of system performance. Both the extension of the SIS to new facets and the reduction of data measures would be pragmatically determined.

Shifts in focal point for academic science may be caused by changes in social, funding, or publication systems. Kuhn (1970) describes some of these shifts in focus in the academic community. These changes could be quite similar or quite distinct from the changes driving business science. Therefore, the theoretical and empirical coverage of a business SIS may diverge from the coverage of a traditional academic discipline over time.

Generalization Given these different goals and approaches, the question becomes whether the results from a SIS constructed and operated for business science would match the results from separate-sample academic science research where they would overlap in some domain. Answering this question depends in part on three major facets that could differentiate SIS research from separate-sample research: sample adequacy, situational generalization, and measurement reactivity.

The most important goal of business science is to generalize to systems within the organization, while the most important goal of academic science is to generalize across organizations (Table 12.3). If a SIS includes all of the target systems in the population, than by definition the SIS has a more inclusive and representative set of cases than any sample. If the SIS results were different from separate-sample academic research, the results for the SIS would be more likely to be true of the population as defined by the target systems within the organization. Conversely, the multiple samples for different organizations covered by academic research may make the research results more generalizable to the broader population of all such target systems in all organizations. Therefore, the business development of a SIS may lead to an extremely good theoretical account of how the target systems function in that organization, but the account may not be entirely generalizable to other organizations.

If the SIS collects data from target systems in the natural environment, the results inherently are based on, and generalize to, system behavior in a natural environment. To the extent that academic research is also based on the natural environment or an effective simulation of that environment, the results should be the same. To the extent that the academic research uses an artificial environment that elicits different processes and behavior of the target systems, the results may differ. In this case, however, the results from the SIS would be more likely to generalize to target system behavior in the natural environment.

Table 12.3 Differential foci in business and academic science

		Focus for validity of theories	
		Within organization	**Across organizations**
Type of science	**Academic science**	less important	critical
	Business science	critical	less important

The databases included in a typical SIS will include much more information and involve more measures than is true of a typical separate-sample academic study. If these measures are part of the normal work routine, they should not be reactive or contaminate the results. If these additional measures are reactive and change the basic processes in the target system, then the results from a SIS-based research could be less valid than separate-sample studies. Since this is an important issue, the possibility of reactive measurement in a SIS should be empirically evaluated using a three-step approach.

The first step would be to establish a core of absolutely necessary and non-reactive measures. Some measures are absolutely necessary and inherent parts of the job. In the aviation domain, for example, the annual review of a pilot's physical condition and flying proficiency is an inherent and necessary part of the job in the natural environment. Therefore, regardless of whether these measures are reactive or not, taking these measures does not degrade generalization to the natural environment. To this set of necessary measures can be added trace or archival measures that are very low in reactivity. For example, the log of computer activity of a worker using a networked workstation can be recorded as a measurement with no reactivity to the worker in the natural system. The combined core set of necessary and non-reactive measures will be the basis for evaluating the reactivity of any potential additional measures.

The second step would be to implement the set of potentially reactive measures in a random subsample of target systems in the SIS while keeping a separate random sample that has just the core measures. Essentially this step is creating an experimental group that has the reactive measures and a control group that has just core measures. If a new set of measures is being considered for the SIS, they may be implemented as a small group tryout in this manner. The important thing in this implementation would be to have a sufficient sample size in both the experimental and control groups to have good statistical power on the final evaluation step.

The third step would be to empirically compare the results of the core measures sample with the sample that includes the potentially reactive measures. This comparison should include all relevant aspects of the core data. For example, most multivariate statistical analyses depend on the mean, variance, and covariance among a set of measures. Therefore, the experimental group should be compared to the control group on the means, variances, and covariances among the set of core measures. If there are no differences between the experimental and control group on these aspects of the data, there is effectively no difference that would affect the statistical results and conclusions about the target systems. This empirical evaluation indicates that the additional measurements are not reactive. In this case, the additional measurements in SIS can be safely used to yield additional information while preserving the validity of the basic results.

Other forms of SIS analysis may use different aspects of the core SIS data such as timing information, the shape of certain distributions, and so forth. The principal of comparison would remain the same; that is, the experimental and control groups would be compared on all aspects of the core data that would impact on relevant analyses. If no differences are found, the additional measures are effectively non-reactive and can be used for SIS analysis.

If significant differences are found, however, then one or more of the additional measurements might be reactive. Subsequent cycles of this process can be used to isolate the subset of reactive measures. Once the subset is isolated, the reactive measures can either be eliminated from the SIS or suitable non-reactive proxy measures substituted for the reactive ones. A side-benefit of this process would be the potential information gained from the analysis of the reactive measures. The reactivity may show, for example, some point at which the processes of the target system are uniquely susceptible to influence. Such sensitive points in the target system may point the way to important underlying system processes such as chaotic processes that are extremely sensitive to initial conditions. Using this three-step procedure, the issue of measurement reactivity can be addressed in the same manner as any other theoretical hypothesis, and results of the evaluation used for theoretical development in the same way as other information from the SIS.

Chapter summary

This chapter has covered three basic aspects of using a SIS in the business context. First, the SIS is a primary source of information about the performance of the target systems. The set of theories and data in the SIS gives a wide variety of scientifically-validated information about the target systems. This information can be used to diagnose system problems as well as to extrapolate the effects of changes or interventions.

Secondly, the SIS is a source of feedback for all levels of management and information stakeholders. The variety of data in the SIS allows for the design of feedback relevant to the information needs of each stakeholder. Careful consideration of principles for the design and dissemination of this information facilitate the communication of the underlying meaning of the results.

Thirdly, the ultimate test of the SIS in a business context is the pragmatic usefulness of the theories and information contained in it. Changes in employ selection, job training, or the work context purchase some examples of interventions that should be guided by relevant theory and data. The SIS approach ensures that the guiding theory and data are scientifically reliable and validated. This maximizes the potential for optimal management decisions and minimizes the potential for errors. The results of business science conducted in this manner should be comparable to academic scientific research in the same domain and potentially better in some respects.

Book summary

This book has introduced the concept of a Scientific Information System, an integrated structure of computer databases containing information that has been scientifically developed and validated. In most domains, the development of such a system must be a bootstrap operation from rough initial approximations of relevant theory and data for a target system to more complete and refined information. The long-term value of complete and accurate information about a target system is to convincingly explain, accurately predict, and precisely control the processes and outcomes of such systems. The scientific evolution of a SIS ensures that the database information will ultimately be sensitive enough to detect important system results, reliable and error-free, and theoretically valid. The development of such systems is justified wherever the benefits of better information outweigh the development costs.

Bibliography

Alliger, G.M. and S. Katzman (1997) *When training affects variability: Beyond the assessment of mean differences in training evaluation.* In *Improving training effectiveness in work organizations*, J.K. Ford and Associates, Editors., Lawrence Earlbaum: Mahwah, NJ. p. 223-246.

American Psychological Association (1992) *Ethical principles of psychologists and code of conduct.* American Psychologist, **47**: p. 1597-1611.

Bandura, A. (1992) *Aggression: A social learning analysis.* Englewood Cliffs, NJ: Prentice-Hall.

Banks, J. (1998) *Handbook of Simulation: Principles, methodology, advances, applications, and practice.* New York, NY: John Wiley & Sons, Inc.

Barquin, R., and Edelstein, H. (1997) *Planning and designing the data warehouse.* The data warehousing institute series, Upper Saddle River, NJ: Prentice Hall PTR.

Bennett, B.S. (1995) *Simulation Fundamentals.* Prentice Hall International Series in Systems and Control Engineering, M.J. Grimble, ed. London: Prentice Hall.

Berkowitz, L. (1989) *Frustration-aggression hypothesis: Examination and reformulation.* Psychological Bulletin, **106**: p. 59-73.

Boehm Davis, D.A., R.W. Holt, and T.L. Seamster (2001) *Airline experiences with resource management training programs*, in *Applying research management in organizations: a guide for training professionals*, E. Salas, C.A. Bowers, and E. Edens, Editors. Lawrence Erlbaum & Associates: Mahwah, NJ.

Boehm-Davis, D.A., R.W. Holt, and J. Hansberger. (1997) *Pilot abilities and performance.* in *Proceedings of the Ninth International Symposium on Aviation Psychology.* Columbus, OH.

Borman, W.C. and S.J. Motowidlo (1993) *Expanding the criterion domain to include elements of contextual performance.* In *Personnel Selection in Organizations*, N. Schmitt and W.C. Borman, Editors. Jossey Bass: San Francisco, CA. p. 71-98.

Buros, O.K. (1978) *Mental Measurements Yearbook.* Highland Park, N.J.: Gryphon Press.

Campbell, D.T. and D.W. Fiske (1959) *Convergent and discriminant validation by the multitrait-multimethod matrix.* Psychological Bulletin, **56**(2): p. 81-105.

Campbell, D.T. and J.C. Stanley (1966) *Experimental and quasi-experimental designs for research.* Chicago, IL: Rand McNally.

Cartwright, D. and F. Harary (1956) *Structural Balance: A generalization of Heider's theory.* Psychological Review, **63**: p. 277-293.

493

Casti, J.L. (1997) *Would-Be Worlds*. New York, NY: John Wiley & Sons, Inc.

Cattell, R.B. (1965) *The scientific analysis of personality*. Baltimore, MD: Penguin Books.

Cliff, N. (1993) *What is and isn't measurement*. In *A handbook for data analysis in the behavioral sciences: Methodological issues*, G. Keren and C. Lewis, Editors. Lawrence Erlbaum Associates, Publishers: Hillsdale, NJ.

Cohen, J. (1977) *Statistical power analysis for the behavioral sciences*. Revised edition. New York, NY: Academic Press.

Cohen, J. and P. Cohen (1983) *Applied multiple regression/correlation analysis for the behavioral sciences*. 2nd ed. Hillsdale, NJ: Earlbaum.

Cortina, J.M. (1993) *What is coefficient alpha? An examination of theory and applications*. Journal of Applied Psychology. **78**(1): p. 98-104.

Cronbach, L.J, G.C. Gleser, H. Nanda, and N. Rajaratnam (1972) *The dependability of behavioral measurements: Theory of generalizability for scores and profiles*. New York, NY: Wiley.

Dalkey, N.C. (1969) *The Delphi method: An experimental study of group opinion*. Management Science, June.

Darley, J.M. and P.H. Gross (1983) *A hypothesis-confirming bias in labeling effects*. Journal of Personality and Social Psychology, **44**: p. 20-33.

De Greene, K.B. (1970) *Systems and Psychology*. In *Systems Psychology*, K.B. De Greene, Editor. McGraw Hill Book Company: New York. p. 1-50.

Fishwick, P.A. (1995) *Simulation Model Design and Execution: Building digital worlds*. Englewood Cliffs, NJ: Prentice Hall, Inc.

Flanagan, J.C. (1954) *The critical incident technique*. Psychological Bulletin, **51**: p. 323-355.

Franta, W.R. (1977) *The Process View of Simulation*. New York, NY: North-Holland Publishing Co.

Gardner, H. (1983) *Frames of Mind*. New York: Basic Books, Inc., Publishers.

Garland, D.J., J.A. Wise, and V.D. Hopkin, eds. (1999) *Handbook of Aviation Human Factors*. Lawrence Erlbaum Associates, Publishers: Malwah, NJ.

Gigerenzer, G. (1993) *The Superego, the Ego, and the Id in Statistical Reasoning*. In *A Handbook for Data Analysis in the Behavioral Sciences*, G. Keren and C. Lewis, Editors. Lawrence Erlbaum Associates, Inc.: Hillsdale, NJ. p. 199-228.

Gilbert, N. and K.G. Troitzsch (1999) *Simulation for the Social Scientist*. Philadelphia, PA: Open University Press.

Gordon, G. (1969) *System Simulation* Prentice-Hall Series in Automatic Computation, G. Forsythe, ed. Englewood Cliffs, NJ: Prentice-Hall, Inc.

Gray, W. and D.A. Boehm-Davis (in press) *Millseconds matter: an introduction to microstrategies and to their use in describing and predicting interactive behavior*. Journal of Experimental Psychology: Applied.

Greenwald, A.G., A.R. Pratkanis, M.R. Leippe, and M.H. Baumgardner (1986) *Under what conditions does theory obstruct research progress?* Psychological Review, **93**: p. 216-219.

Greenwood, D.A., R.W. Holt, and D.A. Boehm-Davis, (submitted) *Training instructor pilots to evaluate aircrew performance in a workshop setting*.

Hansberger, J., R.W. Holt, and D.A. Boehm-Davis. (1999) *Instructor/Evaluator evaluations of ACRM effectiveness*. In *Proceedings of the Tenth International Symposium on Aviation Psychology*. Columbus, Ohio.

Harman, H.H. (1976) *Modern Factor Analysis*. Chicago, IL: University of Chicago Press.

Hathaway, S.R. and J.C. McKinley, ed. (1943) *Minnesota Multiphasic Personality Inventory*. revised edition. New York, NY: Psychological Corporation.

Hays, W.L. (1981) *Statistics*. 3rd ed. New York, NY: Holt, Rinehart & Winston.

Helmreich, R.L. and A.C. Merritt (1998) *Error and error management. Technical Report #98-03*.

Holt, R.W., J. Hansberger, and D.A. Boehm-Davis (submitted). *Improving rater calibration and performance in aviation*.

Holt, R.W., K.A. Fitzgerald, M.M. Matyuf, W.A. Baughman, and D.C. Littman (1991) *Behavioral validation of a hazardous thought pattern instrument*. In *Proceedings of the Human Factors and Ergonomics Society Annual Meeting*.: HFES.

Holt, R.W., E. Meiman, and T.L. Seamster (1996) Evaluation of aircraft pilot team performance. In *Proceedings of the Human Factors Society 40th Annual meeting*. (p. 44-48). Philadelphia, PA.

Holt, R.W., D.A. Boehm-Davis, and J. Hansberger (1998) *Evaluation of advanced crew resource management*. unpublished manuscript.

Holt, R.W., D.A. Boehm-Davis, J.H. Hansberger, J.M. Beaubien, and M. Diaz (1999) *Semi-Annual project report, FAA Grant 94-G-034*. George Mason University: Fairfax.

Holt, R.W., D.A. Boehm-Davis, and J.M. Beaubien (2001) *Evaluating Resource Management Training*. In *Applying research managment in organizations: a guide for training professionals*, E. Salas, C.A. Bowers, and E. Edens, Editors. Lawrence Erlbaum & Associates: Mahwah, NJ.

Huff, D. (1954) *How to lie with statistics*. New York, NY: W. W. Norton and Co., Inc.

Hunter, D.R. and E.F. Burke (1995) *Handbook of Pilot Selection*. Aldershot, England: Ashgate Publishing Limited.

Janis, I.L. (1982) *Victims of groupthink*. 2nd ed. Boston, MA: Houghton Mifflin.

Johnson, P.J. and T.E. Goldsmith (1998) *The importance of quality data in evaluating aircrew performance*. Federal Aviation Administration Office of the Chief Scientific and Technical Advisor for Human Factors (AAR-100): Washington, D.C.

Jonassen, D.H. and B.L. Grabowski (1993) *Handbook of Individual Differences, Learning, and Instruction*. Hillsdale, NJ: Lawrence Erlbaum Associates, Publishers.

Kelley, H.H. (1972) *Attribution in social interaction*, in *Attribution: Perceiving the causes of behavior*, E. E. Jones et al., Editor. General Learning Press: Morristown, NJ.

Kelly, G.A. (1955) *The Psychology of Personal Constructs*. New York: Norton.

Kirkpatrick, D.L. (1976) *Evaluation of training*. In *Training and development* handbook: *A guide to human resource development*, R.L. Craig, Editor. McGraw-Hill: New York, NY.

Kuhn, T.S. (1970) *The structure of scientific revolutions*. Chicago, IL: University of Chicago Press.

Law, A.M. and D.W. Kelton (1991) *Simulation Modeling and Analysis*. New York, NY: McGraw-Hill, Inc.

Likert, R. (1932) *A technique for the measurement of attitudes*. Arch. Psychol., **140**: p. 1-55.

Lorenz, K. (1966) *On Aggression*. New York, NY: Harcourt, Brace, & World.

McClelland, D.C. (1980) *Motive dispositions: The merits of operant and respondent measures*. In *Review of Personality and Social Psychology*, L. Wheeler, Editor. Sage: Beverly Hills, CA. p. 10-41.

McCrae, R.R. and P.T. Costa, Jr. (1990) *Personality in Adulthood*. New York: The Guilford Press.

McGraw, K.O. and S.P. Wong (1996) *Forming inferences about some intraclass correlation coefficients*. Psychological Methods, 1(1): p. 30-46.

Morrison, F. (1991) *The Art of Modeling Dynamic Systems: Forecasting for chaos, randomness, and determinism*. New York, NY: John Wiley & Sons, Inc.

Muchinsky, P.M. (1997) *Psychology applied to work: an introduction to industrial and organizational psychology*. 5th ed. Pacific Grove, CA: Brooks/Cole Publishing Company.

Mulaik, S.A. (1972) *The Foundation of Factor Analysis*. New York, NY: McGraw-Hill.

National Transportation Safety Board (1994) *A review of flightcrew-involved, major accidents of U.S. air carriers, 1978 through 1990*. Washington, D.C. National Transportation Safety Board.

Noy, Y.I., ed. (1997) *Ergonomics and safety of intelligent driver interfaces*. Lawrence Erlbaum Associates, Publishers: Malwah, JN.

Nunnally, J.C. (1978) *Psychometric theory*. 2nd ed. New York, NY: McGraw-Hill.

Pedhazur, E.J. (1982) *Multiple regression in behavioral research: Explanation and prediction*. 2nd ed. Fort Worth, TX: Harcourt Brace.

Perhazur, E.J. and L. Pedhazur Schmelkin (1991) *Measurement, design, and analysis: An integrated approach*. Hillsdale, NJ: Earlbaum.

Presser, S. and L. Stinson (1998) *Data collection mode and social desirability bias in self-reported religious attendance*. American Sociological Review, **63**: p. 137-145.

Reason, K.J. (1991) *Human Error*. Cambridge, England: Cambridge University Press.

Romney, A.K., R.N. Shepard, and S.B. Nerlove, eds. (1972) *Multidimensional Scaling: Theory and applications in the behavioral sciences*. Volume 1: Theory, and Volume 2: Applications. Seminar Press: New York, NY.

Rubinstein, R.Y. (1981) *Simulation and the Monte Carlo Method*. New York, NY: John Wiley & Sons.

Russell, E.C. (1983) *Building Simulation Models with SIMSCRIPT II.5*. Los Angeles, CA: C.A.C.I.

Schmitt, N., W.C. Borman, and Associates (1993) *Personnel Selection in Organizations*. San Francisco, CA: Jossey-Bass Inc.

Schriber, T.J. (1974) *Simulation Using GPSS*. New York, NY: John Wiley & Sons.

Schvaneveldt, R.W., D.W. Dearholt, and F.T. Durso (1988) *Graph theoretic foundations of Pathfinder networks*. Comput. Math. Applic. **15**(4): p. 337-345.

Schvaneveldt, R.W., F.T. Durso, and D.W. Dearholt, (1989) *Network structures in proximity data*. In *The psychology of learning and motivation: Advances in research and theory*, G.H. Bower, Editor. Academic Press, Inc.: San Diego, CA. p. 249-284.

Scientific American (1999) *Revolutions in science.*, ed. J. Rennie.

Scientific American (1999) *Special Report: Revolution in Cosmology*. Scientific American, **280** (1) p. 45-59.

Seamster, T.L., R.E. Redding, and G.L. Kaempf (1997) *Applied Cognitive Task Analysis in Aviation*. Ashgate Publishing Limited: Aldershot, England.

Serlin, R.C. (1987) *Hypothesis testing, theory building, and the philosophy of science*. Journal of Counseling Psychology, **34**(4): p. 365-371.

Serlin, R.C. and D.K. Lapsley (1993) *Rational appraisal of psychological research and the good-enough principle*. In *A Handbook for Data Analysis in the Behavioral Sciences*, G. Keren and C. Lewis, Editors. Lawrence Erlbaum Associates, Inc.: Hillsdale, NJ. p. 199-228.

Simon, H.A. (1983) *Alternative visions of rationality*. In *Reason in human affairs*, H.A. Simon, Editor. Stanford University Press: Stanford, CA.

Stevens, S.S. (1951) *Mathematics, measurement, and psychophysics*. In *Handbook of Experimental Psychology*, S.S. Stevens, Editor. Wiley: New York.

Stevens, S.S. (1959) *Measurement, psychophysics, and utility*. In *Measurement: Definitions and theories*, C.W. Churchman and P. Ratoosh, Editors. Wiley: New York, NY.

Tabachnick, B.G. and L.S. Fidell (1996) *Using multivariate statistics*. 3rd ed. New York, NY: Harper Collins College Publishers.

Tefler, R.A., ed. (1993) *Aviation Instruction and Training*. Ashgate Publishing Limited: Aldershot, England.

Telfer, R. and J. Biggs (1988) *Psychology and Flight Training*. Ames, IA: Iowa State University Press.

Telfer, R.A. and P.J. Moore, eds. (1997) *Aviation Training: Learners, Instruction and Organization*. Ashgate Publishing Limited: Aldershot, England.

Torgerson, W.S. (1958) *Theory and Methods of Scaling*. New York: John Wiley and Sons, Inc.

Tufte, E.R. (1983) *The visual display of quantitative information*. Cheshire, Connecticut: Graphics Press.

Tverskey, A. and D. Kahneman (1982) *Judgment under uncertainty: Heuristics and biases*. In *Judgment under uncertainty*, D. Kahnamen, P. Slovic, and A. Tversky, Editors. Cambridge University Press: New York, NY. p. 3-20.

Ullman, J.D. (1982) *Principles of Database Systems*. 2 ed. Computer software engineering series. Rockville, MD: Computer Science Press.

Van De Ven, A.H. and A.L. Delbecq (1971) *Nominal vs interacting group processes for committee decision-making effectiveness.* Academy of Management Journal, **14**(2): p. 203-212.

Waller, N.G. (1993) *Software review: seven confirmatory factor analysis programs: EQS, EzPATH, LINCS, LISCOMP, LISREL 7, SIMPLIS, and CALIS.* Applied psychological measurement, **17**(1): p. 73-100.

Webb, E.T., D.T. Campbell, R.D. Schwartz, L. Sechrest, and J.B. Grove (1981) *Nonreactive measures in the social sciences.* Boston, MA: Houghton Mifflin.

Williams, D.M., R.W. Holt, and D.A. Boehm-Davis (1997) *Training statistical skills to non-statisticians: A case study of inter-rater reliability training for pilot instructor/evaluators.* Unpublished manuscript.

Yerkes, R.M. and J.D. Dodson (1908) *The relation of strength of stimulus to rapidity of habit-formation.* Journal of comparative neurology of psychology, **18**: p. 459-482.